Biological Periodicity
Its Molecular Mechanism and Evolutionary Implications

BIOLOGICAL PERIODICITY
A Single-Volume Treatise

Editor and Author: **A. LIMA-DE-FARIA,** *Institute of Genetics, University of Lund, Lund, Sweden*

Biological Periodicity Its Molecular Mechanism and Evolutionary Implications

By **A. LIMA-DE-FARIA**
Institute of Genetics
University of Lund
Lund, Sweden

JAI PRESS INC.

Greenwich, Connecticut *London, England*

Library of Congress Cataloging-in-Publication Data

Lima-de-Faria, A.
Biological periodicity : its molecular mechanism and evolutionary implications / by A. Lima-de-Faria.
p. cm.
Includes bibliographical references (p.) and indexes.
ISBN 1-55938-412-3
1. Biological rhythms. 2. Molecular evolution. I. Title.
QH527.L55 1995
574. 1'882--dc20 95-43953
CIP

55 Old Post Road, No. 2
Greenwich, Connecticut 06836

JAI PRESS LTD.
The Courtyard
29 High Street
Hampton Hill, Middlesex TW12 1PD
England

ISBN: 1-55938-412-3
Library of Congress Catalog No.: 95-43953

Manufactured in the United States of America

CONTENTS

PART II. PERIODIC DISTRIBUTION OF FUNCTIONS IN LIVING ORGANISMS

PART IX. MACROMOLECULAR AND ATOMIC MECHANISMS RESPONSIBLE FOR PERIODICITY

INTRODUCTION

Biological periodicity was not recognized until recently because its revelation demanded a novel approach. That is, one based on a link between several areas of knowledge: evolutionary biology, molecular cytogenetics, atomic organization of crystals, chemistry of the elements, and elementary particle physics.

In the first place, one had to await the availability of evidence that evolution is a phenomenon inherent in matter and energy, and that biological evolution is a terminal event in this autoevolutionary process. In the absence of such evidence, phenomena recognized as occurring at the biological level, could not be directly correlated with the evolution of atoms and elementary particles. The recognition of autoevolution turns out to have been the crucial factor that paved the way to this finding.

Secondly, if the periodicity characteristic of the chemical elements was not so well established as it is at present, biological periodicity would not have been readily recognized. This is because it could not fit into a coherent scheme.

Thirdly, the mechanism underlying atomic periodicity, has been elucidated as the result of research carried out in particle physics and chemistry. This serves then as the starting-point of an examination of the atomic and molecular mechanisms that are responsible for biological periodicity.

And fourth, contemporary molecular cytogenetics, as well as molecular biology offer the unique opportunity of allowing to propose a mechanism, occurring at the level of the split gene, that is partly responsible for the emergence of biological periodicity. Thus, without these four types of information the phenomenon by itself would have remained unrecognized, as it has been until now.

Admittedly, a phenomenon that seems at first sight, to be similar to periodicity is convergence, which has been described by many authors both in fossils and living organisms. Convergence is understood as being "The formation of similarities in unrelated organisms living in the same environment" (Webster, 1976). It has also been defined as "The evolution of unrelated organisms which occupy similar niches and which bear superficially similar characteristics" (Taylor, 1983). Thus, what emerges is that what has been called convergence is considered to be: (1) accidental, (2) without any regular emergence, (3) resulting solely from the action of the outer environment, and (4) representing a superficial similarity not anchored in any well-defined biological or chemical process. Moreover, convergence (or parallelism, as it is sometimes called) has been considered a marginal event in evolution, as it does not agree with the prevailing interpretation of this process.

The biological periodicity that has come to light is an entirely different phenomenon from convergence in that it displays the following characteristics: (1) It occurs repeatedly, (2) it emerges at relatively regular intervals, (3) it appears as the intrinsic result of cellular processes, (4) it is mainly determined by internal molecular and atomic mechanisms, (5) it is displayed by both the functions and forms of living organisms, (6) these functions and forms appear suddenly as "ready-made" events, (7) the structural and functional complexes emerge as well integrated packages, (8) the emergence of the periodicity is not necessarily connected with the phylogenetic position of the organism or its degree of complexity, and (9) the periodicity is not necessarily related to the general environment.

When chemical elements were aligned after their increasing atomic weight they turned out to exhibit properties that recurred at relatively regular intervals. It was this unexpected periodicity that led to the establishment of the periodic chart of the elements. Biological periodicity is characterized by an equally unexpected recurrence of groups of properties that emerge at relatively regular intervals, as the biological complexity increases throughout the evolutionary scale. An example is flight. This function originates in the insects, does not show up in any other invertebrate group, reappears in the fossil reptiles, later becomes well established in birds, it occurs in certain fishes, and finally reappears suddenly in the bats but not in any other mammalian species. Many other functions and structures follow similar patterns of periodicity.

The mechanism responsible for the emergence of biological periodicity seems to depend on several processes. Chief among these are:

1. Atomic mimicry seems to occur at every level of organization. This is a phenomenon in which different atoms and different molecules produce the same final pattern or function. It occurs in simple molecules, minerals and living organisms. The emergence of the same solution, be it structural or functional, turns out to be mainly due to the presence of groups of pertinent atoms. However, in the

last analysis, the similarity of properties rests on the electronic configuration of single atoms or of key atom groups. Minerals with different chemical compositions display the same properties and proteins with different amino acid sequences result in the same structure, and even the same function. The cell does not need to synthesize the same chemical to produce an identical structural and functional solution.

2. Novel combinations of DNA sequences occur regularly at the split gene level. These involve rearrangement of the intervening sequences (introns) and the sequences coding for proteins (exons). Exon-intron recombination at the DNA level leads to the production of mosaic proteins. New genes are created by this process and old ones may reappear since the phenomenon is reversible. A similar situation occurs at the messenger RNA level. Alternative RNA splicing changes the position of the sequences and results in different proteins, without changing the genetic constitution. Since this splicing is also reversible specific proteins may reemerge. These events, as well as other ordered DNA rearrangements, are expected to be dominant factors in the production of the periodic reappearance of structures and functions.

3. Single molecules lead to molecular cascades that affect different organs. A single polypeptide can cleave itself into six different hormones that have separate functions. A prime example is the melanocorticotropic hormone that by a series of cleavages produces six hormones. Each of the final amino acid sequences acts on different organs. Many cell mechanisms are known to be directed by molecular cascades in which the activated form of one protein catalyses the activation of as many as 10 proteins, the outcome of which is the formation of distinct packages of structures and functions that are interrelated and that become "ready-made" solutions.

4. The role of the environment. The contribution of the environment appears to be mainly confined to unleashing specific molecular cascades that lead to the emergence of the packages of periodic properties. However, some types of periodicity are not directly related to the predominating environment and are associated with adverse situations.

5. The periodicity of the chemical elements has given rise to the periodicity of the minerals. These have in turn conditioned the construction of the cell and transferred some of their periodic properties to the living state. It may be recalled that the primary function of biologically essential macromolecules such as hemoglobin and chlorophyll, is not localized to their long molecular chains, but in their iron and magnesium atoms, respectively. Moreover, the function of many enzymes depends on the presence of zinc atoms and cobalt is the main component of some vitamins. These metal atoms cannot be produced by the cell and are not part of the genetic code. They were inherited from the minerals. This implies that the minerals have served as the autoevolutionary bridge between the level of the atoms and that of cellular organisms. This is also evidenced by the fact that biological periodicity

follows most of the features that characterize the periodicity of the chemical elements. Although the gap between the two levels is enormous, the essential features remain conspicuous. The reason is actually a natural one, since the atoms building both levels are the same.

Thus, the conclusion reached is that the properties of living organisms mirror the periodic order that the single atoms and the crystals had acquired before the evolution of the living cell, and that they impinged on its formation.

ACKNOWLEDGMENTS

I would like to thank several colleagues at the University of Lund for references to sources of information: Lars Ivar Elding, Per-Åke Albertsson, Anders Liljas, Roland Gorbatchev, Rune Grubb, and Ulfur Arnason. Thanks are also due to Dr. José Lima-de-Faria, Head of the Center of Crystallography and Mineralogy, Lisbon, Portugal, for information on some sources of mineral illustrations.

Excellent technical assistance was provided by Agneta Persson, Institute of Anatomy; Jan Larsson and Thomas Almén of the University Library Media Section, and Med. kand. Johan Essen-Möller, Institute of Genetics, University of Lund, during the preparation of the graphs, figures and typescript respectively.

I also wish to acknowledge a publication grant from the Swedish Natural Science Research Council. Thanks are also due to the Crafoordska Foundation, Sweden, and the Erik Philip-Sörensens Foundation, Sweden, for grants that supported the preparation of the manuscript.

Several illustrations and tables appearing in this book have been obtained from other publications. In all cases reference is made to the original publications. The full sources can be found in the reference lists. Permission for the reproduction of this material is gratefully acknowledged.

A. Lima-de-Faria

PART I

PERIODIC DISTRIBUTION OF PROPERTIES IN CHEMICAL ELEMENTS AND MINERALS

Chapter 1

PERIODICITY IN CHEMICAL ELEMENTS

THE ORDER IN CHEMICAL ELEMENTS TOOK OVER 100 YEARS TO ESTABLISH

The properties of the chemical elements are by no means arbitrary since they depend on the structure of the atom, and vary with the atomic number in an ordered way. The significant feature is that there is a periodic recurrence of characteristic properties.

Until 1817 randomness was thought to characterize the chemical elements. When order started to be perceived they were placed in groups of three, and afterwards in groups of seven. Subsequently, the similarities of properties were so evident as to allow predictions to be made concerning missing elements. This was of course the contribution of Mendeleev who, in 1869, established a relationship between the atomic weights of the elements and their physical and chemical properties. In fact he predicted the existence of six elements that had not yet been discovered. Three (scandium, gallium, and germanium) were soon found while the other three were discovered later. They had the properties predicted by Mendeleev. But it was not until 1922 that Bohr interpreted the Periodic Table in terms of the electronic structure of the atoms (Pauling, 1949).

More than a century elapsed before a coherent explanation of elemental order was made available. It became quite apparent that the mechanism beyond periodicity resided in the electronic organization of the atom. At present, however, much remains to be discovered about this atomic order.

THE PERIODICITY OF PROPERTIES

The properties of elements change in a systematic way through a period. This results in the formation of groups which consist of the elements with closely related physical and chemical properties (Figure 1).

The elements on the left side and center of the Table are metals, i.e., they have high electrical and thermal conductivity and metallic luster. Those on the right side are nonmetals. The metalloids, which have intermediate properties occupy a diagonal position in the Table.

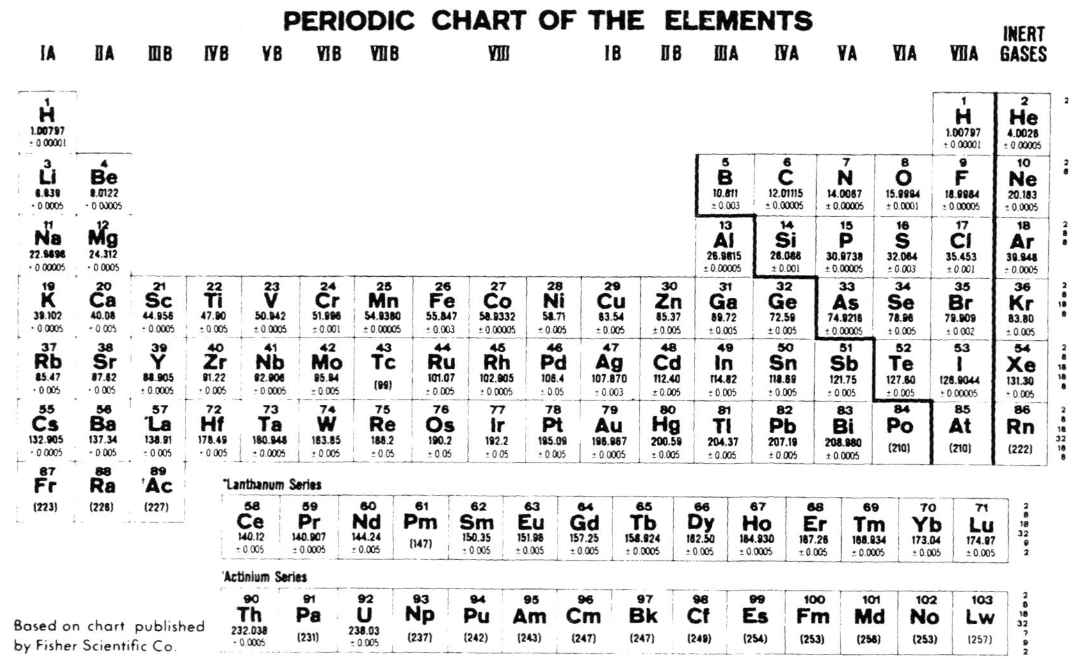

Figure 1. Periodic table of the elements. Compare the figure with Figure 2 in Chapter 11. (From Sanderson, 1967.)

The best example of periodicity is provided by the noble or inert gases: helium (He), neon (Ne), argon (Ar), krypton (Kr), xenon (Xe), and radon (Rn). All build a single group and are chemically inert due to the great stability of their electronic structures, which results from their completed electron orbitals.

The elements of the first group of the Periodic Table (IA) viz., lithium (Li), sodium (Na), potassium (K), rubidium (Rb), and cesium (Cs) are alkali metals characterized by being soft, having a silvery-white color, and great chemical reactivity. Francium (Fr) also belongs to this group but it has a transient existence with a short radioactive half-life.

The halogens are the nonmetallic elements; they differ most from the metals. As can be seen, they occupy group VIIA and are: fluorine (F), chlorine (Cl), bromine (Br), and iodine (I). Many of their compounds are salts. All are volatile and form diatomic molecules. Their color increases steadily with increase in atomic number.

The members of group VA (mainly nonmetallic)—nitrogen (N), phosphorus (P), arsenic (As), antimony (Sb), and bismuth (Bi)—tend to form hydrides and oxides that are of the same type.

The alkaline-earth metals—beryllium (Be), magnesium (Mg), calcium (Ca), strontium (Sr), barium (Ba), and radium (Ra)—belong to group IIA. They are harder and less reactive than the alkali metals and all form oxides.

The Periodic Table contains, among other sequences, two short periods of eight elements each, two long periods of 18 elements, followed by a very long period of 32 elements (Pauling, 1949; Greenwood and Earnshaw, 1989). This means that already at the chemical level the periods have different lengths.

Significant for the understanding of biological periodicity is the finding that the periodicity, at the initial level of the chemical elements, is independent of atomic complexity. The simplest atoms such as helium (He) display the same properties as the most complex such as radon (Rn). The first has an atomic weight of 4 and the second of 222. In between: neon (Ne), argon (Ar), krypton (Kr), and xenon (Xe) have atomic weights of 20.17, 39.95, 83.80, and 131.29, respectively. Simplicity as well as complexity are not a hindrance to the display of the same basic properties.

THE MECHANISM UNDERLYING THE PERIODICITY IN THE CHEMICAL ELEMENTS

The question to be asked is: What does the periodicity of the chemical elements actually mean? The simple answer is that it is essentially the manifestation of their ordered evolution. To grasp the repetition of properties at the biological level, it is important to understand this order.

Periodicity in the elements can be displayed in many ways. The spiral table (see Chapter 11, Figure 2) emphasizes that the simplest element, hydrogen, is the nucleus out of which all other elements were derived. The fusion of four hydrogen atoms yields helium and all the other 105 elements are formed from this simple chemical (Sanderson, 1967). This is a clear demonstration of the creation of complex forms by pure combination of not only the simplest but also the same component, namely hydrogen. Such simplicity was possible because the order followed was laid down by discrete energy levels. Thus far, 106 chemical elements have been identified and these give rise to a little over 1500 variants which are called nuclides but only 269 of them are stable. They represent variations in the numbers of protons and neutrons, as in the well known isotopes, isobars, and isotones.

The picture which emerges is that similarity in chemical behavior results from similarity in electronic structure. The elements of each group have usually the same outer electron configuration, it is the electrons of the outer shell that determine the basic properties (Greenwood and Earnshaw, 1989; Asimov, 1992; Masterton and Hurley, 1993).

GRAPHIC DISPLAY OF CHEMICAL PERIODICITY

More than 700 forms of the periodic table have been proposed, each one emphasizing different types of relations between the elements (Mazurs, 1974). Though the graphic display indicated in Figure 2 is uncommon, it shows in a distinct way the relationship between the atomic volume (molar volume) of the elements in the solid state and their atomic number (Z). The alkali metals appear at the peaks of the curves: lithium (Li), sodium (Na), potassium (K), rubidium (Rb), cesium (Cs), and francium (Fr), whilst the elements that occur near the center of each period are at the troughs: boron (B) and carbon (C), aluminum (Al) and silicon (Si), manganese (Mn), ruthenium (Ru) and osmium (Os). These positions are interpreted on the basis of the electronic theory. The alkali metals (at the peaks) have only 1 electron in their outer-shell, whereas the elements near the center of each period have the maximum number of electrons available for bonding. A line unites magnesium (Mg), calcium (Ca), strontium (Sr), barium (Ba), and radium (Ra). All of them belong to the same group in the periodic table, which is characterized by atoms with 2 electrons in the outer-shell.

NUMEROUS PROPERTIES EXHIBIT PERIODIC TRENDS

Not just one, but a whole series of properties show periodicity. This is why the table has proven to be so important in chemistry. Many phenomena became suddenly intelligible when their periodicity was well established. The various types of atomic behavior that frequently recur are: (1) valence (oxidation state), (2) basicity, (3) electropositivity, (4) bond type, (5) stability of compounds, and (6) stability of coordination complexes and others.

There are not less than 10 periodic trends.

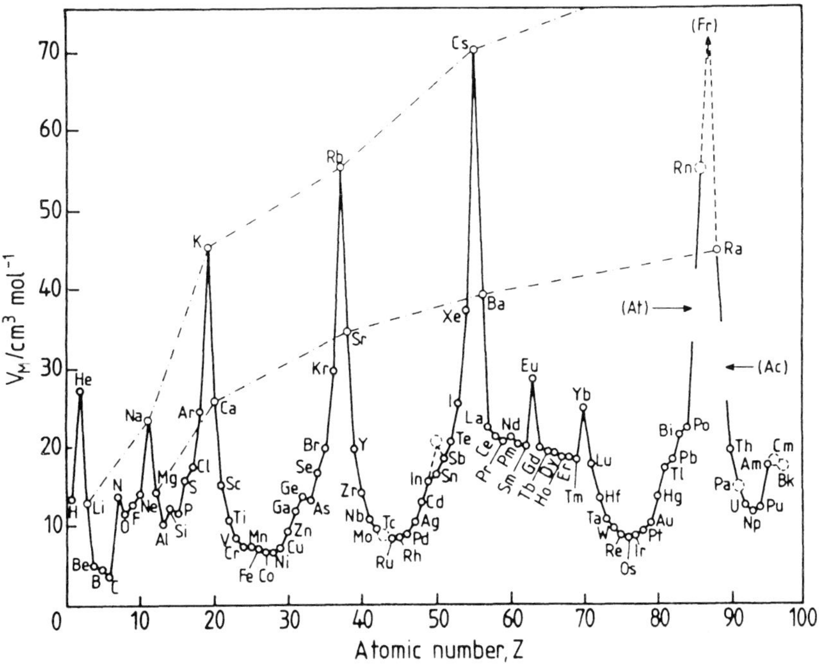

Figure 2. Periodicity in chemical elements. The periodicity found among the little over 100 atoms known is usually displayed as a periodic table of the elements. The periodicity refers to the fact that the atoms in the same group have similar chemical properties. More recent representations depict far better the repetitive character of these properties. The atomic number of an element is the number of protons contained in the nucleus of an atom of the element. Graphic representation of the variation in volume per mol, as the atomic number (Z) increases. The periodicity is revealed by the peaks that occur at *Na* (sodium), *K* (potassium), *Rb* (rubidium), *Cs* (cesium) and *Fr* (francium). The different troughs that emerge, at relatively regular intervals, are also an expression of the periodicity. (Redrawn from Greenwood and Earnshaw, 1989.)

ANOMALIES ALREADY EXIST AT THE LEVEL OF CHEMICAL PERIODICITY

No phenomenon reveals total regularity; exceptions and anomalies are known to exist in the most ordered systems. One tends to think that such deviations are characteristic of higher levels of complexity, such as the biological, but they are found already at the atom level. Several examples can be cited. Helium (He) is an element with an anomalous position in the periodic table (Figure 2). The same is true of europium (Eu) and ytterbium (Yb) that belong to the lanthanide series. Other irregularities include chromium (Cr) and copper (Cu) atoms that carry 1 rather than 2 electrons in their 4s orbitals (Jaffe, 1988).

It is important to realize that every phenomenon is a dynamic process. Although order implies high rigidity, a certain margin of plasticity is present which results in novel and sometimes exceptional solutions or else no evolution could have taken place.

Chapter 2

PERIODICITY IN MINERALS

MINERAL CLASSIFICATION IS BASED ON CHEMICAL HIERARCHY

Approximately 3000 minerals are recognized as valid mineralogic species whose chemical composition is not accidental, for it leads to 12 classes based on chemical similarity. These are: (1) Native elements, (2) Sulfides, (3) Sulfosalts, (4) Oxides, (5) Halides, (6) Carbonates, (7) Nitrates, (8) Borates, (9) Phosphates, (10) Sulfates, (11) Tungstates, and (12) Silicates. As pointed out by Klein and Hurlbut (1985), the classification is mainly the result of the presence of a dominant anion or anionic group. Those minerals that have the same anion or anionic group dominant in their composition have "unmistakable family resemblance." By contrast, a dominant cation does not have the same effect.

That the mineral classes are based on the occurrence of similar properties is best shown by only 20 of the 106 chemical elements occurring in the native state (with the exception of the gases). Apparently, this restriction forced many minerals to have similar properties. Another source of similarity is the mere fact that minerals with the same dominant anion occur generally in a similar geologic environment. An example of this are the sulfides that have a tendency to occur in close association in the same geological formation.

It is an interesting fact that mineral classes are divided into families, groups, species and varieties by mineralogists, just as biologists divide animals or plants according to their degree of similarity. The evolutionary fact that emerges from this classification is that there is already a dominant chemical component (the anions) among the minerals that determines their properties. This is what later nucleic acids do in living organisms. Not all chemicals in the mineral, nor all chemicals in the cell are equally important but some have ended up carrying the critical information that governs the main properties of the cell. Expressed in another way, hierarchy among specific chemicals was present before the cell emerged.

THE PERIODICITY OF THE ELEMENTS HAS DETERMINED THE PERIODICITY OF PROPERTIES IN MINERALS

Arsenic, antimony, and bismuth are considered as both chemical elements and minerals. They are classified as semi-metals because they have an atom packing that differs from that of metals. They build covalent bonds between their atoms. This covalency is related to the position of arsenic, antimony and bismuth (As, Sb, Bi) in group VA of the Periodic Table. This strong bonding results in the layered structure of the atoms, which is the source of the good cleavage of the resulting three minerals. It also conditions their physical properties which are the same, *viz.* poor conductivity of heat and electricity, and lower symmetry.

Another example which comes to mind are the minerals gold, silver and copper, which constitute the so-called gold group within the native metals. The three minerals consist of atoms that belong to the same group in the Periodic Table (Cu, Ag, and Au, Group IB, see Chapter 1, Fig. 1), i.e., these atoms have similar properties. This is also the case when they form minerals. The three have properties in common that arise from the atomic structure. They are rather soft, malleable and ductile, and all are good conductors of heat and electricity. Moreover, the three have hackly fracture, metallic luster and a low melting point. These properties are known to be the result of the metallic bond of their atoms. As Jaffe (1988) remarks: "Electrons govern the chemical behavior of atoms, and how they may be packed into minerals."

Thus, it is not unreasonable to hold that the periodicity of the properties of the elements was transferred to, and perpetuated in the properties of the minerals as a consequence of the original electronic mechanism.

STRUCTURAL AND FUNCTIONAL PERIODICITY—EMERGENCE OF THE SAME PATTERN AND PROTO-FUNCTION IN DIFFERENT MINERAL CLASSES

It is well established that minerals crystallize into seven specific crystallographic systems: triclinic, monoclinic, orthorhombic, tetragonal, trigonal, hexagonal and cubic. All native metals (gold, silver, copper, platinum, and iron) crystallize only in the cubic system. However, the capacity to produce cubic crystals arises independently outside this mineral class. Pyrite (FeS_2) which is a member of the sulfide class forms cubic crystals. Magnetite (Fe_3O_4) which belongs to the oxide class is also cubic. It is thus clear that the same pattern arises in totally different mineral classes. This is a manifestation of structural periodicity and it reinforces the view that the same final pattern can be produced by different molecular arrangements (Whitten and Brooks, 1988).

Mineral proto-functions, such as luminescence occur in minerals with very different chemical compositions. Not only fluorite (CaF_2) but others show this phenomenon: scheelite ($CaWO_4$), willemite (Zn_2SiO_4), calcite ($CaCO_3$), eucryptite ($LiA1SiO_4$) and diamond (pure carbon) (Klein and Hurlbut, 1985).

Already in the minerals, different chemical compositions lead to the reemergence of the same property.

PART II

PERIODIC DISTRIBUTION OF FUNCTIONS IN LIVING ORGANISMS

Chapter 3

PERIOD FLIGHT

THE PREPARATION OF THE GRAPHS REVEALING BIOLOGICAL PERIODICITY

Several reasons can be given for the time of emergence of a given group to be considered as being approximate. One is that it differs considerably among authors. Another is that it sometimes corresponds to the period of greater abundance in the fossil record. A third reason is that there is no accurate information on the emergence of several invertebrate phyla. The figures given are based on data obtained from Bell and Woodcock, 1971; Freeman, 1972; Schopf, 1978; Colbert, 1980; Barnes, 1980; McFarland, 1981; Margulis and Schwartz, 1982; Macdonald, 1984; Whitten and Brooks, 1988. The time of emergence is not drawn to scale to avoid making the graphs too large. The figures refer to million of years ago.

The degree of complexity is a crude estimate. It follows an arbitrary scale divided into six units that represent an increase in number of tissues participating in a given organ or group of organs. These last ones, as well as their functional interactions, cannot be easily measured and this is why the estimate is a rough approximation.

Not all phyla or classes are represented in each graph to avoid an overloading. Further, the number of species that display a certain function is not always easy to establish. Hence, the terms "most" and "some" species are used. In other cases no information is given in the graph on account of the lack of information on how widespread the phenomenon is (Figure 1).

FLIGHT IN INSECTS AROSE FROM NOWHERE

No invertebrate phylum is known to include animals with wings except the Arthropoda within the class insecta, yet there are as many as 30 phyla of invertebrates.

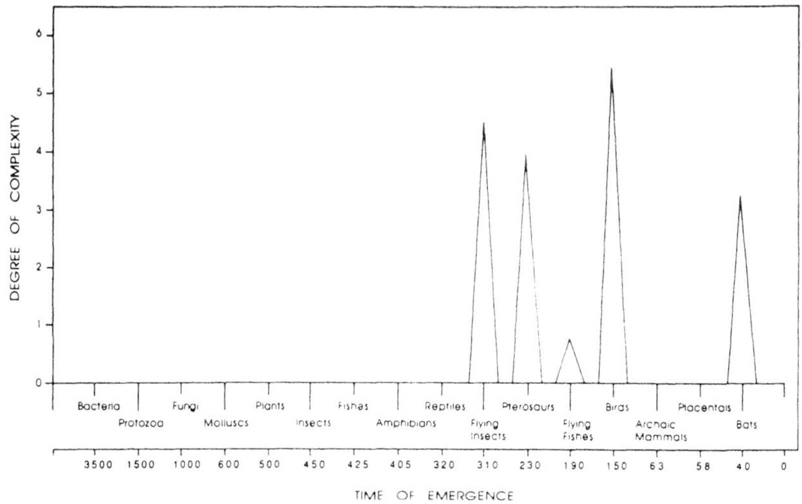

Figure 1. Period Flight. Flight is a phenomenon that seems to have arisen from nowhere. There are 30 phyla of invertebrates among which in only one has the ability to fly been developed. In particular, in insects, one finds this capacity to have been seldom exceeded in evolution. It appears again suddenly in the reptiles called pterosaurs (pteron = wing), in the flying fish, the birds and the bats (which are mammals). None of them is directly related to the preceding group that was able to fly; they have originated from non-flying relatives. Flying fish are teleosts, whose flying ability developed after their emergence some 190 million years ago. The degree of complexity is a crude estimate based on the organs involved in a given function. The time of emergence recorded here is in millions of years. It is not drawn to scale simply to avoid making the graphs too large. Moreover, this time record is only approximate and it is not exactly the same among various authors.

According to the fossil record, the first insects had no wings; they were similar to the wingless species of today. However, wings evolved, and the dragonflies and the mayflies started to fly over the forests of ferns and conifers of the Paleozoic age. Wings and the capacity to fly arose without any predecessors, and with few intermediate forms. Remarkably, they emerged as gigantic structures. For example, the earliest dragonflies are known to have had wings spanning about 70 cm, as evidenced by the available fossil records. This size of a wing is larger than that of many birds and bats (Figure 2).

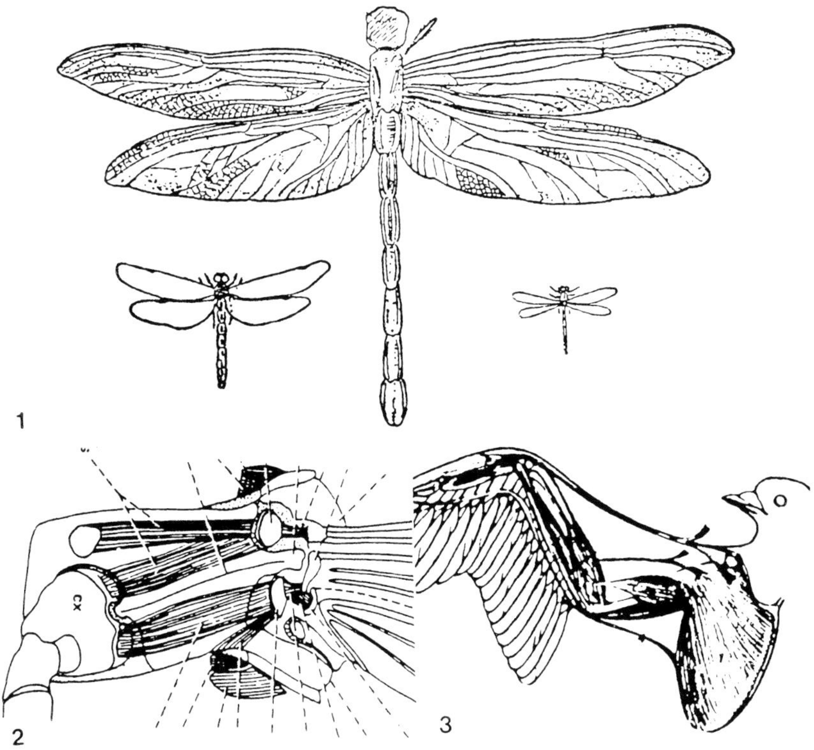

Figure 2. Flight in Insects and Birds. (1) The fossil of the giant dragonfly *Meganeura* that inhabited the forests of the Carboniferous period (350 to 270 million years ago). The wings from tip to tip measure 73 cm. For purposes of comparison two of the largest living dragonflies are depicted here: one from South America (left) and another from Europe (right). (2) The wing muscles of an insect. Four different types of muscles are involved in the wing's movement: basalar, dorsal longitudinal, subalar and tergosternal. Chitinous plates and connections lying between the muscles and the wing base are additional structures that coordinate flight movements. (3) The wing muscles of a bird. The three main types of muscles that move the wing are the: pectoralis major, triceps, and biceps. Several tendons are auxiliary structures that hold feathers in position.

Figure 3. Periodicity of flight. Structural similarity. (1) Insect *Chrysops discalis.* (2) Extinct flying reptile *Rhamphorhynque.* (3) The bird, giant kingfisher, *Megaceryle maxima.* (4) Flying fish *Dactylopterus orientalis* (gurnard). (5) A bat (flying mammal).

FLIGHT DEVELOPED INDEPENDENTLY AT FIVE DIFFERENT TIMES IN BIOLOGICAL EVOLUTION

Flight, with all its complex structures and functions, developed independently, and not less than five times during evolution.

The first event occurred in the invertebrates. Insects emerged 450 to 425 million years ago but winged insects such as the dragonflies, cockroaches and grasshoppers appeared about 310 million years ago (McFarland, 1981) (Figure 1).

The second event was witnessed when the reptiles became aerial vertebrates some 230 to 195 million years ago (Barnes, 1980). In-between the insects and the pterosaurs (flying reptiles) there emerged other invertebrate species, and the fishes, amphibians and the first reptiles that did not fly. Thus, the aerial reptiles are regarded as an isolated event.

On the basis of the evolutionary scale, fishes evolved 425 million years ago, but when present day flying fishes developed from the main stock is not yet known. Since the species that fly belong to the teleosts, which appeared 190 million years ago, it is inferred that flying fishes had emerged after this date (Freeman, 1972) and as such without any connection with the pterosaurs. Moreover, the pterosaurs died out in the Cretaceous period leaving no surviving relatives (Walker, 1974).

Birds are considered to have originated from flightless reptiles such as thecodonts; they emerged about 180 to 135 million years ago (Whitten and Brooks, 1988).

Bats are the last, and suddenly make their appearance in the Eocene age, only 40 million years ago. There are no known intermediate stages between bats and other mammals such as the insectivores (Colbert, 1980). No other mammals fly. Only some others glide such as the squirrels (Figure 3).

FLIGHT IS BOTH A STRUCTURAL AND A FUNCTIONAL PROCESS

All functions are based on structures. It is therefore essential when one describes a given function that the supporting structure be considered. In the case of flight, this is usually solely associated with the existence of a wing. Yet the fact remains that a wing by itself would never allow any animal to lift its weight into the air and move in it efficiently. Clearly, there are birds and insects with wings but they do not fly. Prime examples among birds are the ostrichs and the rheas. The ant *Myrmica scabrinodis* and the fly *Phasmidohelea wagneri* have large wings but cannot fly (Nachtigall, 1974). It is not without significance that a series of other structures and functions are associated with the appearance of a wing and that only when all of them are present does flight become possible.

In treatises on evolution the wing of a bat is considered to be homologous with that of a bird because both have an internal bony skeleton but the wing of an insect is classified as an analogous structure because it has no bones. If the wing is viewed as an isolated structure, then the difference between the invertebrates and the vertebrates is obviously great. When all the other structures and functions associated with it are analyzed in detail, it turns out that the solutions are essentially the same, namely acquiring only different forms in the two groups of animals. Once all the features involved in flight are taken into account the analogy becomes transformed into a prominent homology. Only the degree of homology varies between the insect, the bird, the bat, the pterosaur, and the fish.

FLIGHT DEMANDS MANY MORE STRUCTURES AND FUNCTIONS THAN THE EXISTENCE OF A WING

Let us now examine the main obligatory components of the flying process. The wing is one of them. Flight requires in addition four main specializations that involve: (1) the skeleton, (2) the muscles, (3) blood circulation, and (4) the nervous system. Without the development and coordination of all of them there is no flight.

1. The wing is the bearing structure that functions as an aerofoil and a propeller. Its structure and shape provide the uplift necessary to counteract the action of gravity.
2. The wing possesses a skeleton that maintains form and gives it strength.
3. Articulations are present without which it could not move in different directions.
4. The muscles are of different types, and have antagonistic functions that result in the creation of a variety of movements.
5. Extra energy is necessary to support the great effort spent during flight.
6. Air and blood flow are also needed in greater quantities during flight.
7. The nervous system is involved through sensory nerves and receptors. This allows control of direction, stability and speed.

These different components are obligatory parts, in both invertebrates and vertebrates that are able to fly.

A SERIES OF SIMILARITIES BETWEEN THE FLIGHT OF INSECTS AND THAT OF BIRDS

The flight of a bird and that of an insect are strikingly similar both structurally and functionally. The following features stand out: (1) Flying animals have a fusiform and stream-lined body both in birds and insects. (2) Feathers in birds and bristles in insects smoothen the exterior of the body, thereby adding to the aerodynamic shape. (3) The wings are located above the center of gravity on the dorsal surface of the body in both the invertebrates and vertebrates. (4) Stability during flight is improved by adding a long tail. This is present in birds and mayflies (Ephemeroptera). It is used in gliding, and in turning operations. However, the tail is not absolutely necessary. A sparrow and the insect *Calliphora* can fly well without tails

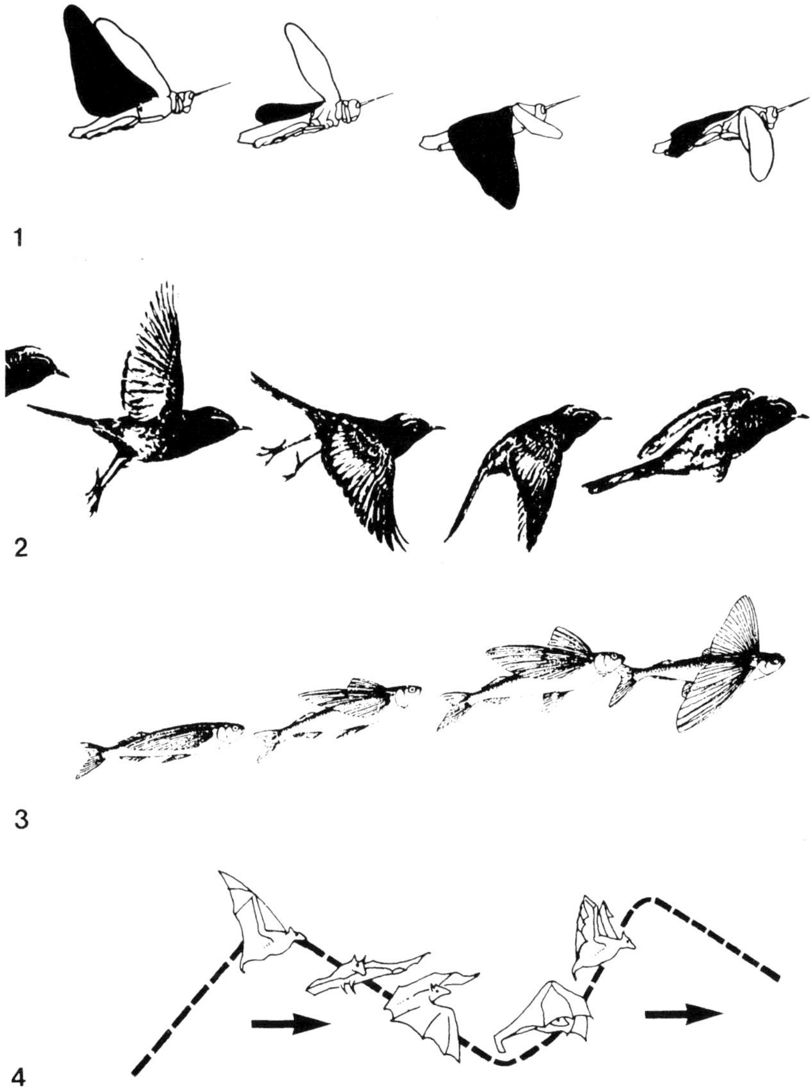

Figure 4. Periodicity of flight. Functional similarity. (1) The course of wing movements in the locust (insect). (2) High-speed cine photography of bird flight. Take off is usually achieved by jumping into the air. (3) The flying fish *Exocoetus callopterus* which uses the propulsion of its tail for forward speed and then opens its pectoral and pelvic fins for flight. (4) Sequential positions of the bat wings during flight.

due to their good balance (McFarland, 1981). (5) The wings of a bird have air-filled proximal bones and the more distal bones are also air-filled in birds of prey. The skeleton of insect wings consists of veins. These are fine tubes of toughened cuticle which also contain air tubes (tracheae), nerves and blood—just like the bones of mammals and birds (Abercrombie et al., 1951). (6) Muscles are involved in the movement of both types of wings. The wings of birds are raised and lowered by two main different muscles. In insects, there are two main groups of muscles which also raise and lower the wings (Figure 2). (7) The bird wing would not be able to fold were it not for articulations in its skeleton. The same holds for insects since articulation of the wing is due to the presence of several skeletal plates (sclerites), as well as the protein resilin that has a rubber-like flexibility (Romoser, 1973). (8) Without the availability of high amounts of energy and its rapid supply, flight cannot occur. In birds and insects the energy source is glycogen. This is stored in the liver of birds and in the fat body of insects. Moreover, both insects and birds increase their fat reserves before starting long migrations. (9) Flying birds have extensive air sacs that allow a continuous air-flow through the lungs. A similar system of air sacs exists in insects. The air circulates through them during flight and is expelled through the abdomen. The high respiratory rate of insect flight muscles is reflected in their huge mitochondria (Lehninger, 1975). (10) The body temperature in birds rises to 35°–40°C, thus allowing the muscles to act efficiently. Many insects including moths and bees maintain temperatures of 30°–40°C in their flight muscles, even when the environmental temperature is only 5°C (McFarland, 1981). (11) Speed in flight in birds is detected through the nostrils while in insects it is detected by hairs in the head and by deflection of their antennae. (12) Birds use four main types of flight: gliding, soaring, flapping and hovering (Beazley, 1974). Insects are able to perform such functions equally well (Figure 4). They also glide, hover and dart. These flight movements are highly complex (Romoser, 1973; Barnes, 1980). (13) Mating may take place in flight both in birds (swifts) and insects (bees). (14) The nervous system participates all the time in the flight process in both the invertebrates and the vertebrates. Nerve impulses are reaching the muscles, which may function in an antagonistic way, and the senses of balance and of vision are in permanent use (Barnes, 1980).

This list of similarities could be extended, but it seems sufficient to show how the two phenomena, though separate in time, and marked by a large gap consisting of non-flying animals, depend on essentially the same solutions, both at the structural and functional levels.

COMPARISON BETWEEN THE FLIGHT OF BATS AND BIRDS

Bats are mammals, yet they fly extremely well.

1. "Nothing is known about the details of the evolution of flight in bats" (McFarland, 1981). This is due to the fact that today's bats are similar to the

1

2

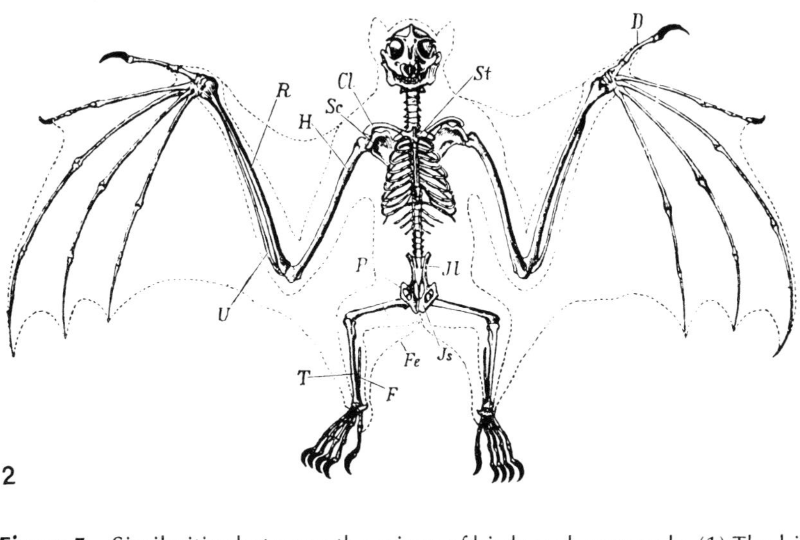

Figure 5. Similarities between the wings of birds and mammals. (1) The bird *Opisthocomus hoazin* as an adult (*left*), and as a young (*right*) climbing by means of clawed wings. (2) Skeleton of the bat *Pteropus edulis* showing the two upper fingers stretching out of the wing and having claws resembling those in the bird.

first fossils, which are 40 million years old. Moreover, a poor fossil record is not the reason. On the contrary, these fossils are so well preserved that components such as membranes are visible and stomach contents are available, indicating that the first bats were already insectivorous (Macdonald, 1984). The same situation is found in birds. The birds are considered to have evolved from a reptilian stock that included the dinosaurs and the crocodiles. The *Archaeopteryx* lived in the Jurassic, about 160 million years ago, but its relationship to present day birds is not well established. By the Cretaceous period, birds showed most modern features such as reduction and fusion of finger bones, a sternum with a deep keel, but still retained teeth, which are not formed in present day birds. From the Eocene onwards the fossils of modern birds appear without obvious forebears. The transition seems to have been sharp. There is no fossil record of the transformation from reptile scales to bird feathers (Beazley, 1974).

2. The bones of the wings of bats are not air filled like those of birds. However, the lungs of bats provide a flow of air that has a rate of exchange similar to that of birds.
3. In bats the bones of the arms are fused and the wrist is much reduced in size. The last four fingers are elongated and account for about half of the wing surface. In birds both the wrist and "palm" bones are fused, and they have three shortened digits (Figure 5).
4. The wing of bats is formed by a membrane supported by the bones of fingers, arms and legs. The membrane is formed from skin. The wing pattern is essentially similar in all species. The wing of birds is also built by the skin supported by the fingers and arms (not the legs), and the feathers are skin modifications formed by highly cornified epidermal cells.
5. The upward and downward movement of the wings is controlled by muscles lying within the wing in both cases.
6. Hinged bones are present in the wings of both birds and bats.
7. The sternum is keeled in both groups.

COMPARISON BETWEEN THE FLIGHT OF PTEROSAURS AND BIRDS

The pterosaurs were flying reptiles that appeared in the Triassic era and disappeared about 70 million years ago.

1. Pterosaurs probably moved their wings by raising and lowering them as birds and bats do, using large pectoral muscles.
2. The sternum of some pterosaurs was keeled as it is in birds and bats.
3. The shoulder girdle was fused both to the vertebrae and sternum. Bone fusion also occurs in birds and bats.
4. The vertebrae of higher forms of pterosaurs were fused together as in birds.

5. The wing bones were hollow, a solution similar to that of the bones of the wings of birds.
6. The pterosaurs had a wing membrane supported by only the fourth finger which was highly elongated. In this respect, they resemble bats more than birds (Carroll, 1987).

THE EMERGENCE OF FLIGHT IN FISH DOES NOT APPEAR TO BE DIRECTLY RELATED TO THE ENVIRONMENT

Fish are not usually included among the flying animals because such species are few and because they fly for relatively short stretches. In reality, their large pectoral fins are activated by strong muscles such as in the flying gurnard *Dactylopterus orientalis*. Some become aerial for more than 100 meters, just as in most birds when moving from one tree to another. It is their type of respiration that obliges them to go back to the ocean water; otherwise, they could continue their flight in air.

Fish differentiated into several 1000 species that were unable to fly but subsequently a few aerial species emerged. It is noteworthy that they were compelled to do so in an environment that was physiologically adverse to their normal respiration in water. Hence, the capacity to fly appears to have arisen without there being a direct connection with the environment. In lakes some fish jump out of water to catch insects, in contrast to the gurnards that live in the ocean and do not prey on insects. It is supposed that they became aerial to avoid predators but they could have sought deeper waters as other fish do. They fly, because like the insects, they got well developed wings that move efficiently. Moreover, they resemble the insects in another respect: there are 2-winged and 4-winged fish. In some species the pectoral fins were transformed into wings while in others, the pectoral and pelvic fins were transformed into a four wing system. In insects, one finds that flies have two wings, whereas butterflies have four wings. It is thus clear that the vertebrates had the possibility of producing a four-winged system, a situation not seen in birds but achieved in a fused form in bats where a single wing unites the arms and the legs.

FLIGHT IN FISH

The 4-winged fish show the most advanced form of flight (Figure 6). The evidence is as follows: (1) The flight starts under water by pressing the wings against the body. (2) In the air they open the wings rapidly. (3) The caudal fin remains in contact with the water surface and its strong muscles allow it to rotate at a high speed, thereby producing a propulsive force (70 contractions per second). (4) The flight is sustained mainly by the pectoral wings that open out. (5) But the flight is not one of passive gliding. It involves strong muscle contractions that oblige the body and the wings to vibrate as in the insects. Moreover, there is some wing beating. (6) The pelvic wings are used to change the direction of flight. (7) These fish show a

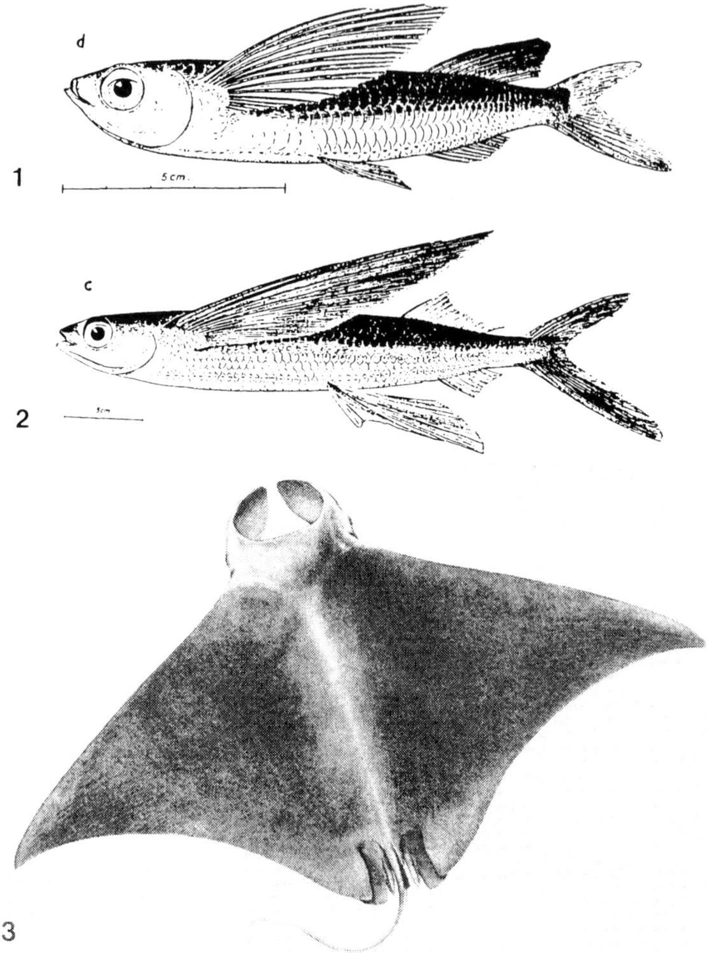

Figure 6. Flight in fish. (1) The Atlantic two-winged flying fish *Parexocoetus mento atlanticus.* (2) The four-winged flying fish *Cypsilurus heterurus* also from the Atlantic Ocean. (3) A ray moves through the water using the wing-beat motions of birds and for this reason is considered to fly but in a different medium than air, which in this case is water.

high capacity to direct their flight. They can circle around a boat, fly at a specific distance from the water surface, as well as suddenly stop their motion in air, just as birds do (Hanström and Johnels, 1962; Beazley, 1980).

Other fish, such as the large rays, have solved the problem by "flying" in water; that is, their huge fins allow them to swim by performing the same type of wing-beat motions that birds of prey use when they move in air (McMahon and Bonner, 1983). Hence, flying is not necessarily correlated with the existence of air.

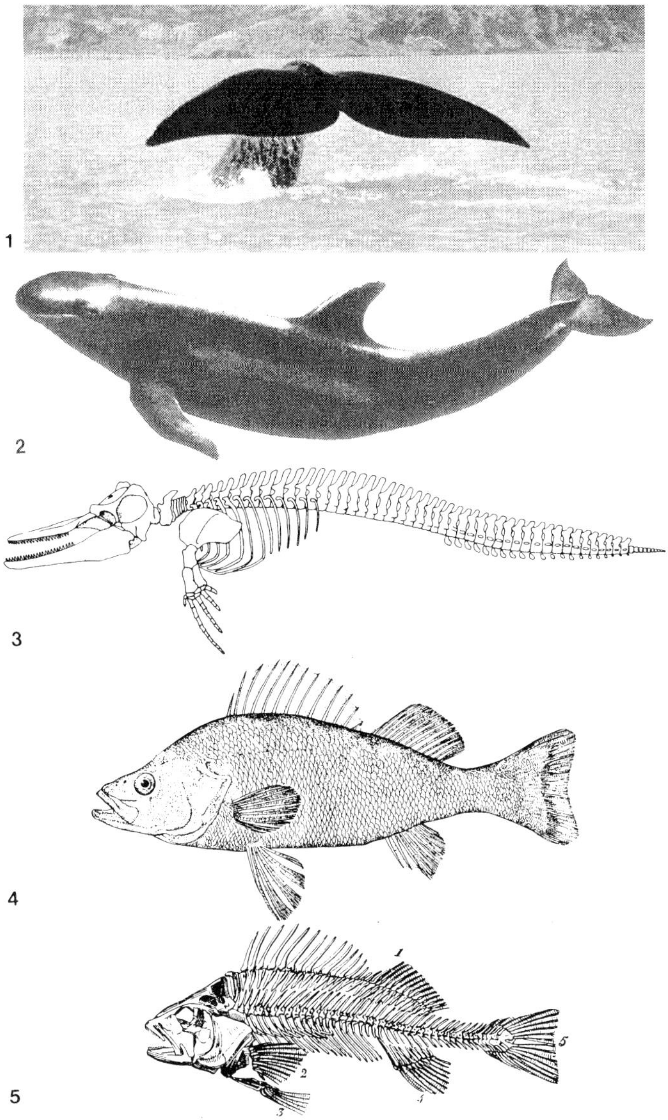

Figure 7. Locomotion organs of vertebrates formed without a skeleton. (1) The enormous tail of the right whale *Balaena glacialis.* (2) False killer whale *Pseudorca crassidens* showing its large dorsal fin and tail. (3) Skeleton of toothed whale which shows that the dorsal fin and the tail do not contain bones; the same is true of baleen whales. (4) The teleost fish *Perca fluviatilis* (perch). (5) The skeleton of the same species showing the bones of the dorsal fins of the tail and of the other fins.

Table 1. Period Flight

Phylum	*Class*	*Group*	*Emergence of Flight Millions of Years Ago*	*General Distribution of Wings*	*Number of Wings Per Animal*	*Limbs Involved in Wing Formation*	*Participation of Tail in Flight*	*Derivation of Flight*
Arthropoda. Inverte-brates	Insects	Flying insects	310	Many insects without wings	2 or 4	No limbs involved. Legs additional	Tail involved in some insects	No direct relationship to other invertebrates. Insects are the only invertebrates that fly.
Chordata. Verte-brates	Reptiles	Pterosaurs	230	Most reptiles without wings	2	Arms	Long tail present allowing stability	No direct relationship to invertebrates that fly. Derived from nonflying reptiles
	Bony fish	Flying fish	After 190	Most fish without wings	2 or 4	Pectoral and pelvic fins corresponding to arms and legs	Tail is additional center of propulsion	Flying fish do not derive directly from insects or flying reptiles
	Birds	Flying birds	150	All birds have wings. A few are flightless	2	Arms	Tail used in gliding and turning	No direct relationship to flying reptiles. Derived from non-flying reptiles
	Mammals	Bats	40	All bats have wings. All bats fly	2 Fusion of 4	Arms and legs	Tail is part of wing in many species	No obvious relationship to other mammals. Derived from non-flying mammals

A WING AND A FIN CAN BE MADE WITH OR WITHOUT BONES

Surprisingly, the very large tail of a giant whale has no skeleton that supports it. Its form and function are solely due to muscles and other types of tissues but not bones. The same is true of its dorsal fin that is also boneless (Macdonald, 1984). However, they perform the same function, have the same structural arrangement and the same form as the tail and the fin of a fish which is made of bones (Figure 7).

The tail of a whale assumes a horizontal position but this is also the case in flatfishes such as the sole *Solea vulgaris*. A comparable situation occurs in the insect wing: it has horny tubes but no bones, yet it performs the same function, has the same structural arrangement and the same form as the wing of a bird consisting of a bony skeleton.

THE WING OF AN INSECT AND THAT OF A BIRD TURN OUT TO BE BUILT BY THE SAME GENES

The wing of an insect has been considered by zoologists to be analogous, to that of a bird, but not homologous. The reason being that the two structures were so different that they seemed to have no relationship. Among other things one had no bones but the other possessed them. The recent work on embryonic development in *Drosophila* and vertebrates (including birds and mammals) has disclosed that the genes, called homeotic genes, not only control the formation of similar structures but have also maintained the same order in the chromosomes of the distantly related species. It has been found, by molecular analysis, that the genes responsible for wing formation in the chicken are homologous to the genes that lead to the production of the wings in *Drosophila* (Affolter et al., 1990; De Robertis et al., 1990; Holland, 1992; Lawrence, 1992; McGinnis and Kuziora, 1994).

CHARACTERISTICS OF FLIGHT PERIODICITY

Flight periodicity is characterized by the following features: (1) Flight arises in every major group from nowhere. It is not exhibited by any immediate predecessors. (2) It emerges with few intermediate forms or suddenly. (3) It appears in a well defined and restricted group. (4) The types of structures and functions that are an obligatory component of the flight process are essentially the same in the groups that exhibit it. (5) It occurs at relatively regular intervals in the evolutionary scale (see Table 1 and Figure 1). (6) It exhibits a degree of complexity that is not necessarily correlated with the evolutionary location of the organism. (7) The number of wings per animal is also independent of the phylogenetic position. And (8) its emergence is apparently not related to a direct relationship with the general environment.

Chapter 4

PERIOD VISION

LIGHT SENSITIVITY IS AN INTEGRAL PART OF THE ORIGINAL CELL CONSTRUCTION

Protozoa, such as infusoria, which do not possess primordial eyes are highly sensitive to light. Experiments have shown that they move in the direction of a light beam or away from it, depending on the circumstances. This is also the case in amoebae. Light sensitivity is an intrinsic property of protoplasm (Kudo, 1971).

PLANT LEAVES ARE MOSAICS OF MICRO-LENSES

With the exception of their roots, plants can be aptly described as living carpets of light-sensitive cells. Most parts of the plant function as light antennae; the leaves in particular are a mosaic of light-receiving cells.

The phenomenon is not without special features. For example, the cells of the leaf show several differentiations that do not appreciably differ from the light-sensitive cells of protozoa or invertebrates (Figures 1 and 2).

It has been known for a long time that in *Fittonia*, and other species, the cells of the leaf epidermis are convex with a biconvex lens at their tip (Niklitschek, 1943). The cytoplasm of these cells is transparent and lacks pigment. This allows the transmission of light to the lower palisade tissue, where the chloroplasts, with their chlorophyll pigment, receive the light and transform it into chemical energy. *Fittonia* plants have broad and thin leaves; these form a ground cover in tropical forests where light is rather dim.

Vogelmann et al. (1989) and Martin et al. (1989) have discovered that, in *Medicago sativa* (alfalfa) which occurs in the fields of Europe, the same efficient light traps are present in the leaves. The epidermal cells in *Medicago* act as lenses that concentrate light, which is focused onto the chloroplasts of the palisade tissue. These cells are plano-convex, as shown by electron microscopy, and they concen-

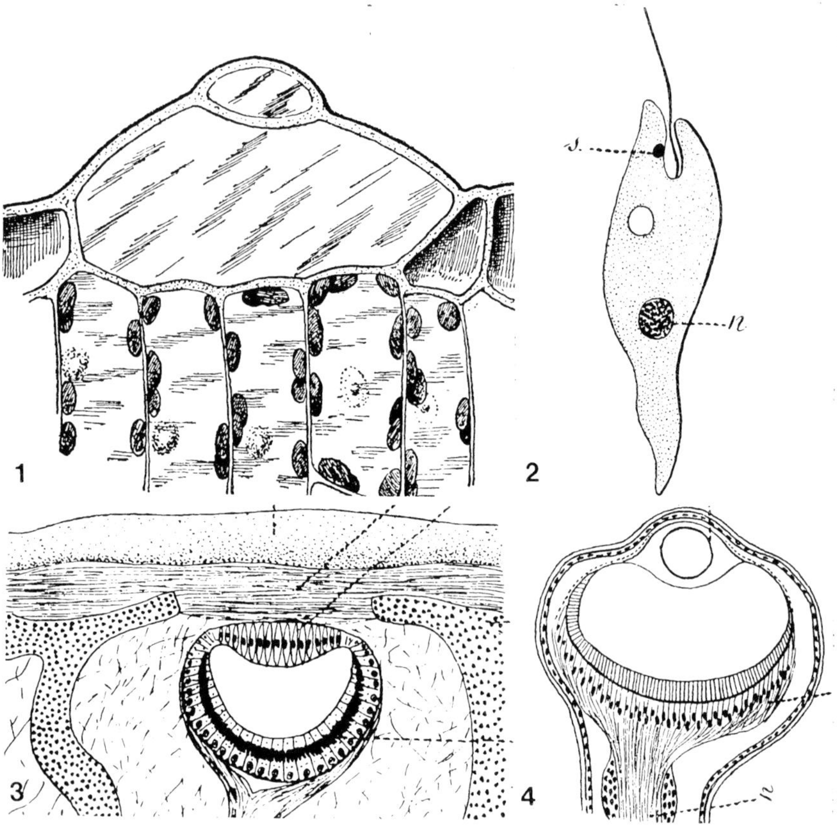

Figure 1. Periodicity of vision. (1) The "eye" of the plant *Fittonia* viz. the epidermal cells of the leaf, function partly as a lens by concentrating the light reaching the chloroplasts situated below (compare with Figure 2). (2) The protozoa *Euglena viridis* showing the eye spot. (3) Eye situated at the top of the head of a lizard (Sauria) consisting of transparent skin (like a cornea) and crystalline surrounded by cells sensitive to light. (4) Eye of an annelid worm (Alciope) with cornea, crystalline and retina.

trate an incident light beam up to 20-fold. The whole leaf epidermis is "a mosaic of micro-lenses that focuses light." As these authors point out, the difference in refractive indices of air (n = 1.000), cell wall (n = 1.425) and cell sap (n = 1.360) converts each epidermal cell into a lens. Actually, the photographs of the surface of the leaf epidermis, with its multitude of convex lenses, are not very different from photographs of the surface of the compound eyes of insects.

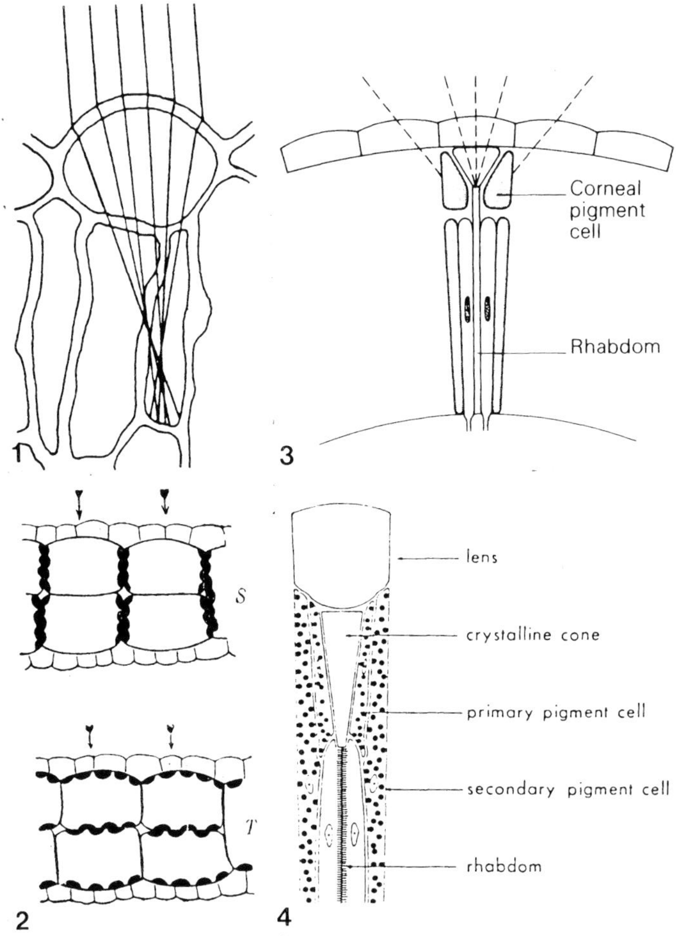

Figure 2. Image formation and movement of pigment in plants and insects.—(1) Image formation in the epidermal cells of the leaf of *Medicago sativa.* The lines represent rays of light converging on lower cells with chloroplasts in which chlorophyll is located; the epidermis of this plant functions as a mosaic of micro-lenses. (2) The chloroplasts (which contain the pigment sensitive to light) change position in the cells of the plant *Lemna trisulca,* depending on whether strong or weak light falls on them (arrows). (3) Image formation in the compound eye of insects (apposition type). The compound eye is also a mosaic of microlenses. The dashed lines represent rays of light converging on pigment cells. (4) One of the elements of the compound eye of a butterfly in which the position and movement of the pigment granules depends on the amount of light.

COMPARISON BETWEEN THE COMPOUND EYES OF INSECTS AND THE LIGHT-SENSITIVE CELLS OF LEAVES

The comparison seems at first sight somewhat far-fetched because we regard leaves as having nothing to do with eyes. But recent studies have shown that there are basic similarities between the light-sensitive cells of plants and those of animals. This includes the following:

1. The leaf consists of several thousand light-sensitive cell units maintained together. The compound insect eye also consists of several thousand light-sensitive cell units which are tightly coupled.
2. In *Medicago* there is both a lens and receptor component. In insects the chitin is transparent and each unit consists of a lens (that is also convex) and a receptor where the light absorbing pigment (retinene) is located.
3. A phenomenon established long ago is that chloroplasts, which contain the light-sensitive pigments, are able to migrate in cells. This movement occurs in response to changes in the intensity of light that reaches the leaf. The chloroplasts of moss cells, and of other plant species, occupy a central position in diffuse light but move to the cell wall when strong light is applied (Figure 2) (Denffer et al., 1971). The cells of the compound eyes of insects (flies) contain pigment granules which are light sensitive. These pigments also migrate and do so when light intensity changes. They are in the middle of the cell or move towards one of its ends, thus attenuating the amount of light and changing its spectral composition (Stavenga and Kuiper, 1977; Stavenga et al., 1977; Stavenga, 1989).

FEATURES OF PERIODICITY IN VISION

Closely related species should have similar types of eyes. This expectation is based on the assumption that eye formation and complexity are mainly decided by two factors: one is the phylogenetic position, the other the environment.

A detailed analysis of vision in invertebrates reveals that the situation is quite different. Within species of the same group there are highly complex eyes, rudimentary eyes and even an absence of eyes. Moreover, this phenomenon is repeated in the most representative invertebrate phyla from the coelenterates to the echinoderms. The two main observations that provide evidence of periodicity of vision are: the repetition of this complexity pattern, and its recurrence at relatively regular intervals.

THE TYPE OF EYES PRESENT FROM THE PROTOZOA TO THE EARLY CHORDATES

The base line recorded in the graph given in Figures 3 and 4 is not at zero because sensitivity to light is a phenomenon which is inherent in the cell. This means that in addition to their eyes animals have light sensitive cells distributed over the body. The scale 1 to 5 is a crude one, but it indicates the increase in complexity from the simplest eye spots, such as those of protozoa and the first organized eyes, to the most differentiated organs of vision that are composed of several tissues, such as those of octopods (Molluscs). Though the systematic classification of the invertebrates is not the same among authors, what is followed here is that adopted by Barnes (1980). Some minor groups are not represented because of lack of detailed information on their vision. However, the Ascidiacea, that belong to the Chordata, and are closest to the vertebrates, are included in the graph, the object being to check on whether the complexity of visual organs correlates with evolutionary complexity.

Dinoflagellates are single celled organisms. Several species such as *Protopsis* have one of the simplest forms of eye. This eye spot consists of a layer of light-sensitive bodies containing carotenoid pigments covered by a clear zone that functions as a primitive lens. Other species, such as *Erythropsidinium*, have already a relatively complex eye. The ocellus consists of: (1) a lens, (2) a fluid filled chamber, and (3) a cup covered by light sensitive pigment. It is most remarkable that although one is dealing with a single cell organism the following can happen: (1) the lens can change its shape, (2) the pigment cup can move freely in relation to the lens, and (3) the whole ocellus can protrude from the cell, thereby allowing it to point in different directions. It is thus not surprising that the dinoflagellate uses its eye to detect the presence of prey just as later invertebrates and vertebrates are able to do. This type of eye occurs in many other protista such as *Euglena*. However, such eye spots are completely absent in other protozoa e.g., *Amoeba* and *Paramecium* (Friday and Ingram, 1985).

The Cnidaria (or Coelenterata) and the Ctenophora are closely related phyla. The classes and orders consist of intimately related animals, yet some have most complex eyes, whereas others have no eyes at all. This makes it seem a paradigm of the periodic emergence of eye structures. On the one hand, not even a trace of eyes is found in both the sea anemones (Cnidaria) and comb jellies (Ctenophora), and on the other, well-developed ocelli occur in hydromedusae (Cnidaria). They consist of patches of pigment and photoreceptor cells which form a pit located on the outside of the tentacles. Some medusae are attracted by light, while others avoid it. Other Cnidaria have developed highly complex eyes. All cubomedusae have visual organs containing a cornea, a lens and a retina. This consists of three layers composed of sensory, pigmented and nuclear cells in contact with nerve fibers. The eyes can orient to points of light.

Within the flatworms the Turbellaria and the Trematoda have primitive eyes but the Cestoda have no eyes or very rudimentary ones.

The Rotifers are little aquatic animals with their heads covered with a crown of cilia. They have an eye-spot which contains a lens and the photoreceptors consist of membranes disposed in layers.

In the Acanthocephala and Gnathostomulida the eyes are again absent.

The molluscs show one of the largest variation in eye complexity. This is significant because it occurs within the same phylum and as such cannot be easily correlated with the degree of evolution attained by the phylum. (1) In species, such as *Patella*, the eye is a simple pit containing photoreceptor and pigment cells. (2) Chitons have ocelli with a lens and a vitreous body. (3) *Murex* represents a more complex type in which the pit is closed, and the eye contains a cornea and lens. (4) Prosobranchs have large telescopic eyes which are more elaborate than those of many fishes. (5) The cephalopods have both a highly complex nervous system, with giant neurons, and the most developed eyes. The eye of the *Octopus* consists of a cornea, a lens, an iris, a ciliary muscle, a retina, and optic nerves. The cephalopod eye is thought to be more efficient than the vertebrate one, since it is not developmentally inverted. Experiments involving training of *Octopus* have shown that they are capable of image-formation and shape-discrimination. (6) Other species within the molluscs have no eyes at all; they belong to the Monoplacophora, the Aplacophora, and the Scaphopoda.

The Annelida, like the Coelenterata and the Mollusca, have species exemplifying these three situations: (1) complex eyes which have lenses capable of accommodation and two types of retinas (Polychaeta, Alciopidae), (2) simple eyes (Hirudinea) and (3) lack of eyes, except a few aquatic forms that have a pigment-cup (most Oligochaeta).

Again, most Araneae have eyes but some lack them.

In the Crustacea there are also: (1) highly developed compound eyes consisting of as many as 14,000 ommatidia (Decapoda, Malacostraca; crabs, lobsters), (2)

Figure 3. Period vision. (1) Part 1, *Protozoa to Gnathostomulida*. Light sensitivity is inherent in the cell. This is why in the graphs the base line, representing the degree of complexity, is not set at zero but rather at the value of 0.2. Simple eyes already appear in the protozoa. Before reaching the vertebrates, whose eyes are complex, a periodic eruption of eye complexity is found among the invertebrates. Moreover, the complex eyes appear within classes or phyla that also consist of animals without eyes. This is the case in the Cnidaria, in which the cubomedusans have complex vision but the sea anemones are eyeless. The same is the case in the Molluscs; for example, the octopods have eyes as complex as those of vertebrates, yet there are many species within the phylum which have rudimentary eyes or no eyes at all. This situation repeats itself in the Annelida, the Crustacea and the Insects. Most significant is that the Hemichordata and the early Chordata, such as the Ascidiaceae, which are evolutionarily more advanced than the invertebrates, have no eyes or very rudimentary ones. This implies that periodicity is not directly related to the degree of evolutionary complexity. (2) Part 2, *Patella to Araneae*.

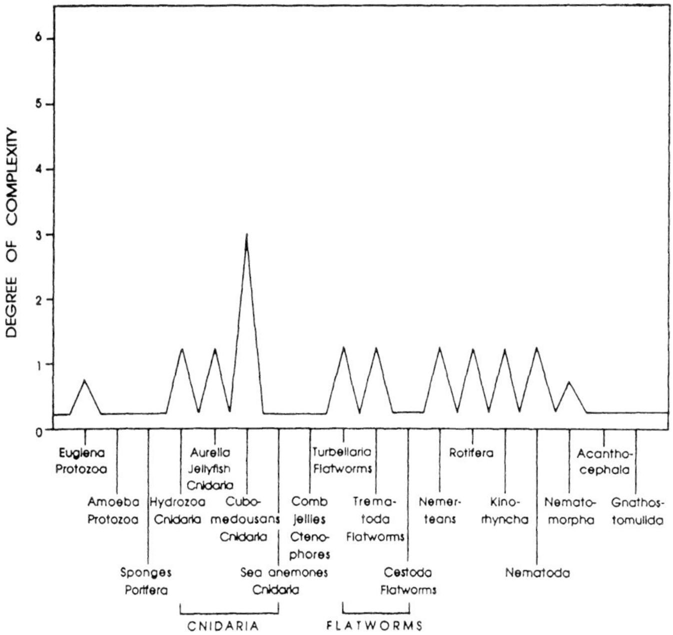
DEGREE OF COMPLEXITY
0
1
2
3
4
5
6
Euglena Protozoa
Amoeba Protozoa
Sponges Porifera
Hydrozoa Cnidaria
Aurelia Jellyfish Cnidaria
Cubo-medousans Cnidaria
Sea anemones Cnidaria
Comb jellies Cteno-phores
Turbellaria Flatworms
Trema-toda Flatworms
Cestoda Flatworms
Nemer-teans
Rotifera
Kino-rhyncha
Nematoda
Nemato-morpha
Acantho-cephala
Gnathos-tomulida
CNIDARIA
FLATWORMS

1

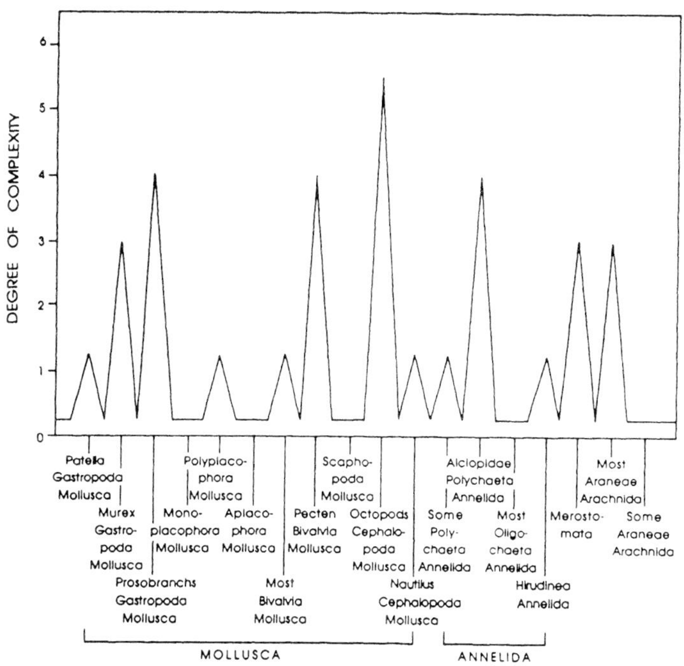
DEGREE OF COMPLEXITY
0
1
2
3
4
5
6
Patella Gastropoda Mollusca
Murex Gastro-poda Mollusca
Prosobranchs Gastropoda Mollusca
Mono-placophora Mollusca
Polyplaco-phora Mollusca
Aplaco-phora Mollusca
Most Bivalvia Mollusca
Pecten Bivalvia Mollusca
Scapho-poda Mollusca
Octopods Cephalo-poda Mollusca
Nautilus Cephalopoda Mollusca
Some Poly-chaeta Annelida
Alciopidae Polychaeta Annelida
Most Oligo-chaeta Annelida
Hirudinea Annelida
Merosto-mata
Most Araneae Arachnida
Some Araneae Arachnida
MOLLUSCA
ANNELIDA

2

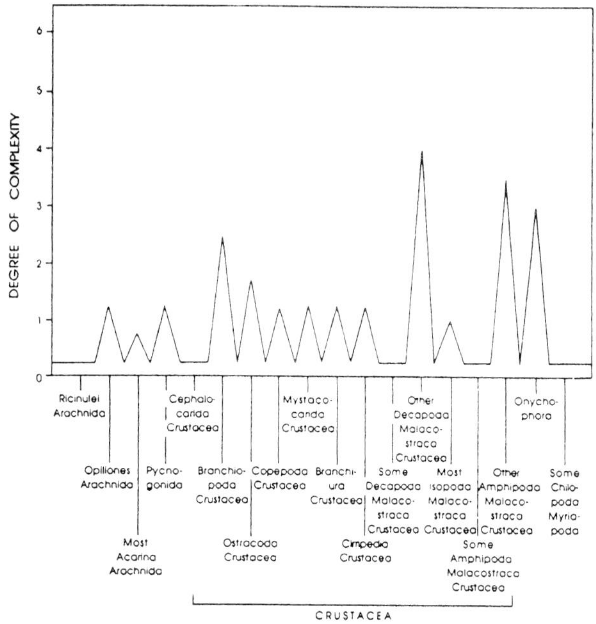

1

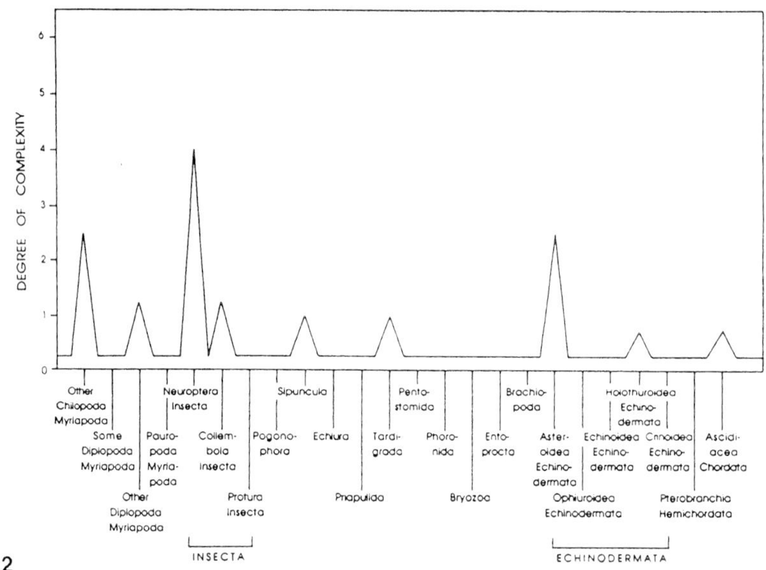

2

Figure 4. Period vision (1) Part 3, *Ricinulei to Chilopoda* (2) Part 4, *Other Chilopoda to Ascidiaceae.* (For details about the preparation of the graphs see Chapter 3.)

simple eyes (Branchiura and Cirripedia), and (3) absence of eyes (Cephalocarida, some Decapoda, and some Amphipoda).

In the Myriapoda there are simple eyes but several groups of centipedes lack eyes.

The situation repeats itself in the insects: (1) Two orders lack eyes, (2) others have simple lateral and dorsal eyes, and (3) most of them have the well-known compound eyes in which the number of ommatidia may reach 25,000.

A close look at Figure 4 reveals that between the insects and the echinoderms there are not less than eight distinct phyla that are characterized by the absence of eyes: Pogonophora, Echiura, Priapulida, Pentastomida, Phoronida, Bryozoa, Entoprocta, and Brachiopoda.

These phyla are followed by the Echinodermata in which the general pattern recurs: (1) elaborate eyes, with the cuticule acting as a lens and pigment cells forming an optic cup in the Asteroidea; (2) simple eyes in the Holothuroidea, and (3) absence of eyes in Ophiuroidea, Echinoidea, and Crinoidea.

One would suppose that as one leaves the invertebrates, and enters the advanced chordates (which include the vertebrates), one would find highly complex vision, as it occurs in the vertebrates. However, this is not the case. The Hemichordata (Pterobranchia) and the Ascidiacea, which are true Chordata, have been included in Figure 4, to call attention to the fact that the first have no eyes and the second possess only dispersed photoreceptor cells, i.e., both classes consist of blind species.

COMPARISON BETWEEN THE EYES OF HUMANS AND CEPHALOPODS

The eyes of cephalopods are, as many zoologists have noted, "strikingly similar" to those of mammals, humans included. The form, the organization and the number of parts is about the same (Figure 5). There are certain differences, such as a fixed focal lens that results in the image formed being of another type.

The similarity becomes more striking when one realizes that one is dealing with such unrelated groups as invertebrates and vertebrates. Moreover, the two types of eyes have different embryonic origins, different tissues being involved in their formation. Finally, the cells of the retina have not the same orientation. In squid eyes, these cells point toward the light, whereas in vertebrates they point away from light; this is what Romer and Parsons (1978) call "the wrong way." Millions of years of evolution have not changed their "wrong way" (Tansley, 1965).

VISION WITHIN INSECTS DISPLAYS PERIODICITY

The insects are the largest group of living organisms that evolved on Earth comprising over 750,000 species. They are also very old going back to the Silurian Period (450 million years ago). Their eyes have been studied extensively (Nilsson,

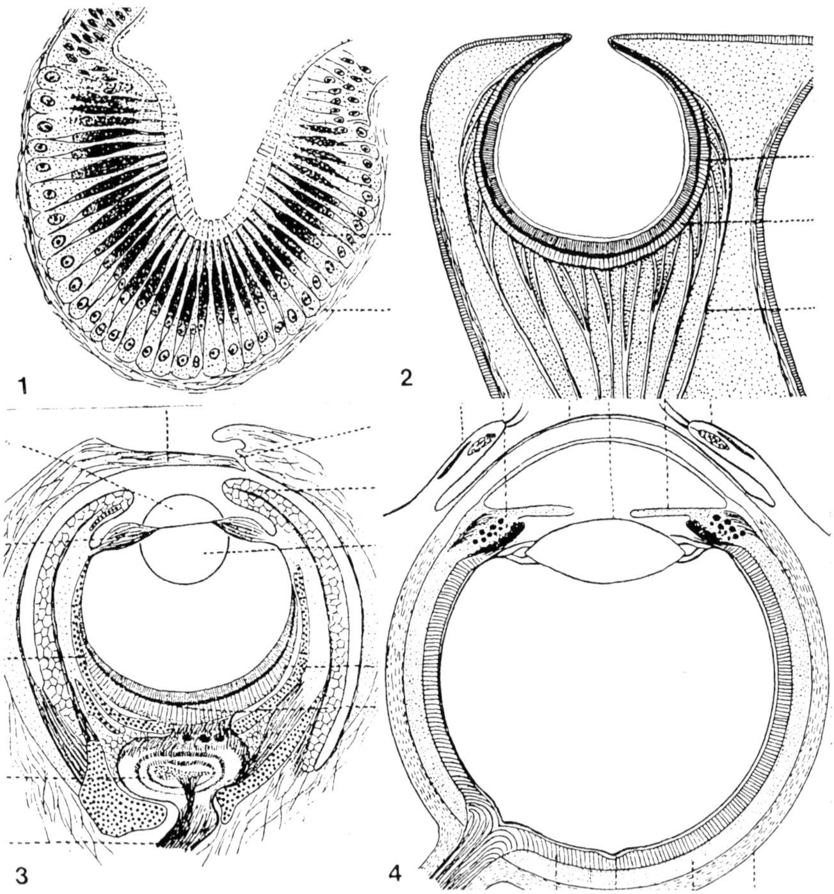

Figure 5. Periodicity of vision. (1) Eye of the mollusc *Patella* consisting of a simple depression in contact with the exterior and retina cells. (2) Eye of the mollusc *Nautilus* with a deeper depression, pigmented zone, retina, and nerves. (3) Eye of the mollusc *Sepia* (cephalopods) consisting of 11 differentiated parts: retina, pigment cells, optic ganglion, optic nerve, cartilaginous capsule, epithelial body, two juxtaposed crystallines, iris, cornea and a small opening to the surface. (4) Human eye, formed by cornea, iris, crystalline, retina, and optic nerve.

1989, 1990). For these reasons the analysis of their vision organs should give some clues to the evolutionary processes that have been at work for such a long period of time.

When the types of eyes present in all the orders are compared, the periodicity of eye formation becomes evident. Opinions about the relationships between the

Table 1. Dorsal Eyes In Insects. Presence and absence of these eyes form recurrent groups in the 26 orders known. Numbers refer to the phylogenetic position of the orders

	Dorsal Eyes	
	Group I Absent	*Group II Present (in some or most species)*
PERIOD 1	1, 2, 3	4, 5, 6, 7, 8
PERIOD 2	9, 10	11, 12, 13
PERIOD 3	14, 15	16, 17, 18, 19, 20, 21, 22
PERIOD 4	23, 24*, 25	26

Note:* Generally absent.

Key: The orders have the following names: 1 Protura, 2 Entotrophi, 3 Collembola, 4 Thysanura, 5 Ephemeroptera, 6 Odonata, 7 Plecoptera, 8 Orthoptera, 9 Dermaptera, 10 Embioptera, 11 Isoptera, 12 Zoraptera, 13 Psocoptera, 14 Mallophaga, 15 Anoplura, 16 Thysanoptera, 17 Hemiptera, 18 Neuroptera, 19 Mecoptera, 20 Trichoptera, 21 Lepidoptera, 22 Diptera, 23 Siphonaptera, 24 Coleoptera, 25 Strepsiptera, 26 Hymenoptera. The sequence of the orders is according to Romoser (1973).

orders are not the same. That of Romoser (1973) is followed in Table 1. The 26 orders form groups just like the chemical elements. This is especially evident in the case of the dorsal ocelli. What is remarkable is that the phenomenon has the following characteristics: (1) The orders with similar properties (absence or presence of dorsal ocelli) are not dispersed; rather, they form well defined groups consisting of related orders. (2) These groups occur at relatively regular intervals; four groups with the property are intercalated by four without it, thus establishing in this way four natural periods. (3) The lateral ocelli also display a recurrent pattern; they are present in orders No. 3, 18, 21, 23, 24, and 26 but are absent in the remaining 20 orders. They do not build large clusters of orders, presumably because they are not so widespread. The compound eyes exhibit a distribution that seems to be the reverse of that of the lateral ocelli. They are absent in four orders (1, 2, 3, and 23) and present in the remaining 22.

THE INDEPENDENT EVOLUTION OF THE EYE

Virtually all zoologists who have written about the organization and evolution of visual organs, have emphasized the independent origin of this structure in different animal groups. Among them is Hertwig (1929b) who pointed out that complex eyes appear repeatedly and independently. More recently, Romer and Parsons (1978) stated that well-developed eyes have many common features, though "surely independently evolved."

Barnes (1980) considers two different lines of origin of receptors types in the Animal Kingdom. These are: (1) A *ciliary line* in which the photoreceptor is associated with a cilium (as in protozoa) and the microvilli are evaginations of the ciliary membrane. This line includes the coelenterates, the Echinoderms and the vertebrates. And (2) in the *rhabdomeric line* the photoreceptor does not appear to be derived from a cilium, this is found in the flatworms, annelids, and molluscs. This is taken to mean that the origin of the highly complex eyes of cephalopods is different from the complex eyes of vertebrates and humans.

A third eye medially situated on the forehead and directed upward has been recorded in fossil fishes such as the placoderms, ancient amphibians and paleozoic reptiles. It disappeared by Triassic times but is still found in lampreys and some lizards. The skin lying over this third eye is transparent to light and the organ consists of a miniature cornea, lens and retina (Rabaud, 1932). As for the origin of its parts, it is not the same as that of the median eyes. The retina of the median eyes not only corresponds to the lens of the third eye, but also has an orientation to the light that is inverse in both types of visual organs.

VISION AND ENVIRONMENT

There is both a relationship and an absence of correlation between vision and the environment. This means that several mechanisms are involved in this process. (1) *Relationship.* Invertebrates which are parasites, such as the Acanthocephala, have no eyes. Among the Arachnida the species of Ricinulei, that live in caves, lack eyes. There are also blind decapodes (Crustaceans) among those that inhabit the deep-sea waters. (2) *Absence of correlation.* No eyes are present in animals of the phylum Echiura, yet the majority live in shallow waters. Most evident is the situation in the phylum Bryozoa that have no eyes, but are found mainly in coastal waters. This is also the case among the echinoderms such as the Echinoidea, which live near the water surface but have no eyes. Even in the chordates, such as Ascidiacea, although they inhabit shallow waters, they have no eyes. In general, the insects living in caves have reduced eyes but there are insects living under these conditions that possess normally developed eyes (Aron and Grassé, 1939).

Hence, there is no obligatory relationship between vision and environment.

THE INSECT EYE AND THE HUMAN EYE ARE PRODUCED BY THE SAME TYPE OF GENES

No zoologist would take seriously the postulate that an insect eye would be the product of the same type of genes that are found in humans and which are responsible for eye formation in this species. The two types of eyes are anatomically totally different. One is built by the association of thousands of ommatidia (which may be as many as 25,000), the other results from the association of several tissues into a single spherical unit. The difference could not be more extreme in structural and anatomical terms. One would simply deny any genetic homology between the two organs which allow vision in both groups of animals. Yet the results that have been obtained recently by Walter Gehring and his collaborators at Basel University, change our notion on this subject.

The *Drosophila* gene *eyeless* (*ey*) is a mutation that leads to the reduction or complete absence of compound eyes in the fly. An experiment was carried out in which multiple copies of this gene were inserted into fly embryos, after the *ey* gene was coupled, by genetic engineering, with a segment of DNA that can activate any gene, obliging it to carry out its normal transcription. The result was most unexpected. Compound eyes appeared in the wings, all six legs and in the antennae of the flies. The fine structure of these eyes was analyzed by scanning electron microscopy and they turned out to be formed like the eyes that normally appear at the top of the head.

Of importance from the point of view of evolution is that when the *ey* gene of *Drosophila* was replaced by the mouse gene *Small-eye* (*Pax-6*) it was found that the mouse gene also produced flies with multiple eyes. These results implied that the mammalian and the insect gene were closely related and were derived from a common ancestor.

A subsequent study disclosed that the human gene *Aniridia* (whose mutations cause defects in the iris, retina and the lens of human beings) is homologous to the gene *ey* of *Drosophila* and to the *Small-eye* of the mouse. The molecular analysis revealed that the proteins encoded by these genes share 94% sequence identity in one domain and 90% in the homeodomain, and that they contain additional similarities in the flanking sequences. Furthermore, the gene split sites have been conserved from the invertebrates to humans. The genes are also similar in their DNA sequences and in the expression pattern that occurs during embryonic development. The cloning of the homologous genes from ascidians, cephalopods and nemerteans (ribbon worms) has led these authors to suggest that this gene may be present in all metazoa. The *ey* gene is considered to be a master control gene, determining eye morphogenesis, that is shared by invertebrates and vertebrates including humans (Quiring et al., 1994; Halder et al., 1995).

Hence, the inclusion of all types of eyes, as a source of evidence for the establishment of the periodicity, is now supported by the genetic and molecular results.

GENERAL FEATURES OF VISION PERIODICITY

Periodicity in vision is expressed by the following features: 1. The same combination of basic structures and functions is found in the most diverse species in the form of a functional eye. 2. Complex eyes, simple eyes and absence of eyes occur in closely related species. This is a pattern that recurs in major invertebrate groups. 3. There is a recurrence of phyla whose species are blind. 4. Dorsal eyes in insects appear in clusters of orders. 5. There is a recurrence of highly complex eyes in unrelated groups of invertebrates, and in still more distant groups, such as higher mammals. 6. The eye has had an independent origin in several animal phyla. 7. There is no obligatory relationship between vision and environment. 8. The recurrence of some of these phenomena tends to be relatively periodic. 9. The eyes of the simplest invertebrates such as ribbon worms (nemerteans), those of insects, cephalopods, mouse and humans, have been shown, by genetic and molecular analyses, to be determined by the same genes. The homology between these genes has been demonstrated both at the DNA and protein levels.

Chapter 5

PERIOD PLACENTA

DEFINITION OF PLACENTA

The placenta of eutherian mammals is defined as "a double vascular spongy structure formed by interlocking of fetal and maternal tissue in the uterus, and in which maternal and fetal blood vessels are in close proximity, allowing nutritive, respiratory and other exchange" (Holmes, 1979), and Holmes adds that the placenta in plants is the ovule bearing part of the carpel, which is also a sexual organ that belongs to the female part. Denffer et al. (1971) remark that "the ovules of flowering plants are provided with nutrients through vascular bundles from their point of origin on the placenta."

This means that the application of the same term to animals and plants is not a loose connotation but results from the fact that the same type of structure and function is involved.

PLACENTA IN FLOWERING PLANTS

The placenta of vertebrates is well known. What is less well known is that similar organs occur both in invertebrates and plants that perform essentially the same functions.

The placenta of mammals is an organ richly supplied with blood vessels that transports oxygen and the entire food supply to the embryo.

Plants have sex organs but one thinks of them as structurally and functionally very different from those of humans. Such a view is not necessarily correct. In flowering plants the stigma and the ovary (with its egg cells) maintain a configuration which resembles the vagina, the uterus and the ovary of humans. But this does not end the similarity. Nagl (1973, 1976) has pointed out that the suspensor of several plants is most similar to the trophoblast of mammals and humans. The trophoblast is the tissue surrounding the embryo, from which the placenta is formed. Both the suspensor and the trophoblast have comparable anatomical locations in relation to their embryos. Both are involved in the transport of nutritional substances necessary for the embryo and both show DNA amplification and polytene chromosomes.

THE PLACENTA IN INVERTEBRATES

Scorpions, which are arachnids, usually give birth to young. The yolkless eggs produce embryos which become attached to the maternal intestine by a structure not different from an umbilical cord.

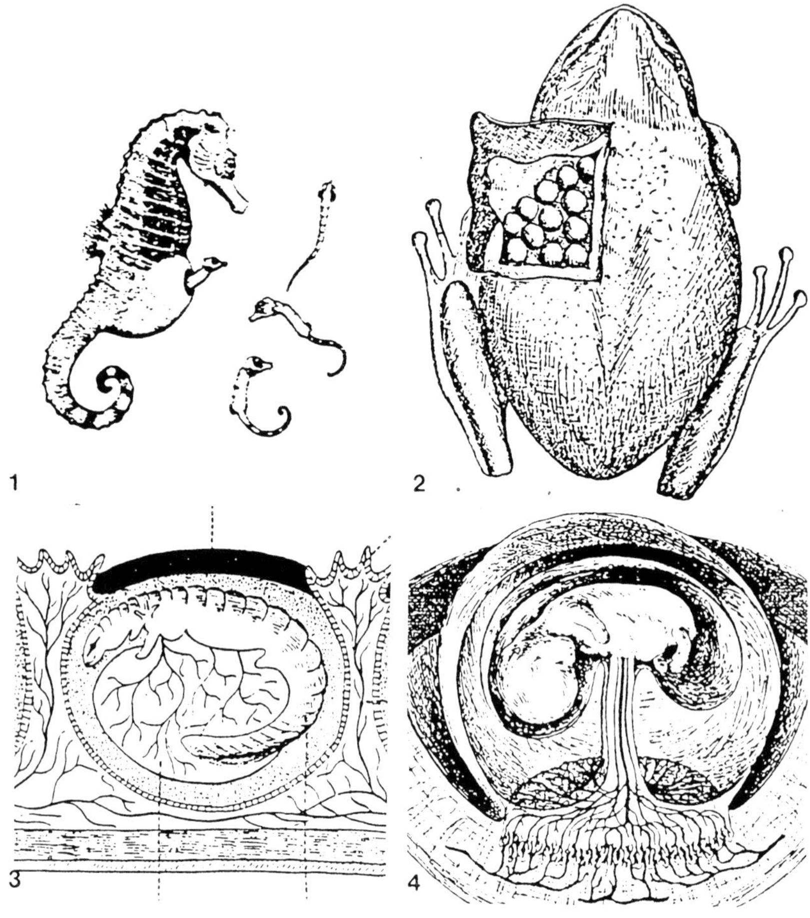

Figure 1. Periodicity of the placenta. (1) The male of the sea-horse *Hippocampus* sp. (a fish) with young that are emerging from its pouch. (2) The dorsal pouch of the frog *Nototrema marsupiatum* with part of the skin lifted to show the developing eggs. (3) Placenta in *Pipa americana* (a frog) showing the embryo with the adjacent tissues and blood vessels. (4) Placenta in the human female with embryo, the surrounding tissues and blood vessels.

Peripatus belongs to the onychophorans, which is a phylum closely related to the myriapods, and known as velvet worms. Such a simple animal has an advanced system of live birth. Each embryo is surrounded by an amniotic membrane and connected to the oviduct by a placenta and an umbilical cord. These two organs have the same functions as those present in vertebrates.

THE PLACENTA IS PRESENT IN FISHES

The male sea horse (a teleost) carries the eggs and young in special pouches and nourishes them as well. The placenta does not develop in a uterus or even female but rather in the pouch of a male. The female deposits her eggs into the male's pouch where they become embedded in a spongy structure irrigated by blood vessels. Oxygen, carbon dioxide and nutrients are exchanged across the placenta. This is the same functional solution appearing in a different sex and connected with a different organ (Figure 1).

In the case of cartilaginous fish e.g., sharks, they have a placenta similar to that of the mammals. The membranous sac, attached to the embryo, which contains the food necessary for its development (yolk sac) fuses with the tissues of the oviduct. The connection actually is so intimate that the blood vessels of the embryo and of the female parent come into contact (Burton, 1987).

THE PLACENTA IN AMPHIBIANS AND REPTILES

The black salamander of the Alps is viviparous. The tadpoles develop within the mother and not in the water, as with frogs. The gills of frog larvae are respiratory organs while the gills of salamander tadpoles function as a placenta.

The use of other body organs as a placenta becomes more apparent in the Surinam toad, *Pipa pipa*. The eggs are enclosed in pockets on the back of the female and each embryo develops into a tadpole. The tail of the tadpole is used as a placenta through which nutrients from the mother's body are absorbed (Figure 1).

In the European lizard (skink) and the Australian snake (copperhead), the membranes of the embryo and the tissues of the oviduct fuse so intimately that the nutrients are easily transferred between mother and embryo, and without difficulty in the opposite direction (Rabaud, 1934).

THE PLACENTA DOES NOT EXIST OR IS RUDIMENTARY IN MARSUPIALS

In marsupials, or pouched mammals, the young are born alive but they emerge from the mother at a very early embryonal stage. They are then kept in a pouch where the newborn is nourished by suckling the mother's milk. The egg and the early development of the embryo in marsupials resembles that of reptiles and birds. There is no placenta in this group or only a rudiment of this organ, as is the case in the bandicoots (Macdonald, 1984).

THE PERIODICITY OF THE PLACENTA

The placenta not only occurs in many different groups but shows features of periodicity (Figure 2).

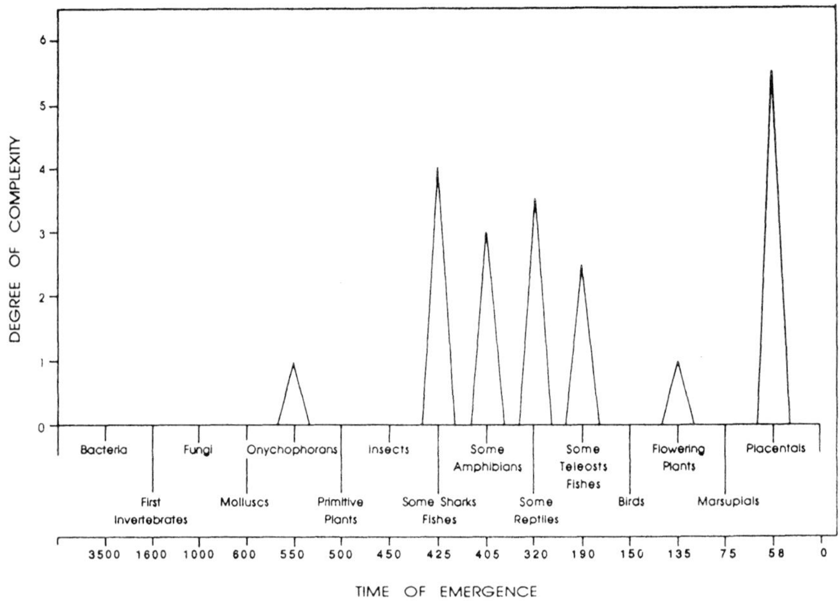

Figure 2. Period Placenta. The placenta is the tissue complex that provides the embryo with nutrients. It is already well developed in the Onychophorans but absent in many other invertebrate phyla, such as the molluscs. Its periodicity is evident in its sudden appearance in some fishes, such as the sharks, where it is highly developed. It recurs in some amphibians, some reptiles, some teleost fish, as well as in flowering plants. Birds, however, do not possess it. It is not without significance from the point of view of periodicity that marsupials, which are mammals, lack it (or only have a rudiment of it), whereas their closest relatives, the placentals, got their name from the presence of this organ in every species of the group. The time of emergence of a given group indicated in this graph is approximate because of the absence of agreement among authors. The time values are not drawn to scale to avoid making the graph too large. The figures refer to million of years ago.

Repetition in Well Defined Groups

The placenta appears in plants, invertebrates, fish, amphibians, reptiles, and eutherian mammals (also called placental mammals), but it emerges only in well defined groups of organisms that belong to a specific class, order, or family.

Absence of Direct Phylogenetic Relationship

Most orders of fishes do not have a placenta but it occurs in two unrelated groups: the sharks which are cartilaginous fishes and the teleosts which are bony fish. Moreover, its degree of complexity is independent of the phylogenetic position. Sharks have a placenta similar to that of mammals, which are the most advanced vertebrates.

Occurrence at Regular Intervals

The first invertebrates can be traced to fossils dated 1600 million years ago. Most invertebrates do not possess a placenta but it is found in the velvet worms which appeared 550 million years ago (Margulis and Schwartz, 1982). It emerges again in the sharks which were present as early as 425 million years ago, but the teleosts only developed 190 million years ago. Hence, the emergence of the placenta in fish may have been separated by over 150 million years (Carroll, 1987). According to Burton (1987), viviparity, which is accompanied by placental function, evolved in lizards and snakes more than 30 times. The primitive plants appeared 500 million years ago but the flowering plants, which have a placenta, emerged 135 million years ago. The marsupials appeared 75 million years ago and like the birds have no placenta (or only a rudiment). Suddenly, it makes its appearance in the eutherians 58 million years ago, only to become the most significant feature of this group.

Absence in Closely Related Species

The frogs and toads belong to the order Anura. Some families, such as the tree frogs (Hylidae) have a placenta, whereas their close relatives, the Ranidae, do not possess this organ. The same is true of fishes and reptiles in which close relatives do not have this organ.

Formation of the Same Organ from Different Tissues

In the seahorse the placenta is formed without any obvious anatomical difficulty using the skin instead of the uterus. In the larvae of amphibians the gills and tails become placentas. The same applies to plants in which this organ arises from tissues which are different from those of animals.

The Same Type of Function is Performed in All Cases

In plants the placenta already provides the ovules with nutrients and does it through vascular bundles. These connections in the invertebrates and vertebrates become the blood vessels. In some teleost fish, such as *Heterandria formosa*, absorption of nutritional materials is so highly developed as to be almost mammalian in type (Needham, 1950).

The Emergence of the Organ is not Directly Related to the General Environment

Most fish live in water, and their eggs are released into this medium, where they are fertilized and where they develop independently of their parents. The placental fishes occur in the same habitat. Hence, the emergence of the placenta is not dictated by the general environmental conditions. The same also holds for amphibians. *Pipa americana* has a well developed placenta, whereas *Xenopus laevis* eggs develop freely, but both species live in water. Moreover, the eggs of fish and amphibians possess large quantities of reserve materials and are not in need of a large supply of organic substances from the bodies of their parents (Rabaud, 1934). Although the environments that marsupials and placental mammals face are about the same, they differ sharply in the possession of this organ.

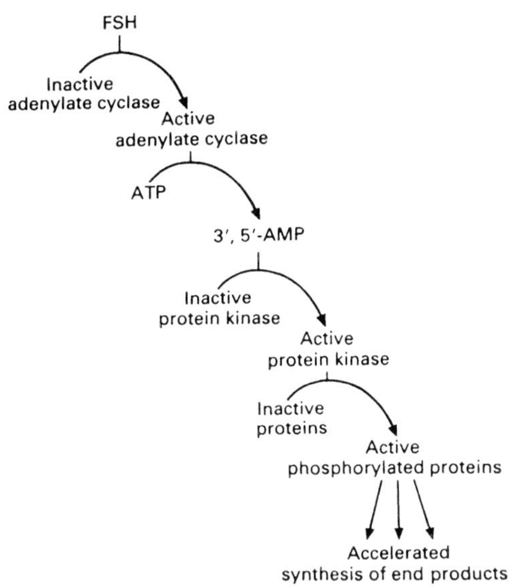

Figure 3. Cascade of chemical events involved in the mechanism of action of the Follicle Stimulating Hormone (FSH). (From Dorrington, 1979).

The Emergence is Decided by Cascades of Hormones

Recent studies in humans and other mammals have shown that the formation of the placenta is the result of the action of two cascades of hormonal signals (Figure 3). The first is produced by the follicle stimulating hormone which increases the intracellular concentration of cyclic AMP. This, in turn, leads to the production of phosphorylated proteins that affect transcription of DNA into RNA. The follicle stimulating hormone promotes the maturation of the egg follicles, which secrete a second hormone, estrogen. This initiates the second cascade. The estrogen prepares the walls of the uterus, where the placenta is formed, and leads to the surge of the luteinizing hormone. In a further step this last hormone provokes the secretion of progesterone that allows the full pregnancy of the fertilized egg. Prostaglandins are also involved in this process (Dorrington, 1979; Challis, 1979; Friday and Ingram, 1985).

This rigid and ordered sequence of chemical signals is the basis of the coherent framework of structures and functions that is the placenta. Once an initial threshold of a given hormone is reached a cascade ensues that cannot be stopped and which results in a novel and harmoniously constructed organ.

Chapter 6

PERIOD BIOLUMINESCENCE

LUMINESCENCE IN MINERALS

Barite ($BaSO_4$) is a heavy mineral that can retain the light from the sun and re-emit it in the dark. Other minerals, e.g., fluorite (CaF_2), have the property of luminescing, and a large number of inorganic and organic compounds also luminesce after being irradiated by ultraviolet or by visible light. The light lasts only a short time but in phosphorescent minerals it persists for many minutes (Figure 1).

Thermoluminescence arises when crystals are usually heated between 100 and 500 °C. This rise in temperature provides the energy necessary to release electronic energy initially absorbed from light or subatomic particles. The presence of extra atoms in crystals favors the phenomenon. In a pure crystal the energy of the particles is dissipated as heat, but the additional atoms create structural boundaries that trap the particles, the result being photon emission (Campbell, 1988; Iyer and Xie, 1993).

CHEMICAL PROCESSES INVOLVED IN BIOLUMINESCENCE

Light emission by living cells requires oxygen, either as O_2 or one of its metabolites. Other chemicals are also required. These include: (1) luciferin which is a substrate in the reaction. (2) luciferase which is the protein catalyst. (3) cofactors, such as a long chain aldehyde. (4) cations that are alkaline earths (Mg^{2+}, Ca^{2+}) or transition metals (Fe^{2+}, Cu^{2+}). (5) fluor, which is bound to a protein. Both cations and fluor are considered regular components of minerals. It is interesting that oxygen is at one end and fluor at the other of the energy transfer cascade that results in the emission of light (Campbell, 1988).

THE OCCURRENCE OF BIOLUMINESCENCE

Bioluminescence occurs mainly in bacteria, fungi, dinoflagellates, jelly fish, worms, shrimp, insects, some squid, and fish. There are thousands of living species which are luminescent, representing 16 phyla and almost 700 genera. They are found on land, sea and air.

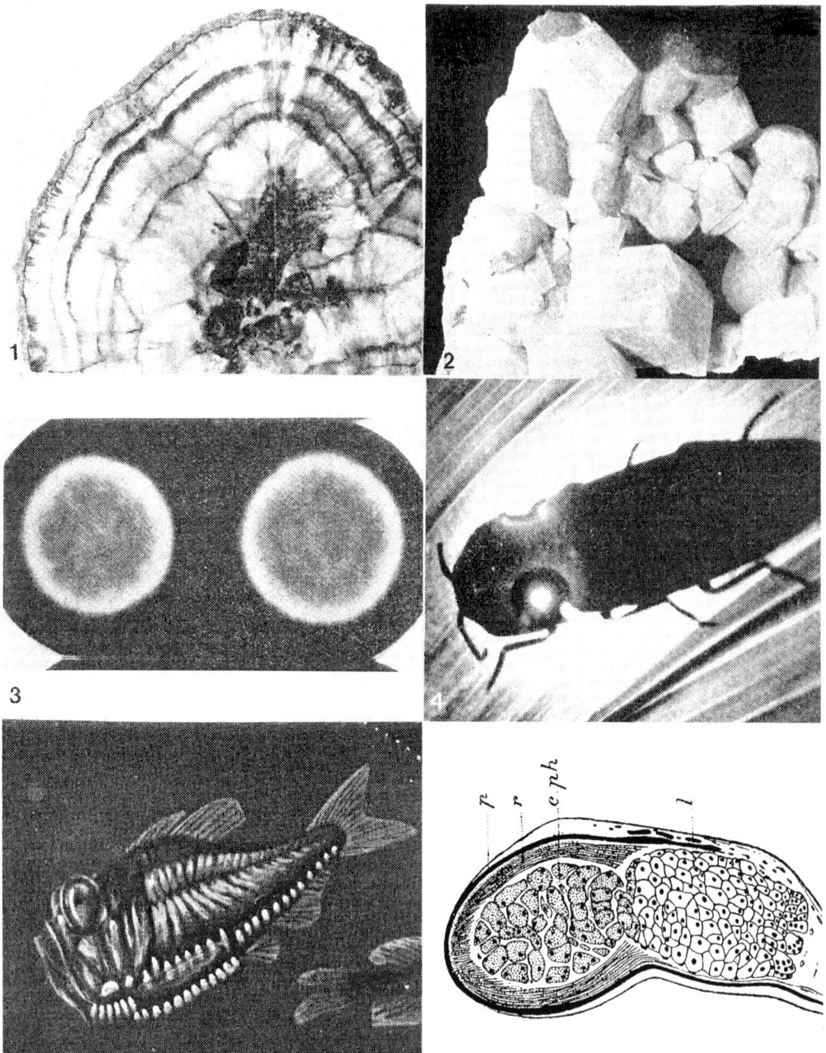

Figure 1. Periodicity of luminescence. (1) The mineral fluorite (CaF_2). The word fluorescence derives from fluorite which emits visible light when activated by invisible radiation such as x-rays. (2) Group of calcite crystals, another mineral that also fluoresces. (3) Two colonies of *Photobacterium* photographed in obscurity. (4) The tropical insect *Pyrophorus noctilucus* which produces light that is strong enough for reading. (5) The abyssal fish *Argyropelecus* with the body covered by luminous organs. (6) The luminous organ of the fish *Argyropelecus affinis* showing the cells that produce light, the lens and the reflecting structure.

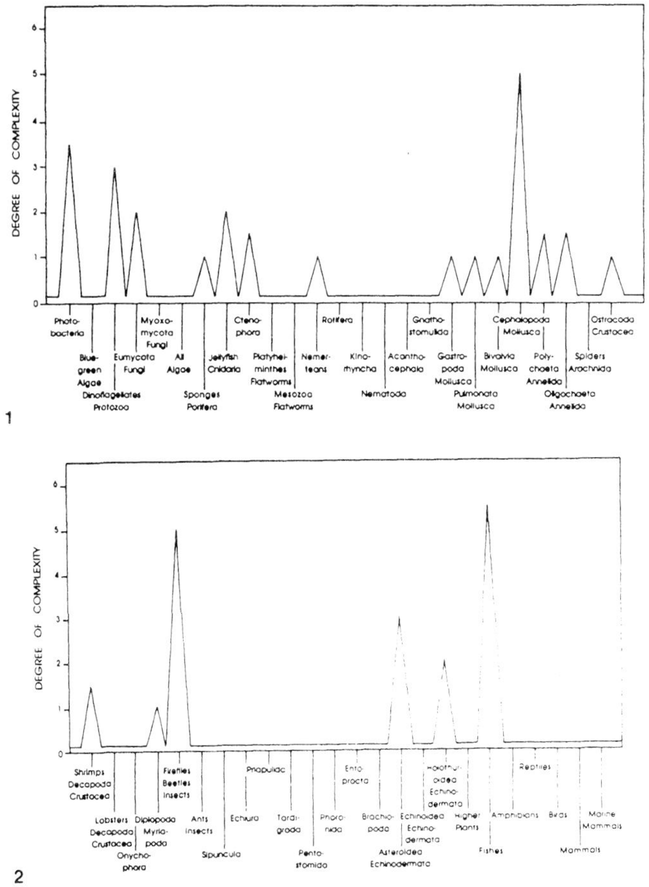

Figure 2. Period luminescence. (1) Part 1, *Photobacteria to Ostracoda.* Very weak luminescence can be detected in most cells. For this reason, as shown in the graph, the degree of complexity is represented by a base line whose value is 0.1. The ability to emit light emerges at the start of life. The photobacteria are strongly luminescent. Periodicity of the phenomenon is evident in several ways: The close relatives of the photobacteria, the cyanobacteria (blue-green algae) could also be luminescent but are not. As the eukaryotes evolve there are whole phyla that do not exhibit this phenomenon but it erupts with high intensity and at regular intervals in the Dinoflagellates (protozoa), the Cephalopods (molluscs), the fireflies (insects) and the fish. But, luminescence is not found in other vertebrates. Even marine mammals such as the whales that inhabit the same dark depths of the ocean as the fishes and the cephalopods have not developed luminescence. Lack of luminescence is also found in all eukaryotic algae living in sea waters. Hence, the emergence of this property is not necessarily correlated with a given environment. (2) Part 2, *Shrimps* to *Marine Mammals.*

About 60 to 80% of all the fish species in deep oceans are luminous. However, whales which live at the same depth do not emit light. Sperm whales have been recorded at below 2,000 meters (Line and Reiger, 1980; Savage and Long, 1986).

Organisms that produce light intracellularly achieve this either by generating their own light in the form of a glandular secretion or as a result of association with symbiotic luminous bacteria. Some organisms combine the two processes. The number of light organs can vary from one to several hundred. These organs occur in almost every part of the body: eyes, jaws, mouth, dorsal parts, tails, and fins.

Most light organs contain accessory structures. These are: (1) filters (containing melanin and other chemicals), (2) light guides (having collagen), (3) reflectors (made of multilayers of refractive material such as chitin), and (4) lenses.

CHARACTERISTIC FEATURES OF BIOLUMINESCENCE

There are marine and fresh water species of bacteria that emit light, but no luminous blue-green algae (cyanobacteria) are known (Figure 2). Moreover, many fungi species are luminous, such as mushrooms and moulds, yet there are no luminous plants.

Not a single mammal including humans has been found to emit light. However, many of the cells of the human body are able to produce ultraweak chemiluminescence, which is invisible to the naked eye but detectable by photomultipliers. Ultraweak chemiluminescence produces less than one photon per cell per second. This means that luminescence is a phenomenon intrinsic to the cell.

Bioluminescence occurs in many common groups such as mushrooms, jelly fish, squid, starfish, worms, beetles and fish. However, other equally common groups of animals are not luminous. These are: spiders, ants, crabs, lobsters, sea urchins, flat worms, amphibians, reptiles, birds, and mammals. Within a given group of animals, the phenomenon tends to be localized in a certain sub-group. Bioluminescence has been described in more beetles than in any other order of insects (Romoser, 1973). Moreover, it may appear in one species but another closely allied may contain no trace of luminosity (Cormier, 1974).

According to Cormier (1974) and Campbell (1988), it is possible that luminescence arose independently more than 30 times during evolution.

Bioluminescence can be measured physically. No other function can be measured with such accuracy. This represents a valuable asset in judging periodicity. The relative intensity in photons per unit wavelength, i.e., the energy emitted by the organism, as well as the wavelength can be recorded. It turns out that the intensity of emitted light and the wavelength, vary with the species (Figure 3). A bacterium, a crustacean and a fish have, however, not very different illumination powers (Wampler, 1978; Withers, 1992). This means that, as in other functions, the re-emergence of a given property, and its degree of organization are not necessarily related to the complexity of the organism.

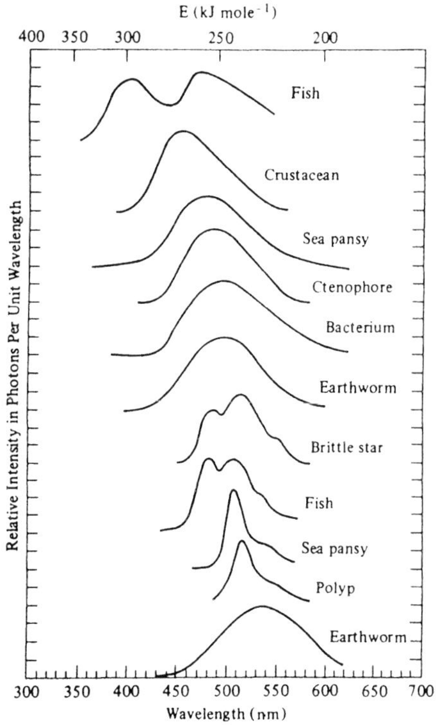

Figure 3. Wavelength (lambda; nm) and energy (kJ mole photons^{-1}) emitted by bioluminescent organisms. Searsid fish exudate, crustacean *Cypridina* reaction *in vitro,* sea pansy *Renilla* reaction *in vitro,* ctenophore *Mnemiopsin* emission *in vivo,* photobacterium emission *in vivo,* earthworm *Diplocardia* emission *in vivo,* brittle star emission *in vivo,* fish *Porichthys* emission *in vivo,* sea pansy *Renilla* emission *in vivo,* hydropolyp *Clytia* emission *in vivo,* earthworm *Diplotrema* emission *in vivo.*

Source: (From Wampler, 1978).

THE PERIODICITY OF BIOLUMINESCENCE

The periodicity of bioluminescence is born out as follows:

1. It emerges as a property of minerals before it becomes a property of the cell.
2. Atoms characteristic of minerals are part of its chemical cascade in living organisms.
3. It occurs in well defined groups of organisms but not necessarily in their close relatives.
4. Light organs, like eyes, can occur as several hundreds in the same individual, and in almost every part of the body.

5. It is absent from very large groups of organisms such as all algae, all flowering plants and the vertebrates, with the exception of the fishes. It is also not found in whole phyla in the invertebrates (Table 1).
6. Its intensity in eukaryotes is independent of the phylogenetic position of the organism. It has about the same strength in molluscs, insects and fishes.
7. The emergence is not necessarily related to the environment. Both sharks (that are fish) and whales (which are mammals), have no luminous organs, yet they prey in the darkness of the abyss. The sleeper shark *Somniosus microcephalus*, which can be as long as eight meters, swims to depths of at least 2,000 meters. Sperm whales (*Physeter catodon*) go below this depth (Line and Reiger, 1980).

Table 1. Period Bioluminescence. Presence and Absence of Bioluminescence

Presence	*Absence*
Photobacterium and other Species	Cyanobacteria (Blue-Green Algae)
Dinoflagellates (Protozoa)	Protozoa (many species)
Eumycota (Fungi)	Myoxomycota (Fungi)
Porifera (Sponges)	All Algae
Cnidaria (Jellyfish)	Flowering Plants
Ctenophora	Platyhelmintha (Flatworms)
Nemerteans	Rotifera
Mollusca — Bivalvia (Pholas)	Gastrotricha (Aschelminths)
— Pulmonata (Land Snails)	Kinorhyncha (Pseudocoelomates)
— Gastropoda (Limpets)	Nematoda (Roundworms)
— Cephalopoda (Squids)	Acanthocephala
Annelida — Oligochaeta	Entoprocta
— Polychaeta	Priapulida
Insects — Fire Flies	Sipuncula
— Beetles	Euchiura
Diplopoda (Millipedes)	Tardigrada
Crustacea — Ostracoda (Vargula)	Pentastomida
— Decapoda (Shrimp)	Onychophora
Echinodermata — Asteroidea (Star Fish)	Phoronida
— Holothuroidea (Sea Cucumbers)	Gnathostomulida
	Loricifera
Hemichordata	Brachiopoda
Chordata — Fish	Chaetognatha
	Chordata — Amphibians
	— Reptiles
	— Birds
	— Mammals

Notes: Only main groups are represented.
"Presence" means that the group contains some or many luminous species. "Absence" means that luminescence has not been recorded.

Source: Based on data from Campbell (1988).

8. It recurs at relatively well defined regular intervals (Figure 2), and it may have arisen independently more than 30 times.
9. The periodicity found at the biological level cannot be expected to be as regular as that of chemical elements, since living organisms are more complex. Yet the regularity is remarkable if one takes into consideration the many factors involved.

Chapter 7

PERIOD PENIS

THE PERIODICITY OF THE OCCURRENCE OF THE PENIS

Gnathostomulids are marine worms that are hermaphroditic but cross-fertilize. They possess a penis that is stiffened by a stylet. During copulation the animal injects sperm packets into its partner.

The Gastrotrichs are worm-like animals belonging to another phylum that are also hermaphrodites. Those that behave as males use a penis to transfer sperm to those that behave as females.

The spiny-headed worms (Acanthocephalans) are bisexual and possess a penis that is introduced into the vagina of the female. These worms are parasites of vertebrates.

The flatworms (Platyhelminthes) have a well developed penis located in a genital atrium (Figure 1).

Among the molluscs *Busycon* species (gastropoda), display a large penis that seats in the frontal part of the body. An equally long penis is present in *Buccinum undatum*, a sea dweller (Margulis and Schwartz, 1982).

There is a sudden occurrence of a well developed penis in the insects. The dragonflies copulate by means of such an organ (Romoser, 1973).

Among the crustaceans, barnacles (*Balanus*) are known to have a long penis. Some of these marine animals are hermaphroditic, but cross fertilization is usually the rule. The testes are located in the cephalic region and the two sperm ducts unite within a penis which upon protrusion out of the body reaches a distant individual acting as female (Figure 1). The penis enters the mantle cavity, where it deposits the sperm (Barnes, 1980).

Unlike human females, marsupials possess a double vagina; the males have a double penis, whereas humans and other placental mammals, have only one (Macdonald, 1984).

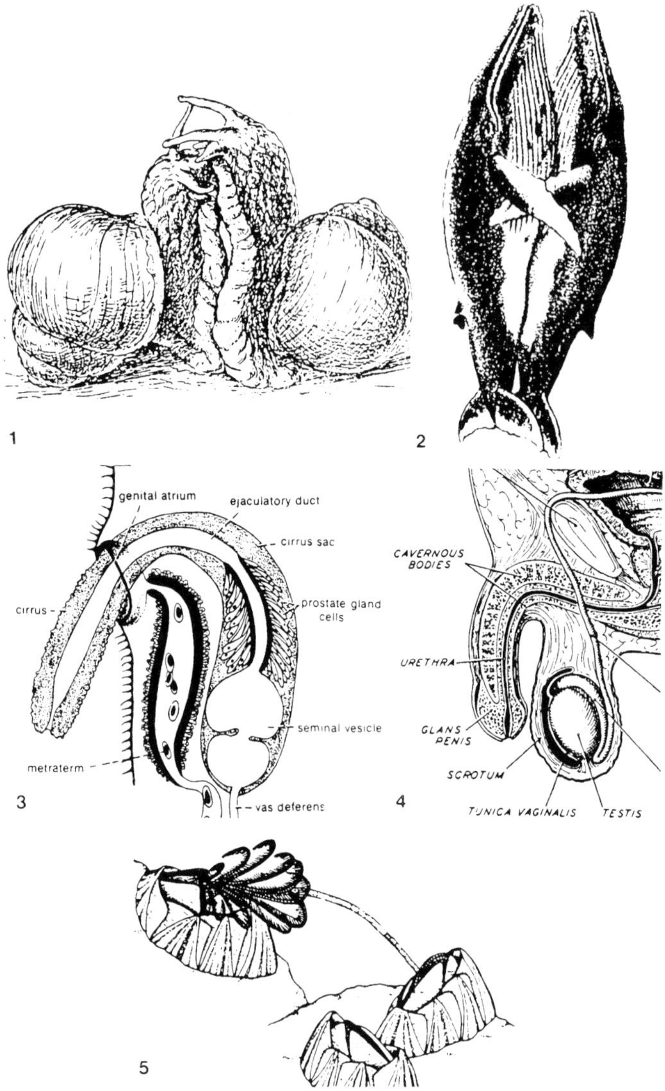

Figure 1. The periodicity of the penis. (1) Copulation in snails; these molluscs are hermaphrodites. However, one may function as male (extruding its penis) while the other functions as female. (2) A pair of mating humpback whales which, like other mammals, have a long penis. (3) The protruding penis (cirrus) of the flatworm *Fasciola hepatica*. This organ has a seminal vesicle, a prostate gland and an ejaculatory duct. (4) Penis in the human male with its ejaculatory canal and seminal vesicle which is also connected with a prostate gland. (5). The penis of the barnacle (an invertebrate) is so long that it extends from one individual (on the left) to another (on the right) during their copulation.

SIMILARITIES BETWEEN THE PENIS OF HUMANS AND INVERTEBRATES

The similarity between the human penis and that of invertebrates is based on morphological, functional, and physiological processes.

Many invertebrates have a hydraulic skeleton. Mammals with their large, heavy, and well articulated bones do not need to have a frame of this type. Although the human penis has no bones, there is no reason why it could not have had them. In some invertebrates, the penis is stiffened by a stylet, while in seals and walruses the penis has a long bone that grows with increase in sexual maturity (King, 1964), but this bone is absent in other mammals. Under the influence of hormones, the boneless penis of most mammals stiffens so much that it becomes as rigid as any arm or leg. In this sexual organ the hydroskeleton creates an equally effective rigidity and articulation as that found in the body of invertebrates.

When a pulmonate snail copulates, both the sensory and locomotory organs of the head and foot are flaccid and crumpled. This is the result of a change in blood flow within the body. A large proportion of the blood volume in the snail is involved in the dilatation of the penis and other genital structures. This is also what happens in humans during copulation. A previously flaccid penis becomes erect as the result of a change in blood flow.

The penis of several species of gastropods (molluscs) is a long cylindrical body arising just behind the right cephalic tentacle. The vas deferens opens at the tip of the penis and there is as in humans, a prostate, that has the same function, i.e., it produces seminal fluid.

In flatworms this organ is as complex as in mammals. It consists of: (1) a cirrus (the protruding part of the penis), (2) genital atrium, (3) ejaculatory duct, (4) prostate gland, (5) seminal vesicle, and (6) vas deferens (Figure 1).

WATER PERFORMS WITH EQUAL EFFICIENCY THE FUNCTION OF BONES AND OTHER SUPPORTING TISSUES

Worms possess a hydraulic skeleton created by the fluid of the body cavity. Consider, for example, velvet worms; their many legs are hollow and unjointed. The rigidity of these legs is not maintained by bones but by water pressure. The muscles present in the body wall oppose the hydrostatic pressure that moves the animal (Margulis and Schwartz, 1982) (Figure 2). It is thus apparent that water alone performs with equal efficiency the function that in vertebrates is accomplished by the mineralized tissues of the bones, and in insects and myriapods by the outer skeleton.

The body of brown seaweeds (the sporophyte) differentiates into: (1) a blade that has the shape of a leaf, (2) a stipe, that resembles a stem, and (3) a holdfast that functions as a root. This allows the algae to grow to sizes of several meters, just as trees do. These algae lack the conducting and supporting tissues characteristic of

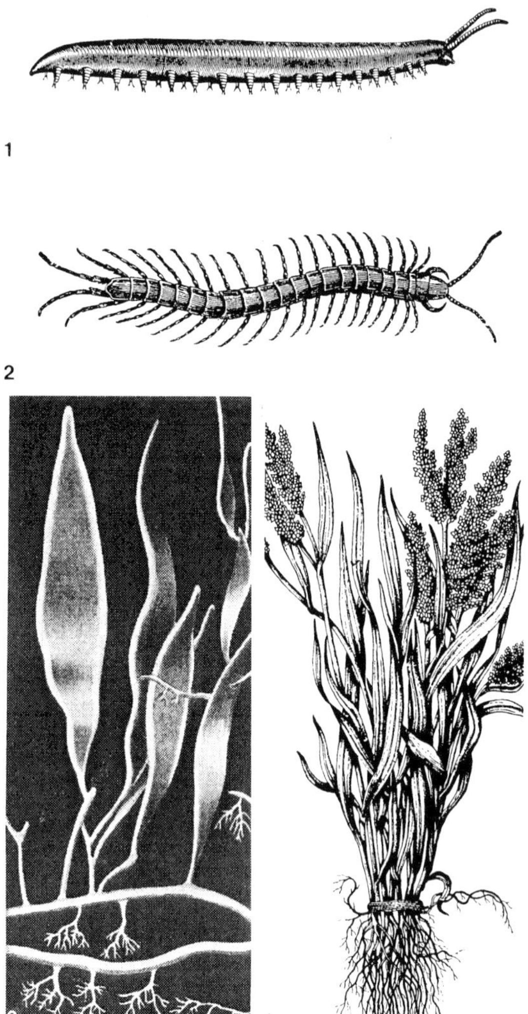

Figure 2. Water skeleton. (1) The velvet worm *Peripatus capensis* has many hollow legs whose rigidity is maintained by water pressure. (2) The myriapod *Scolopendra morsitans* has many legs but these are supported by a rigid exoskeleton. (3) The marine alga *Caulerpa prolifera* resembles a flowering plant with its root and leaf-like structures, yet its gigantic structure measuring several decimeters is formed by a single cell which mainly consists of water. (4) By contrast, a millet plant *Panicum mileaceum,* of comparable size and shape, is built with the help of rigid supporting tissues.

flowering plants, such as the phloem, xylem, collenchyma, and others. They have instead sieve tubes that transport water and photosynthetic products. Though the tissues and organs differ, their shape and function are the same. Moreover, such a parallelism is also found in other groups of algae. The red and the green algae, obey the same construction pattern. Most algae have a water content of 98 percent (Denffer et al., 1971).

An extreme case is provided by the alga *Caulerpa prolifera*, which has leaf- and root-like structures, and attains a size of several decimeters, yet it consists of a single cell (Figure 2).

THE EMERGENCE OF THE PENIS IS NOT DIRECTLY RELATED TO THE GENERAL ENVIRONMENT OR ORGANISM COMPLEXITY

A significant aspect of the evolution of the penis is that it occurs in marine species, such as the molluscs and crustaceans, in which the sperm could be easily released into the water, as is the case in many fish. Besides, the existence of this organ in vertebrates, does not represent an evolutionary solution that is related to life on land. After all, a simple cloaca exists in fish and amphibians. Reptiles living on land or in the sea have rudimentary copulating organs, in addition to the cloaca, that function as primitive penises. In birds the situation is more extreme since most species (though they live on land) do not have a penis but only a cloaca. There are some exceptions in which a rudimentary penis is present.

The fact that primitive penises can occur in reptiles and birds, but have not developed into large and efficient structures suggests that the genetic constitution necessary to produce this organ exists in these vertebrates, but it remains latent. Moreover, the marsupials which are less complex than the placental mammals have two well developed penises instead of one as found in the advanced Hominids and other placentals.

To recapitulate, internal molecular processes seem to have determined the emergence of the penis since it developed under the following conditions:

1. The organ emerged in totally unrelated groups.
2. It appeared in environments where a need for it did not seem apparent.
3. Its emergence failed to occur in situations in which it could have been useful for the propagation of the organism e.g., birds, which have had 150 million years of evolution at their disposal.
4. Its degree of differentiation, as well as its number per individual are unrelated to organism complexity.
5. The tendency for it to occur at relatively regular intervals is apparent (Figure 3).

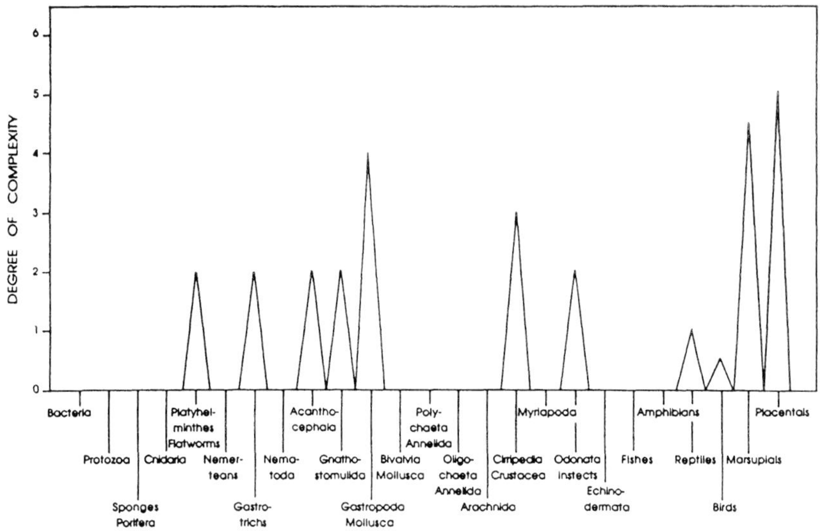

Figure 3. Period penis. The penis in Gastropods (Molluscs) occurs in a highly developed form and is well-developed in other invertebrate phyla, such as the Crustacea (Cirripedia), flatworms (Platyhelminthes) and insects (Odonata). It is absent in many other invertebrates, such as the Echinoderms. In both the Marsupials and the Placentals, the penis is highly developed but nearly absent or rudimentary in birds and reptiles, and totally absent in fishes and amphibians. The emergence of the penis bears no direct relationship to the general environment: it occurs in marine forms, such as the Cirripedia and the Molluscs, in which the sperm could have instead been released into the water, and it is hardly present in birds that live on land.

Chapter 8

PERIOD RETURN TO AQUATIC LIFE

WATER CHANGES THE CONFIGURATION OF MINERALS AND MACROMOLECULES

Different forms arise during the crystallization of bimalate of ammonia, depending on the water conditions under which its crystals are allowed to grow (Sharp, 1988).

The same phenomenon occurs in minerals. Calcium sulphate, in the absence of water, builds anhydrite ($CaSO_4$) which crystallizes in the orthorhombic system. However, in the presence of water it becomes gypsum ($CaSO_4{\cdot}2H_2O$), which produces crystals that belong to the monoclinic system (Figure 1).

The change of configuration with the number of water molecules available is especially critical in large macromolecules, such as DNA, which occurs in forms called: A, B, C, D, E, and Z, depending on the available amount of water and salt concentration. Since DNA is the genetic material, its change of configuration is critical, as it may affect transcription and replication (Saenger, 1988).

THE PLANTS THAT LIVE IN WATER HAVE STREAMLINED FORMS

The algae have a streamlined form that is most evident in many marine species. The same is true of the flowering plants that live completely submerged in the oceans, such as the eelgrass (*Zostera*). These have leaves that are highly elongated, like the body parts of algae and the body of elongated fish, from which their name eelgrass derives. Other flowering plants, such as those that are only partially submerged display hydrodynamic structures only in the part of the plant that is under water. This is evident in the well known species *Ranunculus peltatus* and *Potamogeton gramineus*. The submerged leaves are highly dissected and thin, whereas the aerial ones are thick and entire. The leaves that grow under water assume the algal shape. These changes occur within the same individual.

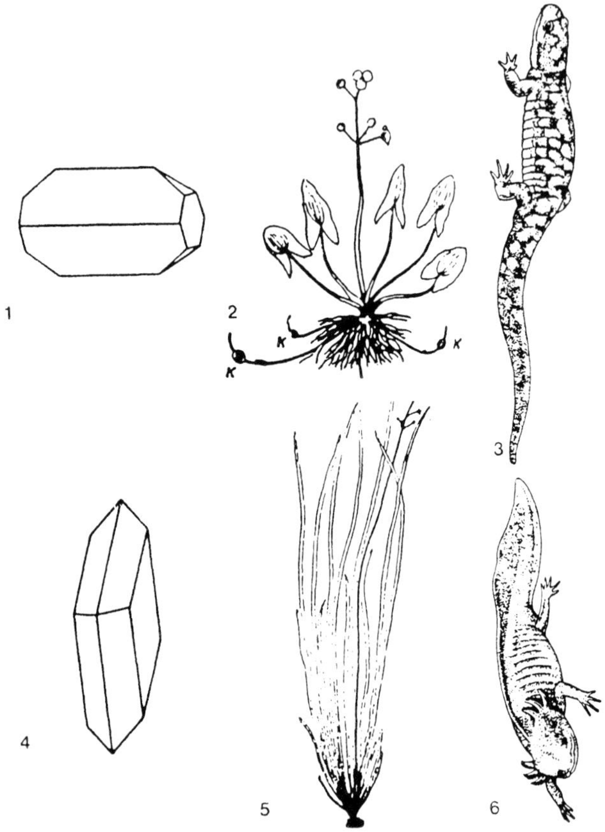

Figure 1. Same change of form in the presence of water. (1) Single crystal of anhydrite ($CaSO_4$) without water molecules, orthorhombic system. (2) The plant *Sagittaria sagittifolia* as a fully developed terrestrial form with its broad leaves. (3) The axolotl is the terrestrial form of the salamander of the genus *Amblystoma*. It has a broad head and tail. (4) Single crystal of gypsum ($CaSO_4 \cdot 2H_2O$) monoclinic system. This is the same calcium sulfate molecule having in addition two water molecules. The presence of the water molecules changes completely the crystal form to another system. (5) Fully developed aquatic form of *Sagittaria sagittifolia* with linear leaves. (6) Aquatic larval form of *Amblystoma* that has become sexually mature. The tail and the head are flat and arrow shaped. The presence of water leads, within the same mineral, plant and animal, to changes of pattern that result in comparable solutions.

Not only the form of the leaves is affected, but their function also changes. The photosynthesis of submerged leaves and body parts is modified, presumably because the plant faces reduced carbon dioxide and oxygen concentrations, as well as less light with different wave-lengths.

THE PLANTS REVEAL THAT NO CHANGE IN GENETIC CONSTITUTION IS NECESSARY TO PRODUCE A NOVEL HYDRODYNAMIC FORM AND FUNCTION

The fact that the parts of the plant that are above and below the water have the same genetic constitution renders the occurrence of the hydrodynamic form and function as being most significant from thc point of view of evolution. It simply tells us that it is not necessary to change the gene constellation, so as to obtain a radically different form and function, when water is the new medium. The clear implication of this evidence is that neither the reptile, the bird, nor the mammal had necessarily to modify their genetic constitution to assume different forms and functions related to the new medium.

Consider first minerals. The presence of water leads to a different crystal growth as the result of a rearrangement of the same atoms. In plants, the presence of water induces the same genetic constitution, i.e., the same sequences of bases in the DNA, to produce different molecular configurations that result in what is usually called a different phenotype.

WATER-AIR AND AIR-WATER TRANSFORMATIONS IN PLANTS

The plant *Polygonum amphibium* occurs in two forms viz. an aquatic *P. natans* and the other terrestrial *P. terrestre*. Both forms can be obtained from fragments of the same original plant which are allowed to develop in different environments. The aquatic form is characterized by long stems terminated by a group of large leaves without hairs, which are oval and have long petioles. By contrast, the terrestrial type has straight stems almost without ramifications, with lanceolate and hairy leaves with short petioles.

Aquatic plants that are moved to places with dry terrestrial conditions develop the terrestrial phenotype. Conversely the terrestrial forms if submerged are found to develop a new stem and leaves; i.e., the water phenotype. No transition types have yet been identified. This explains why botanists had originally divided this species into two subspecies, for they were convinced that the aquatic form was unrelated to the terrestrial one (Blaringhem, 1923).

The fact that the phenotype can be reversed is taken as evidence that it is the water environment that determines the structure of the whole organism. This is because the stems and leaves, as well as the type of ramification are affected.

A plant of *Hedera helix* (ivy) may be divided into two portions (which have the same genotype). These will acquire different phenotypes if cultivated in a moisted and dark atmosphere or in a dry and sunny locale. In conditions where the water concentration is high and the amount of sunlight low, they develop creeping stems with adventitious roots and broad leaves. Sections of the same plant are found to develop into small rounded shrubs with elongated leaves if cultivated in a sunny atmosphere with little moisture. Transitional forms of this species arise when the water concentration in the air is somewhat reduced.

EXPERIMENTAL DEMONSTRATION THAT WATER DECIDES THE LEAF PATTERN

Two types of leaves are produced by *Ranunculus aquatilis* (water buttercup) and other plants that grow in aquatic or semiaquatic habitats. Those submerged in water are narrow and highly dissected, whereas those floating are broad and entire (Figure 1).

Experiments confirm the view that the type of leaf produced is determined by the environment to which the upper top of the plant is exposed. If the plant crown is under water, then dissected leaves are produced but if moved to air, then floating entire leaves appear. If the reverse experiment is carried out, i.e., the crown is moved from air to water, then the production of broad leaves is found to stop and dissected ones are formed. In addition, the amount of light is known to affect the type of leaf formed (Dale, 1982).

Plant hormones are probably involved in the transmission of information from the crown to the leaves. This applies to the differences in pattern found between juvenile and adult leaves, so common in plants, in which the concentration of gibberellin is the main agent responsible for the structural difference.

What such experiments reveal is that pattern differences between the submerged or the floating form are regulated by the direct action of both chemical (water) and physical (light) agents. Under their action the same genotype is able to produce two entirely different solutions, irrespective of the consequences these may have for the plant.

THE TRANSFORMATIONS INVOLVED IN THE RETURN TO WATER IN INVERTEBRATES ARE SIMILAR TO THOSE THAT OCCUR LATER IN HIGHER MAMMALS

Not only did the Gastropoda, which started as marine animals, conquer the land, but they returned to the aquatic medium by invading freshwater habitats (Russell-Hunter, 1979). This return to water resulted in the creation of forms and functions which are repeated in vertebrates that moved to similar conditions later on.

Figure 2. Change from terrestrial to aquatic life. (1) The terrestrial snail *Helix pomatia*. (2) The fresh water snail *Limnea stagnalis* upside down below the water surface. (3) The penguin *Aptenodytes patagonica* on land. (4) A closely related species of penguin *Pygoscelis antarctica* using its wings to swim. (5) The green turtle *Chelonia mydas* on land. (6) The same species swimming.

1. The fresh-water forms of pulmonates such as *Limnaea* come to the surface to obtain air just as the seals and whales do (Figure 2).
2. The lung has an opening through which air is allowed (the pneumostome). In *Limnaea* this opening is closed when the animal is submerged. The whales close their nostrils by means of valves.
3. This pulmonate species can remain submerged for more than an hour. Whales can dive for the same duration.

This phenomenon is widespread since it also occurs in land insects that have secondarily entered an aquatic medium. They developed new organs, such as the tracheal gills, in which the tracheae have ramified and lie under a very thin cuticule. Other insects surface at regular intervals to breathe air, while others use an air bubble as a "physical gill" from which they absorb oxygen.

Newly changed organs also appear in aquatic mammals. The lungs of whales have become highly elastic and extensible, thereby allowing the retention of greater quantities of air in the body.

THE CONQUEST OF THE LAND AND THE RETURN TO WATER IN AMPHIBIANS

The amphibians are known to have been the first vertebrates to conquer the land some 405 million years ago. They started breathing air using lungs, developed limbs that allowed them to move on hard surfaces, and a tough skin that protected them from the drying effects of air. Though terrestrial, they returned, and still return to water, or moist places, where they lay eggs. This change to the original habit was manifested in the extreme form by whole groups of amphibians, such as the stereospondyls, which became specialized aquatic types, thus reversing the general trend of evolution approximately 230 million years ago.

The return to water was accompanied by the following morphological and functional changes:

1. There was a simplification of the vertebral column.
2. They increased in size, becoming the largest amphibians ever to live.
3. Their heads became immense in relation to their body.
4. The limbs and feet diminished in size (Colbert, 1980).

A matter of some significance is that the same type of changes occurred much later in land mammals when they returned to water. Whales are the largest living mammals, whose heads are quite large in relation to body size, and their limbs are greatly reduced.

STRUCTURAL AND FUNCTIONAL MODIFICATIONS IN REPTILES FOLLOWING THE TRANSFER TO AQUATIC LIFE

The reptiles represent the first mode of life that can be regarded as fully terrestrial. They developed a reproductive pattern free of standing water. Their eggs have extraembryonic membranes that retain water, and their scaly skins prevent them from dessication. A series of groups of reptiles invaded land masses, about 320 million years ago. Later, two types of reptiles returned to the marine environment: the ichthyosaurs and the plesiosaurs.

The ichthyosaurs emerged about 200 million years ago and were the most specialized. The body form corresponds very closely to that of a modern tuna fish which can swim at more than 40 kilometers an hour. The ichthyosaurs did not lay their eggs on sandy beaches, but gave birth in the water to their young, just as whales do today. The land reptiles had efficient lungs which were not abandoned when they became marine animals. The gills had long since been lost and were not replaced. The legs and feet were transformed into paddles, achieving the same function as the fins of fish. The tail having the bifurcated form, found in fishes, had also long disappeared in the reptiles. The new tail that developed in the ichthyosaurs is "astonishingly" similar to that of fish.

They also acquired, like the whales did much later, a fleshy dorsal fin. This new structure without any skeleton, took the place of the bony dorsal fin of fish. It appeared suddenly and without antecedent. Especially significant from the point of view of evolution is that, as Colbert (1980) stresses, this reptile group appeared suddenly in the middle of the Triassic. There are no clues in the fossil record as to the possible ancestors of the ichthyosaurs. Moreover, all species known adhere to a single pattern. The bones of the limbs were shortened. The wrist and ankle bones, and the bones of the fingers and toes, were modified into flat hexagonal structures.

The other marine reptiles, the plesiosaurs, were the dominant group between 181 and 135 million years ago. They had large limbs, but were less specialized in their aquatic propulsion than the ichthyosaurs. They resembled the modern sea lions (that are mammals) in that they probably moved the pectoral fins in the horizontal plane (Carroll, 1987).

Other reptiles became mainly or partly aquatic, notably the sea turtles, the crocodiles, and the fresh water turtles (which appeared 63 million years ago).

The return to aquatic life in the reptiles occurred in at least three distinct groups and at different periods of evolution.

THE HYDRODYNAMIC FORMS AND FUNCTIONS OF BIRDS DERIVE FROM THOSE OF LAND RELATIVES

The penguins originated from land birds. They acquired a hydrodynamic shape totally different from that of an ostrich or a pheasant. The wings are paddle-like flippers, the whole body is fusiform in shape and the feet webbed. Interestingly, penguins continue to maintain their lungs but use their wings not to fly but rather to swim. Both in shape and function they are not very different from the flippers of a seal or whale (Figure 2).

The flightless cormorant of the Galapagos Islands uses its powerful legs with webbed toes to move in water instead of its wings (Beazley, 1974). In this respect, it represents a solution similar to that of the hippopotamus that also lives in water, swims with its strong legs, and has slightly webbed feet. Not all swimming birds have webbed toes.

There are several other similarities between swimming birds and the higher mammals that have returned to water. These include:

1. Diving ducks use both their wings and feet under water but the wings have a secondary function. They act as stabilizers in the same way as the forearms of the dolphins (Perrins, 1976).
2. The bones of the wings of penguins are of the same pattern and number as in flying birds. However, they have become very flattened and expanded, and the elbow and the wrist joints have fused. In whales, the arm bones are shortened and flattened, and the wrist bones are also flattened discs.
3. Bird divers, as the long-tailed duck, can remain submerged for 15 minutes and descend to about 60 meters. Although this is what the seals regularly do, they can stay under water much longer and dive to greater depths.
4. When birds dive they are able to draw oxygen from oxyhemoglobin, which is stored in the muscles, and have a high tolerance to carbon dioxide (Perrins, 1976). Consider too seals whose blood contains much oxygen-carrying hemoglobin (three times that of humans). High amounts of the oxygen binding protein myoglobin are present in their muscles. They also tolerate high concentrations of carbon dioxide in their blood (Macdonald, 1984).
5. Penguins have thick layers of fat that insulate them from the cold water. This is also true of whales since they have developed a thick layer of fat beneath the skin that gives them insulation (Colbert, 1980).
6. Penguins are swift swimmers attaining speeds up to 40 kilometers per hour (Line and Reiger, 1980). Seals that prey on them move at about the same speed.

THE RETURN OF MAMMALS TO AQUATIC LIFE OCCURRED SEVERAL TIMES AND FROM DIFFERENT ORDERS

The whales (order cetacea) arose from ancestors of the deer (order artiodactyla, the even-toed ungulates). This is revealed by studies of the action of restriction enzymes on DNA followed by DNA–DNA hybridization experiments, as well as other molecular methods (Janvier, 1984; Goodman et al., 1985; Arnason, 1985; Arnason and Widegren, 1986; Springer and Kirsch, 1993).

The seals (order pinnipedia) comprise three families whose ancestors are the carnivores (order carnivora). The eared seals (sea lions) originated from a dog-like branch (25 million years ago), the walrus from a bear-like stem and the true seals from an otter-like group (15 million years ago) (Macdonald, 1984).

The sea cows (order sirenia) come from the primitive ungulates, i.e., the ancestors of elephants (order proboscidea).

The hippopotamuses (order artiodactyla) derive from the suiforms (which belong to the same order) and have as close relatives the pigs and the peccaries (Colbert, 1980).

THE RETURN TO AQUATIC LIFE IN THE UNGULATES: THE WHALES

The emergence of the whales, about 50 million years ago, was marked by the following events:

1. No gills were formed but the lungs were preserved. One should not forget that rudiments of gills exist even in the human embryo and could be made to develop in the adult.
2. The change was relatively rapid since they appear suddenly in early Tertiary times. Colbert (1980) puts it in the following terms: the whales "enjoyed at the outset a series of extraordinarily rapid evolutionary changes" that suddenly transformed them into aquatic animals. What is significant is that the transformation was not only rapid but also involved several traits at the same time. The first fossils that are recognized as whales already look like whales. This is not due to the lack of fossils because the same happens in the bats and the ichthyosaurs that seem to come from nowhere. The new pattern, which is dependent on a series of simultaneous modifications, emerges from the start as a well-organized structural and functional package (Figure 3).
3. A highly streamlined body appeared, which was accompanied by the emergence of a fishlike tail. The tail of the whale is, in fact, a new feature. It contains no bones, as it does in fish, yet it can be 2 meters long and has a horizontal rather than a vertical orientation. Another new formation is the dorsal fin. This is fleshy (no bones) in the whales and was also fleshy in the ichthyosaurs (Chapter 3, Figure 7).
4. In the ichthyosaurs, the four legs were modified into paddles. The same transformation occurred initially in the whales as attested by the fossil remains of *Pakicetus* (Savage and Long, 1986). Later the hind limbs were suppressed, or became vestigial; only the front limbs functioned as paddles (Figure 3).
5. The arm bones of whales are short and flattened, and the fingers are elongated by the multiplication of the phalanges. Again, this is the same

Figure 3. Air-water transformations in whale ancestors and living whales. (1) A male elk, *Alces alces* (even-toed ungulate) is shown swimming across a Swedish lake; females also swim and have no horns. DNA studies show that the terrestrial even-toed ungulates and the whales have common ancestors. (2) Restoration of *Pakicetus*, which is the most primitive known whale (from early Eocene). (3) Sei whale, *Balaenoptera borealis*, a baleen whale with its characteristic mouth plates that allow it to ingest plankton. (4) Sperm whale, *Physeter catodon*, a toothed whale which has strong teeth and is a carnivore.

solution that occurred earlier in the reptiles. The fore and hind limbs of ichthyosaurs and plesiosaurs consisted of bones that were short and flattened, and the fingers were elongated by the presence of many phalanges (Gregory, 1974).

6. The earliest whales had external nostrils that were located in the front. In present day whales they are shifted to the top and back of the head and can be closed by valves. As for the lungs, they are highly elastic and extensible, thereby enabling great quantities of air to be taken in. Whales can stay under water for periods of over 60 minutes and can swim to great depths.
7. Whales give birth to their young under water. The mother helps the newborn to the surface so that it can breathe.
8. The cetacea are divided into two suborders: baleen and toothed whales. They arose about 45 and 40 million years ago, respectively. Their first fossils were found on separate continents (Asia and North America) (Halstead, 1978).

THE RETURN OF THE CARNIVORES TO WATER: THE SEALS

The seals (pinnipeds) had as ancestors the terrestrial carnivores. Their transformation was never as extensive as that in whales or ichthyosaurs. Seals have retained a flexible neck, and have failed to evolve both a dorsal fin and propulsive tail. However, their body is streamlined. They live partly on land.

In these mammals the four feet are transformed into paddles with webbing between toes similar to that found in aquatic birds. The front paddles of seals are used in propulsion and steering, while the back paddles function as a substitute for a caudal fin when they are in water. On land the flippers and the body are employed in locomotion. The sea lions (eared seals) swim like penguins by making long sweeps of the foreflippers.

The pinnipeds close their nostrils on immersion and can stay under water for as long as 73 minutes. Their young are born on land or on ice (Figure 4).

The walruses were not included separately in the diagram on periodicity, since they represent a single species closely related to the other seal species (Figure 7).

THE SEA COWS ARE DERIVED FROM THE ANCESTORS OF THE ELEPHANTS

The sirenians, or sea cows, are a group of mammals entirely adapted to live in water. They swim with paddle-like forelimbs, have no hind limbs like most whales, and possess a flattened tail. They have descended from an ancestor shared by the elephants, and they still maintain several features in common with these animals. They are non-ruminant herbivores and have a tooth replacement like that of elephants (Figure 5).

Figure 4. Air-water transformations in seal ancestors, seals and walruses (order Pinnipedia). DNA studies disclose that the terrestrial carnivores had the same ancestors as the Pinnipeds: the otter and the bear being their closest relatives. (1) Tiger bathing, (*Panthera tigris,* carnivore); tigers often lie in water to keep cool. (2) and (3) Polar bear (*Ursus maritimus,* carnivore); like the sea lions, the polar bear swims only with its forelimbs (the hind limbs being used only as a rudder) and closes its nostrils when it dives. (4) Swimming Southern sea otter (*Enhydra lutris*), otters are truly amphibious, foraging in water but living also on land. (5) The typical eared seal *Arctocephalus pusillus.* The eared seals comprise the fur seals and the sea lions. (6) A typical true seal *Phoca vitulina.* (7) The walrus *Odobenus rosmarus.*

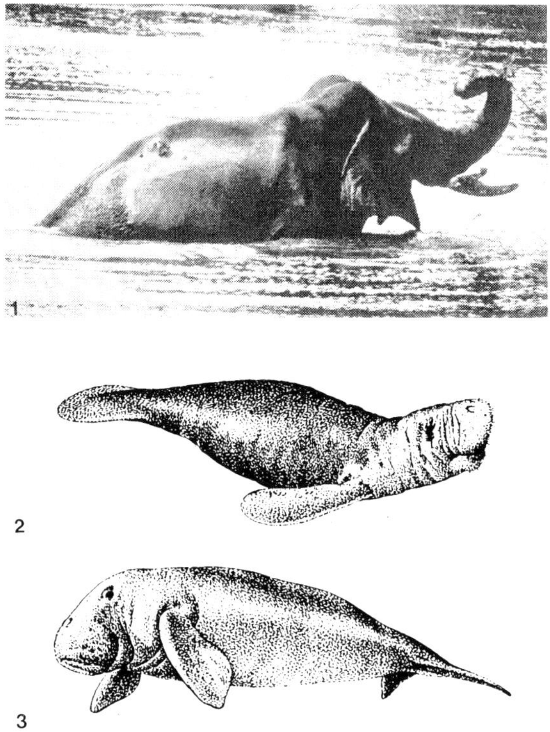

Figure 5. Return to aquatic life—sea cows. (1) An elephant bathing. These animals are good swimmers and eat usually the plants that grow under water. The sea cows are derived from the ancestors of the elephants. (2) The North American manatee *Trichechus manatus.* (3) The Dugong *Dugong Dugon.* The manatee and the dugong are sea cows.

The sea cows emerged as a group 54 to 38 million years ago. They followed two lines of development: the manatees and the dugongs which are fresh-water and sea dwellers, respectively.

Unlike the seals, but like the whales, they never left the water. The nostrils of the sirenians are valved. They remain under water for periods of up to 16 minutes and give birth under water. The sirenians have not developed a dorsal fin resembling that present in the whales (Colbert, 1980; Macdonald, 1984; Dorst and Dandelot, 1988).

THE PIGS AND PECCARIES ARE THE CLOSE RELATIVES OF THE HIPPOPOTAMUSES

These mammals live during the night on land but during day are mainly under water. Their large, heavy, short-legged bodies have a thick layer of fat under the skin. This

Figure 6. Return to aquatic life—hippopotamuses. (1) The collared peccary (*Tayassu tayacu*) is a close relative of the hippopotamus. (2) *Hippopotamus amphibius*: this animal on land shows its strong legs which have not changed into paddles. (3) Although its body form is not as hydrodynamic as that of seals, it swims well.

is a feature found also in sea cows. Their feet are slightly webbed. Females give birth to young on land or under water. The adults can stay immersed for up to 6 minutes. There are no traces of hippopotamuses in the fossil record until 20 million years ago. Their closest relatives are the pigs and the peccaries (Colbert, 1980; Dorst and Dandelot, 1988).

Although it took hippopotamuses a longer time to evolve than the true seals (that appeared only 15 million years ago), they have not changed their feet into paddles, nor acquired a streamlined body, and have only developed partial webbing between their toes. Both groups live on land and in water (Figure 6).

THE PERIODICITY OF THE RETURN TO AQUATIC LIFE

Few phenomena show such a distinct recurrence and periodic emergence as the return from terrestrial to aquatic life (Figure 7):

1. The recurrence is evident in: (a) The land invertebrates such as gastropods and insects. (b) In the land amphibians and reptiles that became stereospon-

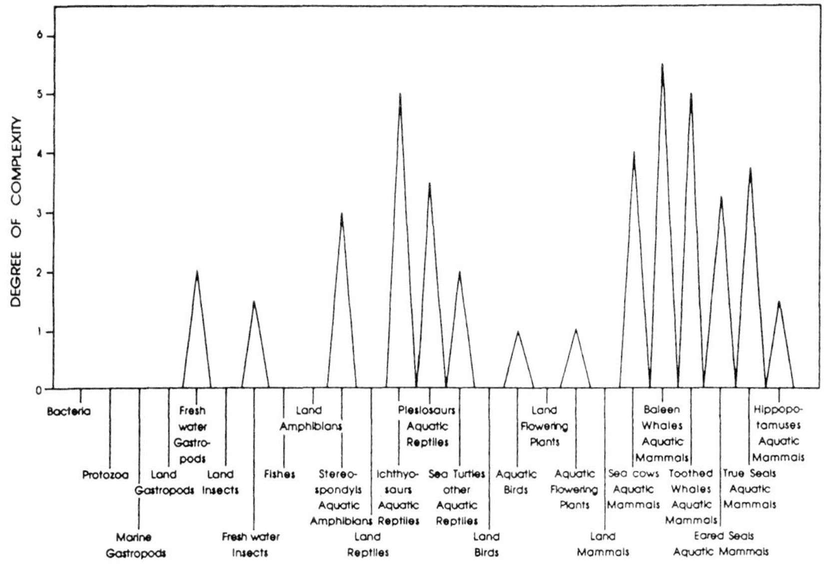

Figure 7. Period return to aquatic life. The invertebrates started living in the sea or fresh-water, but some classes e.g., insects became terrestrial. The early vertebrates were also marine, such as the fishes but became successively terrestrial, starting with the amphibians. In both groups, the terrestrial species returned to the water environment and did this periodically. The land gastropods returned to aquatic life; the terrestrial insects returned to fresh-water; the land amphibians became aquatic as they transformed into the stereospondyls; the land reptiles got transformed into the marine reptiles called ichthyosaurs and plesiosaurs. Some birds and some flowering plants also became aquatic. The periodicity is particularly evident in the return of the land mammals to water which occurred in several orders not closely related: the sea cows, the whales (baleen and toothed), the seals (eared and true), and the hippopotamuses.

dyls, ichthyosaurs and plesiosaurs, respectively. (c) The land birds that became partly aquatic. (d) The flowering plants that also became partly aquatic. (e) The land mammals that returned to water. There are 16 to 19 orders of eutherian mammals (depending on the classification), and in four of them the species became fully aquatic (whales, sea cows) or amphibious (seals and hippopotamuses).

2. The periodicity is evidenced by the fact that the transformation that led to aquatic life occurred only in specific orders, families or genera and emerged at relatively regular intervals from the early invertebrates to the latest mammals (Figure 7).
3. The structural and functional changes that occur in these organisms, relative to aquatic life tend to be of the same type, irrespective of their phylogenetic position. Further, they can be traced back to changes in atomic arrangement occurring in minerals in the presence of water.
4. The structural and functional changes arose as a "package," which is found both in invertebrates and vertebrates. This essentially consists of the following components: — (a) a streamlined body shape, (b) openings connected with the lungs that can be closed during submersion, (c) ability to remain under water for long periods of time, this being due to a higher capacity to bind oxygen in blood, (d) limbs or corresponding body parts, diminished in size, (e) legs and feet, when present, were usually transformed into paddles getting the same function as the fins of fish, (f) new tails and new dorsal fins were created in reptiles and mammals having the same shape as in fishes, (g) toes, became partly webbed in such distant groups as the birds, the seals and the hippopotamuses, and (h) thick layers of fat that insulate the body developed in birds and whales, as well as in other groups.
5. The structural and functional changes arose quite suddenly. In the reptiles and mammals there is no fossil record of long and successive intermediate stages.
6. This is most evident in the case of the formation of new organs. Some have even arisen without the participation of the vertebrate skeleton. The dorsal fin of whales and ichthyosaurs was formed without bones, yet it has the same shape and function as that of fishes, which is part of the skeleton. This new organ, like the rest of the structural and functional "package," seems to have emerged from nowhere, i.e., it has no predecessors in closely related species.
7. No change in genetic constitution needs to occur to transform an aquatic organism into an air dweller or for the reverse air-water situation to occur. This is the case in plants where the same organism can be divided into two parts: one which acquires the terrestrial and the other the aquatic form and function. This is also the case in frogs in which the same animal changes from an aquatic to a terrestrial life by modifying its structures and functions radically.
8. The emergence of new organs has been studied in detail in some species. The "conquest of the land" by adult frogs is achieved by a tenfold increase in thyroid hormone levels in the blood of the swimming tadpole. This small hormone is responsible for the cluster of transformations that include the following: reabsortion of the tail, change from a gill to pulmonary respira-

tion, formation of a complete skeleton with accompanying muscles, reshaping of the intestines and other drastic modifications. Experiments have confirmed this chain of events. Thus, removal of the thyroid gland from a frog embryo stops the transformation, but the animal continues to grow until it becomes a giant tadpole. Injection of this hormone restores the original conversion to terrestrial structures and functions (Fox, 1981; White and Nicoll, 1981). This cascade has now been studied at the molecular level in the frog *Xenopus laevis*. It is found that the hormone binds directly to receptor proteins. The genes coding for these proteins have been characterized and cloned. The amount of messenger RNA, produced by thyroid hormone receptor genes, increases sharply before the adult transformation is initiated and diminishes when this is concluded. Moreover, the thyroid hormone action changes gene expression in every tissue of the tadpole (Yaoita et al., 1990; Shi and Brown, 1990).

9. Water is also an agent in this process. The salamander of the genus Amblystoma can occur as an aquatic larval form and as a terrestrial adult. These were considered separate species and even genera, until they were transformed into one another by the addition of thyroxine (Raff and Kaufman, 1983). Water can also cause indirectly this transformation. When the aquatic larval form is moved to a dry habitat, it changes rapidly into the terrestrial adult type. The absence of water leads to an increase in thyroxine levels (Aron and Grassé, 1939; Freeman, 1972).
10. The critical interference of water in plants such as *Ranunculus* is well established. If the plant crown is under water, then leaves acquire the aquatic shape. If, however, the crown is exposed to air, aerial leaves appear. The action of this chemical is confirmed by a reverse experiment in which this organ is moved from air to water: broad leaves stop being formed while those dissected are replaced again. Plant hormones such as gibberellin are known to be the agents in the interaction between the apex of plants and their leaf primordia (Dale, 1982).
11. The water-air-water transformations in both animals and plants appear to be conditioned by an external agent, namely water, and an internal one, the hormone, which by binding to receptor proteins unleashes a cascade of totally different gene expressions.

Chapter 9

PERIOD PLACENTAL-MARSUPIAL EQUIVALENCE

THE PLACENTALS AND MARSUPIALS ARE "CARBON COPIES" OF ONE ANOTHER

During the Cretaceous period (135–65 million years ago) mammals underwent separation into two distinct groups: The marsupials which carry their young in pouches, and the placentals that retain the young inside the mother's body. The first marsupial fossils are 75 million years old. All placentals are considered to have appeared later. Most of the modern orders were established about 50 million years ago (Halstead, 1978; Macdonald, 1984).

It has been repeatedly pointed out that the two groups are very much alike. However, this finding has not been examined in terms of periodicity. The reason may lie in the fact that the similarities have not been seen as a result of internally coordinated cellular processes. Another difficulty in interpreting this relationship, has been that the marsupials and the placentals, not only evolved separately, but also mainly on different continents and in different climates where most plant species are different, as is the case in Australia.

The similar versions that emerged include species, genera, families and orders. Equivalents are found for members of the following orders: carnivora, primates, rodentia, dermoptera, insectivora, edentata, perissodactyla, and artiodactyla (Table 1).

The evidence coming from fossils already reveals the phenomenon of periodicity. The large sabretooth cats or tigers had very long upper canines that protruded from the sides of their mouths. In these animals the lower jaw had to open to an angle of 90° before it could effectively stab the prey. They emerged in the placentals not only once but several times from independent stocks. Well known genera are *Smilodon* and *Eusmilus*. Further, there existed a fossil marsupial sabretooth *Thylacosmylus* which was practically identical to the placental species; its

Table 1. Equivalence between Placentals and Marsupials

Placentals			*Marsupials*		
Order	*Family*	*Species*	*Species*	*Family*	*Order*
Carnivora	—	Sabretooth *Smilodon* (fossil) and *Eusmilus* (fossil)	Marsupial Sabretooth *Thylacosmylus* (fossil); upper canines had same form and shape as those of placental sabretooth, lived at the same time and had same carnivore habit as placental equivalents.	—	Marsupialia
Carnivora	Felidae	Lion *Panthera leo*	Marsupial lion *Thylacoleo carnifex* (fossil); strong stabbing and shearing teeth, lion-like in body and head shape, well equipped for hunting and carnivore nutrition.	Thylacoleonidae	
Carnivora	Hyaenidae	Spotted hyena *Crocuta crocuta*	Pouched hyena *Borhyaena* (fossil); hyena-like.	Borhyaenidae	
Carnivora	Mustelidae	European otter *Lutra lutra*; Aquatic habit, paws are webbed in most species, catches food with forefeet or mouth, feeds mainly on crustaceans and fish, closes ears and nostrils under water.	Water opossum *Chironectes minimus*; Aquatic habit, hindfeet are webbed, swims with hindfeet, forefeet used to catch food, feeds mainly on crustaceans and fish, closes pouch under water.	Didelphidae	
Carnivora	Canidae	Common wolf *Canis lupus lupus*	Tasmanian wolf *Thylacinus cynocephalus*. Wolf-like in overall form and carnivore nutrition.	Thylacinidae	
Carnivora	Felidae	Wild cat *Felis silvestris*	Tasmanian cat *Sarcophilus harrisii*. Similar to a cat, has powerful jaws and sharp teeth; carnivore.	Dasyuridae	
Primates	Tarsiidae	Spectral tarsier *Tarsius spectrum*. Arboreal habit, large eyes directed forward, flattened face and long tail.	Black-shouldered opossum *Caluromysiops irrupta*. Arboreal habit, protruding eyes directed forward, face similar to that of primates; long tail.	Didelphidae	
Primates	Cebidae	Spider monkey *Ateles paniscus*, Tail prehensile as an extra hand.	White-eared opossum *Didelphis albiventris*. Tail prehensile as extra hand.	Didelphidae	

Rodentia	Muridae	House mouse *Mus musculus*	Marsupial mouse (mulgara) *Darcycercus cristicaudata*. Similar in body shape and habits. Many species in this family resemble placental mice.	Dasyuridae	Marsupialia
Rodentia	Anomalur-idae	Scaly-tailed squirrel *Anomalurus peli*. Has a membrane that extends between forefeet and hindfeet and glides well.	Greater glider *Petauroides volans*. Has a membrane that extends between forearms and hindfeet, same body solutions and gliding efficiency as placental species.	Pseudocheiridae	
Dermoptera	Cynoce-phalidae	Philippine colugo *Cynocephalus volans*. It is also called "flying lemur" because it glides well, has a membrane that extends between forefeet, hindfeet and end of tail	Sugar glider *Petaurus brevicepts*. Has a membrane between forefeet and hindfeet, same body solutions and gliding capacity as placental species.	Petauridae	
Insectivora	Talpidae	Common mole *Talpa europaea*. Subterranean way of life, cylindrical body feeds on insects and worms; long and naked muzzle, eyes covered by skin, no external ears, short tail.	Pouched mole *Notoryctes typhlops*, Subterranean way of life, cylindrical body, feeds on insects, hands used to excavate, horny shield on tip of snout, no eyes and no external ears; tail is a stub.	Notoryctidae	
Insectivora	Soricidae	Common shrew *Sorex araneus*	Red-tailed Phascogale *Phascogale calura*. Has long conical snout, insect eater, resembles the shrew in appearance and habits.	Dasyuridae	
Edentata	Myrmeco-phagidae	Lesser anteater *Tamandua tetradactyla*. Digs with clawed fingers, eats ants and termites with long narrow tongue	Numbat, banded anteater *Myrmecobius fasciatus*. Digs with clawed fingers, eats termites with long narrow tongue.	Myrmecobiidae	
Perissodac-tyla	Rhinoce-rotidae	Rhinoceros *Diceros bicornis*	Marsupial Rhinoceros *Diprotodon* (fossil), A three meter long herbivore with the size, shape, and habits of a placental rhinoceros.	Diprotodontidae	
Artiodactyla	Suidae	Wild boar *Sus scrofa*	Pig-footed bandicoot *Chaeropus ecaudatus*. Forefeet have only two functional digits, thus forming a paired foot pad, like that of artiodactyla.	Peramelidae	

Figure 1. Some of the species that are "carbon copies" of one another in placental and marsupial mammals. (1) The sabretooth large cat *Smilodon* (placental) and (2) the sabretooth cat *Thylacosmylus* (jaguar sized maruspial). Both are fossils. (3) Scaly-tailed squirrel *Anomalurus peli* (placental) and (4) the greater glider *Petauroides volans* (marsupial). (5) The common mole *Talpa europaea* (placental) and (6) the pouched mole *Notoryctes typhlops* (marsupial). See also Figure 1 in Chapter 12.

upper canines had the same form and shape, it lived at the same time and had the same carnivore habit (Figure 1).

The lions are similarly represented by an equivalent species: the living lion *Panthera leo* (placental) and *Thylacoleo carnifex* (marsupial fossil). The marsupial lion had strong stabbing teeth, and a body and head resembling that of a lion. It was well built for hunting and had carnivore nutrition. Other carnivore equivalents include the hyena, the wolf and the wild cat. Among the most interesting ones is the otter. The European otter (placental) and the water opossum (marsupial) have the same aquatic habit: they swim and catch food in similar ways and are able to close certain organs while under water (Table 1).

Among the primates one finds monkeys that resemble several opossum species. And in the rodents several species of mice (placentals) are difficult to distinguish on the basis of shape and habits from the marsupial mice (Chapter 12, Figure 1).

A most striking equivalent is found in the gliding squirrel. The scaly-tailed squirrel (placental) and the greater glider (marsupial) have the same type of membrane and the same gliding efficiency (Figure 1). The squirrel belongs to the order Rodentia.

Gliding mammals are also found in the order Dermoptera. The Philippine colugo (placental) and the sugar glider (marsupial) have the same type of membrane and glide equally well.

The common mole and the pouched mole serve as classical examples of the phenomenon of periodicity. The subterranean way of life, insect feeding, cylindrical body, and reduced eyes and ears make these two species a paradigm of the parallel occurrence of the same structural and functional solutions (Figure 1).

The shrew, anteater, rhinoceros, and wild boar complete this collection of animals whose structural and functional features are "carbon copies."

A COHERENT PACKAGE OF STRUCTURES AND FUNCTIONS ACCOMPANIED THE EMERGENCE OF SABRETOOTHS AND ANTEATERS

The enlargement of the upper canines in carnivores and marsupials was not an isolated structural and functional event. A series of modifications seemed to have occurred simultaneously both in the skull and jaws of the sabretooths: (1) The face is always short, (2) the neck musculature is strong, (3) the lower jaw opens in a wide angle, (4) the canines show a backward curve and serrated edge, and (5) the jaws have powerful muscles. Additionally, these animals digested meat and had the same type of body shape (Savage and Long, 1986).

Anteaters occur in the monotremes, the marsupials and the placental mammals. The species of these three distinct groups have the following in common: (1) elongated snouts, (2) small eyes, (3) long tongue, (4) reduced or absent dentition, (5) large claws on the fore feet, and (6) digestive enzymes that cope with the formic acid coming from the ants. Examples include the *Tachyglossus* (monotreme), the giant anteater *Myrmecophaga* (placental), the scaly pangolin *Manis* (placental) and the pouched *Myrmecobius* (marsupial).

RECURRENCE OF THE SAME STRUCTURAL AND FUNCTIONAL PACKAGE IN GLIDING SPECIES

Gliding placentals and marsupials are characterized by: (1) a large membrane of skin formed between the four limbs, (2) the membrane is attached to the neck and sides of the body, (3) it extends to the fingers and the toes, (4) it also extends to the whole tail or part of it, (5) the membrane is furred, and (6) it forms a large rectangle.

In addition to a membrane, gliding demands: (1) the appropriate muscles, (2) the right amount of energy, and (3) coordination by the nervous system. Some species cover distances of 100 meters when they glide, and all of them do it with efficiency.

The capacity to glide has evolved independently in three families of marsupials during the mid-late tertiary period. They are the greater glider (*Petauroides volans*) in the family Pseudocheridae, the sugar glider (*Petaurus breviceps*) in the family Petauridae and the pygmy glider (*Acrobates pygmaeus*) in the Burramyidae (Macdonald, 1984). These animals are practically indistinguishable from the placental gliding squirrels and colugos.

PERIODICITY IS THE RESULT OF MODIFICATIONS AFFECTING MOST PROPERTIES

When the placentals are compared with the marsupials nearly all body structures and functions are found to have participated in the establishment of their "equivalence." At least the following structures are involved: (1) body configuration, (2) body size, (3) skeletal traits, (4) place of muscle attachment, (5) shape of cranium, (6) shape of mouth, (7) type of tongue, and (8) type of teeth. And at least the following functions are involved: (1) position of the eyes, (2) reduction in eye size, (3) protrusion of the eyes, (4) feeding habits, (5) type of diet, (6) digestive functions, (7) strength of muscles, (8) running capacity, and (9) gliding capacity. Obviously other structures and functions, though not as prominent, were also involved.

In essence, one is witnessing mammal mimicry. The species of marsupials and mammals are as similar to one another as insects from different orders that turn out to be virtually identical (See Chapter 12, Figure 1). The marsupials and the placentals seldom inhabit overlapping areas, whereas the insects that resemble each other usually coexist in the same territory. One has refrained from using the term mimicry because they live mainly on different continents and do not usually compete for food or environmental niches. However, the biological solution has been the same: mammals mimic one another, just as insects do. The "equivalence" appears as being the product of the same internal periodicity.

Chapter 10

PERIOD HIGH MENTAL ABILITY

THE RECURRENT EMERGENCE OF HIGH MENTAL ABILITY

Although it may seem strange, mental ability per se shows periodicity. It is found in most unrelated groups and bears no relationship to a specific environment, but occurs at certain intervals of time. Consider for example cephalopods. They are constructed on the basic molluscan pattern of structure and function — a pattern which produces, on the one hand, such simple animals as bivalves and snails, and on the other, leads to highly complex patterns of behavior. Some cephalopods have the capacity to learn or be trained to perform certain tasks. Both abilities are as developed as those in several species of birds and mammals. Trained octopuses can distinguish between different geometrical figures. They were shown a figure at the same time that they got a reward (e.g., a crab to eat) or subjected to an electric shock if they made an error. This is the same type of experiment performed with dogs and other mammals. The brain of octopuses and of vertebrates has an additional feature in common: short-term and long-term memory (Figure 1) (Russell-Hunter, 1979; Wells, 1983).

Experiments with *Octopus vulgaris* were carried out to test whether octopuses can learn to perform a task by observing other trained octopuses. This is a phenomenon well established in mammals. Untrained octopuses were allowed to visually observe the choices made by octopuses trained to discriminate between two stimuli that were identical in shape and size but not in color. Once the animals saw the behavior of the trained ones, they were able to choose the same object that was picked up by their previous companions. Learning by observation occurred, no matter what the chosen object was (Fiorito and Scotto, 1992). The brain of Cephalopods possesses giant axons and in the squid (*Loligo*) there are giant nervous fibers that form a complex system of three types of very large axons whose conduction velocity is high viz. 5 to 25 $m.sec^{-1}$ (Bullock and Horridge, 1965; Katz and Miledi, 1966; Withers, 1992).

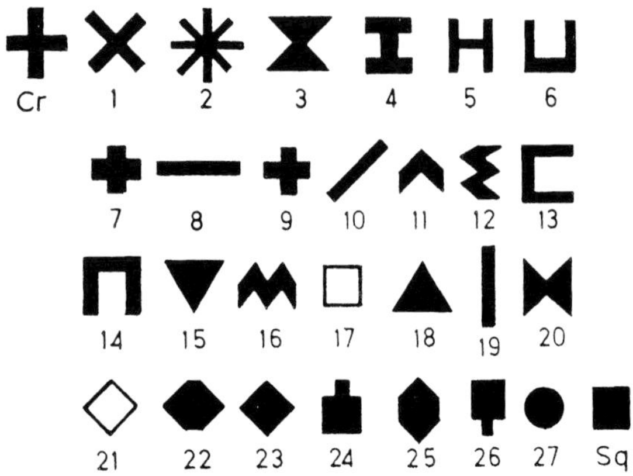

Figure 1. Octopuses can distinguish by sight between the 29 shapes shown above. They can be trained to distinguish between a cross (*above left*) and a square (*below right*) and 27 intermediate forms. They were given rewards of food when they attacked one of the indicated symbols. The numbers refer to the order in which the octopuses ranked the shapes. This turns out to be directly related to the transition from cross to square.
Source: (From Wells, 1983).

Other groups of invertebrates display an equally high mental ability. Bees, ants, and termites are social insects belonging to two different orders; namely Hymenoptera and Isoptera, that have developed: (1) a highly organized society, (2) a well defined language, (3) possess dialects, (4) decide the fate of their progeny by differential fertilization, (5) have an agriculture consisting of the cultivation of fungi, (6) raise other species such as aphids, and use their sugary products in a manner akin to humans raising cattle and collecting their milk. Yet, they are not directly related to mammals, and have a brain that is at least one million times smaller than ours (Figure 2).

The brain of a human male weighs on average 1,400 g and consists of at least 10 billion cells (Changeux, 1985), while the brain of an insect weighs between 1 milligram and 1 microgram. The whole body of the common ant *Lasius niger* weighs about 0.6 mg and the number of cells in its brain is thought to be about 10,000. The difference could hardly have been bigger.

The honey-bees are said to be the only creatures other than man that communicate by a symbolic language (Dreller and Kirchnert, 1993). Moreover, bees and ants, like birds, are known to use celestial navigation. They follow with different degrees of accuracy the movements of the sun and its relationship to the horizon. This type of navigation is a highly complex process since the rate of movement of

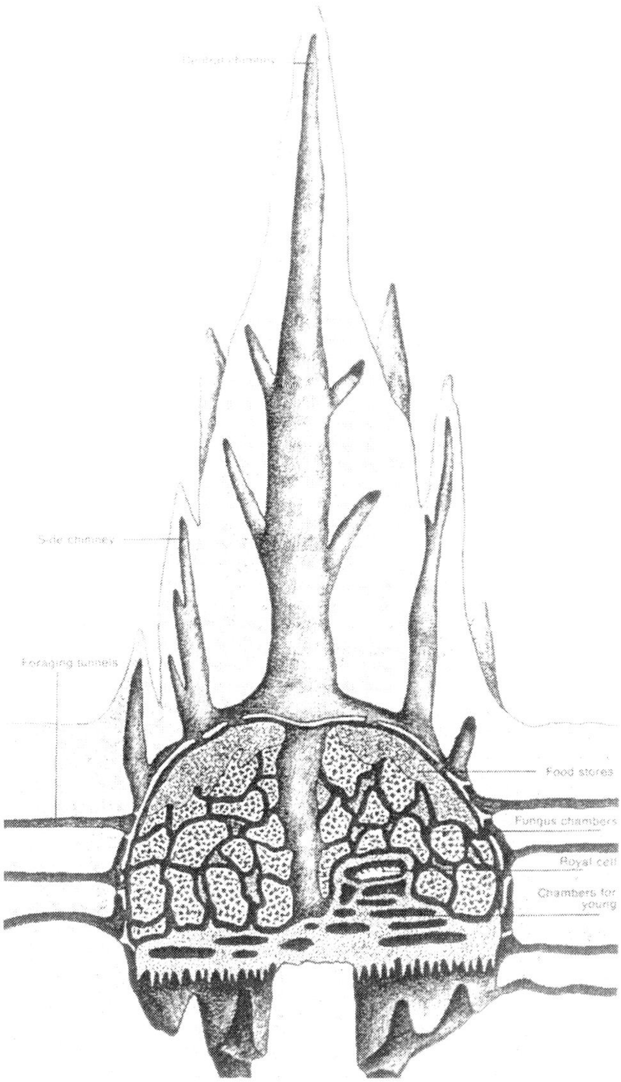

Figure 2. The complexity of the termites' nest is evident from this cross-section. The central chimney is a ventilation shaft that helps to keep the temperature constant. Outside changes of as much as 13°C occurring between day and night do not affect the inside temperature. Other compartments function as: foraging tunnels, food stores, chambers where fungi are cultivated, the royal cell and chambers for the young termites.

Source: From Pope, 1986.

the sun depends on time of day, time of year and geographical latitude (Wehner and Müller, 1993).

It may be pointed out that the cephalopods inhabit the oceans. The social insects and the mammals live on land and have emerged at totally different geological times.

Just as in the case of other periodic functions, the high mental activities are not necessarily displayed by closely related species. Chimpanzees and gorillas are not more intelligent than dolphins, though they are the most closely related species to humans. Both in wasps and bees related species show no social behavior. According to Hinde (1981), social organization has evolved independently 11 times in the bees, wasps, and ants, and only once in the other insect orders (in the termites).

Many other functions show periodicity, but they are either of less importance or there is little information about them.

PART III

HOW THE ORDER INHERENT IN MATTER AND ENERGY PAVED THE WAY TO BIOLOGICAL PERIODICITY

Chapter 11

THE THREE EVOLUTIONS THAT PRECEDED THE BIOLOGICAL ONE

ORDER IN THE ORGANIZATION OF THE SOLAR SYSTEM

Press and Siever (1982) draw attention to the "amazing regularities" occurring in the solar system. These are:

1. The planets revolve around the sun in the same direction and in nearly the same plane.
2. Most of the planets' moons revolve in the same direction.
3. The orbits of the planets are all elliptical and nearly circular.
4. The planets rotate in the same direction as their revolution. The only exceptions are Venus and Uranus.
5. Each planet is situated at a distance from the sun which is approximately twice as far as the next inner planet. This is the *Titius-Bode* rule.
6. The sun has a mass that is 99.9% of that of the whole solar system. However, 99% of the angular momentum is concentrated in the large planets (speed of turning in relation to the mass of the turning object).
7. The inner group of planets, i.e., Mercury, Venus, Earth, and Mars are small, rocky and dense, whereas the outer group of planets, i.e., Jupiter, Saturn, Uranus and Neptune consist of gigantic gaseous bodies with a low density (Table 1).

SIMILARITIES AND DIFFERENCES BETWEEN THE ORGANIZATION OF THE ATOM AND THE SOLAR SYSTEM

When the atom structure was discovered by Niels Bohr in 1913 it was compared to a solar system in miniature — a rather crude simplification then. Evidence which establishes this more clearly is now available. There is a correlation between the ratio of the mass of the atomic nucleus and its electron cloud, and that of the sun

Table 1. Main Dates in Autoevolution

Events	*Billions of Years Ago*
Origin of the universe	15 ± 4
Formation of the solar system	4.6
First self-replicating system	3.5 ± 0.5
Prokaryotic-eukaryotic divergence	1.8 ± 0.3
Plant-animal divergence	1.0
Invertebrate-vertebrate divergence	0.5
Mammalian radiation beginning	0.1

Source: (Based on data from Doolittle et al., 1986).

and its planetary system. The atomic nucleus represents 99.95% and the electron cloud 0.05% of the mass of the atom. The sun represents 99.85% and the planets 0.15% of the total mass of the solar system (Jaffe, 1988).

The main difference between the two levels lies in the fact that the planets describe elliptical orbits that lie in about one plane, whereas the electron moves in an orbital denser in some regions than in others but having a specific shape. Moreover, the electron is restricted in its movements and combinations, and in these processes it follows well-established rules, called Hund's rules (Figure 1).

The positions of the planets around the sun allow the location of celestial bodies between them. This does not hold for the electrons that circle the nucleus. Only a few discrete orbits are allowed. In Bohr's planetary model of the atom, each electron is in a particular orbit and, even if it changes position, it can never be located halfway between two orbits. The atom can only handle energy of certain sizes, the so-called quanta (Von Baeyer, 1992; Asimov, 1992; Masterton and Hurley, 1993).

PRINCIPLES GOVERNING THE AUTOEVOLUTION OF ELEMENTARY PARTICLES

Autoevolution is the transformation process which is inherent in matter and energy (Lima-de-Faria, 1988).

Nature uses only one quark pair and one pair of leptons, that it combines in well defined ways, to construct our world (Mulvey, 1979; Von Baeyer, 1986). This is due to the fact that elementary particles can now be classified into two distinct families: quarks and leptons. The electron is a lepton. Combinations in pairs or groups result in the formation of larger particles. This type of evolution was subject to certain restrictions; for example, only left-handed neutrinos have been found in nature. Restrictions also occur in the association of electrons with other elementary particles. Thus, the electron can not move at random but can only circle on a certain

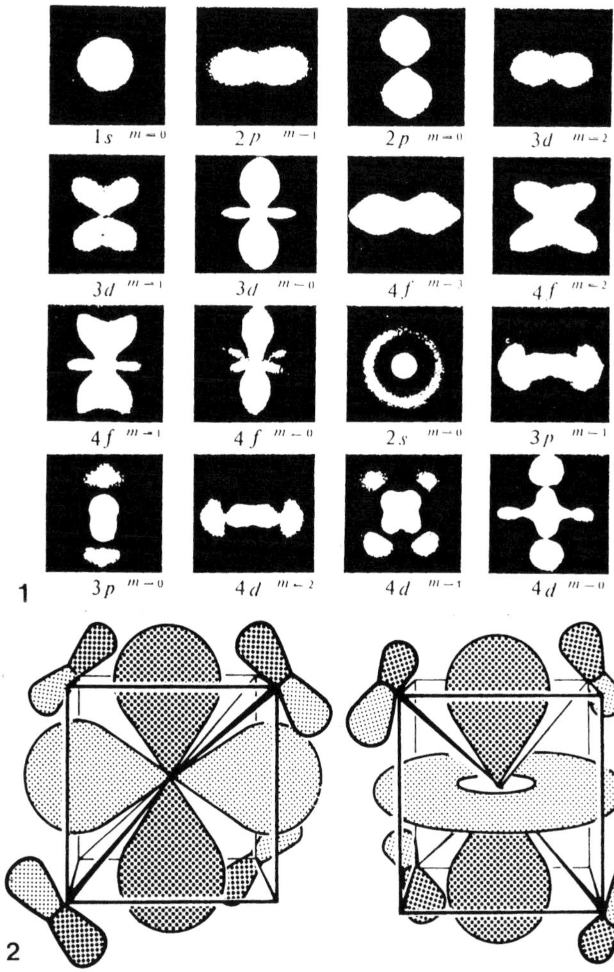

Figure 1. Forms of electron clouds produced by different states of the same atom. (1) Space distribution of the electron cloud in the hydrogen atom in different states. (2) Overlap of electron clouds in phosphorous and oxygen atoms which combine in the phosphate tetrahedron.

Source: (1) From Vainshtein et al. (1982). (2) After Cruickshank (1961), from Saenger (1988).

move at random but can only circle on a certain number of privileged orbits, each of which corresponds to a certain energy (Figure 1) (Amaldi, 1966).

Of the 20 elementary particles hitherto known, only four are fully stable: photon, electron, neutrino, and proton. All other particles eventually decay into these particles (Pitt, 1988). This is to say that self-assembly and self-disassembly already occur at the level of elementary particles. The protons result from the self-assembly of quarks, and mesons self-disassemble into other components. Moreover, it turns out that the formation of the universe has not been a slow and continuous process but rather a rapid and abrupt event (Weinberg, 1977).

What emerges from the present state of particle physics is that elementary particles have had an evolution of their own, and that this evolution already contained restrictions that led to canalization of the process. Structures and anti-structures were involved in the formation of new particles, and this followed symmetry principles. The variation that was produced derived mainly from combination and was not continuous.

A higher degree of order is now known to exist among elementary particles. The principles that this order follows have laid down, at the primeval level, the rules and the type of canalization that directed all future evolution, whence the term "autoevolution" (Lima-de-Faria, 1988, 1991b).

THE EVOLUTION OF THE CHEMICAL ELEMENTS

The chemical elements known to date number a little over one hundred and have structures and properties that originate from a single basic form: the hydrogen atom (Figure 2) (Bynum et al., 1981). Beginning with the universe's formation, hydrogen was mainly synthesized from quarks. Helium, in turn, originated from hydrogen. Carbon, the key element of compounds in living organisms, had a different origin. Its atoms were formed from three helium atoms in the interior of stars (Von Baeyer, 1986).

At present, helium is constantly being formed on the earth's surface through radioactive decay of minerals, and then released into the atmosphere. Hydrogen is also being built anew as a result of photochemical processes, which cleave the water molecules at altitudes above the stratosphere (110,000 meters). The same elements are also being built in the sun and are a main constituent of the solar wind (Allen, 1984).

Since the discovery of radioactivity it is well established that elements can be transformed into other elements. However, not all kinds of forms are possible. The element is transformed not into a loose new configuration but into a stable state, or a transitory state that later becomes stable. When a nucleus of aluminum is bombarded with an alpha particle it emits a proton and becomes a stable nucleus of silicon. Elements that have suffered a transmutation to another element, as in the case of the carbon-nitrogen conversion, can revert to the original element when the energetic conditions favor such a transformation.

Thus, the evolution of the chemical elements is that they have descended from hydrogen and were formed by electrons, protons, and neutrons which combined in different ways. Combination was possible because of the order in the process of element transformation, the result being a small number of elements, namely, 106. Elements can be transformed into other elements by means of self-assembly and

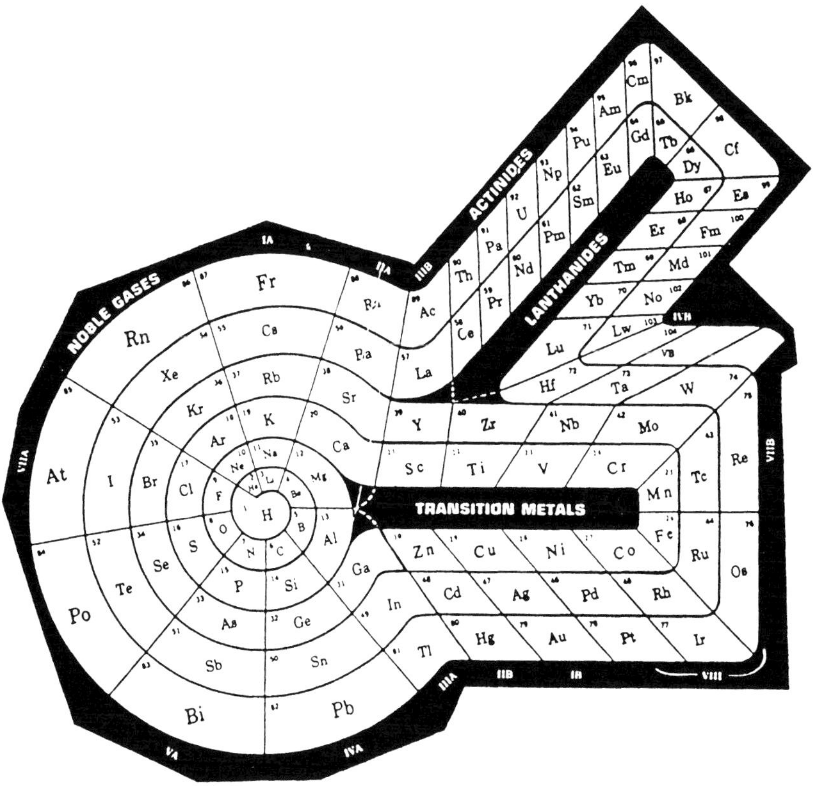

Figure 2. The periodicity of the chemical elements displayed in the form of a spiral chart. The diagram shows strict periodicity in the properties of most elements but at the same time it exposes, equally well partial deviations from this periodicity, chiefly found among the actinides, the lanthanides and the transition metals. Compare this with Chapter 1, Figure 1.

Source: (After Benfy, 1964; From Jaffe, 1988).

self-disassembly processes. However, there are restrictions, since only specific combinations are possible (Segrè, 1980; Asimov, 1992). The crucial factor seems to be the energy state of the structure itself: it determines the element transformation. This particular point is of key significance to an understanding of the stabilization of biological structures during evolution.

THE ELECTRONS OF THE CHEMICAL ELEMENTS CONTROLLED THE SPECIES OF MINERALS THAT WERE ALLOWED TO BE FORMED

Crystal chemistry has advanced so rapidly that it has led to an important conclusion: that the nuclear structure of atoms governed the abundance of chemical elements in the universe, whereas their electronic structure controlled the species of minerals that were allowed to form. Nuclear stability depends on proton and neutron properties, while crystal stability depends on chemical bonding mediated by electrons (Jaffe, 1988).

As few as 18 of the 106 available chemical elements make up 99.98% of the Earth's crust, and as such, are the most abundant among the minerals. All of them occupy the first four rows of the periodic table (Dickerson and Geis, 1979). This is consistent with the view that there has been a strong restriction in the formation of the minerals dictated by the internal properties of the atoms.

THE PROPERTIES OF CRYSTALS RESULT FROM THE PROPERTIES OF THEIR ATOMS

Pauling (1960, 1987) recognized that the bonding of atoms in crystals results from five types of bonds all of which involve different electron behaviour. These bonds are: ionic bond (electron transfer), covalent bond (electron sharing), metallic bond (electron mobility), hydrogen bond (electron orientation), and van der Waals bond (electron synchronization). The joining of atoms into crystals may thus represent a combination of these five different bond types. As for the chemical reactions leading to crystallization, they involve the self-assembly and self-disassembly of atoms — a condition determined by the types of electron behaviour.

Take for example crystalline sodium chloride. Its properties are determined by the way in which sodium and chlorine ions aggregate to form the crystal. The density itself of the crystal results from the equilibrium distance between these ions, that is, the size of the ions; while the crystallization system is determined by the symmetry of the ionic arrangement. In regard to the shape of the crystal, this is primarily determined by the internal ionic arrangement but secondarily by the surrounding medium. Salt is known to form cubes in a water solution, but produces octahedra if the crystals are allowed to grow in urea. The cleavage is independent of shape. Both cubes and octahedra show the same cleavage. Both hardness and high melting point are the result of strong electrostatic attraction between the cations (Na^+) and the anions (Cl^-).

EXAMPLES OF MINERAL EVOLUTION

Pyrite (FeS_2) exemplifies mineral evolution in several respects. (1) It has produced many crystal types which occur as cubes, pyritohedrons, octahedrons, twinned

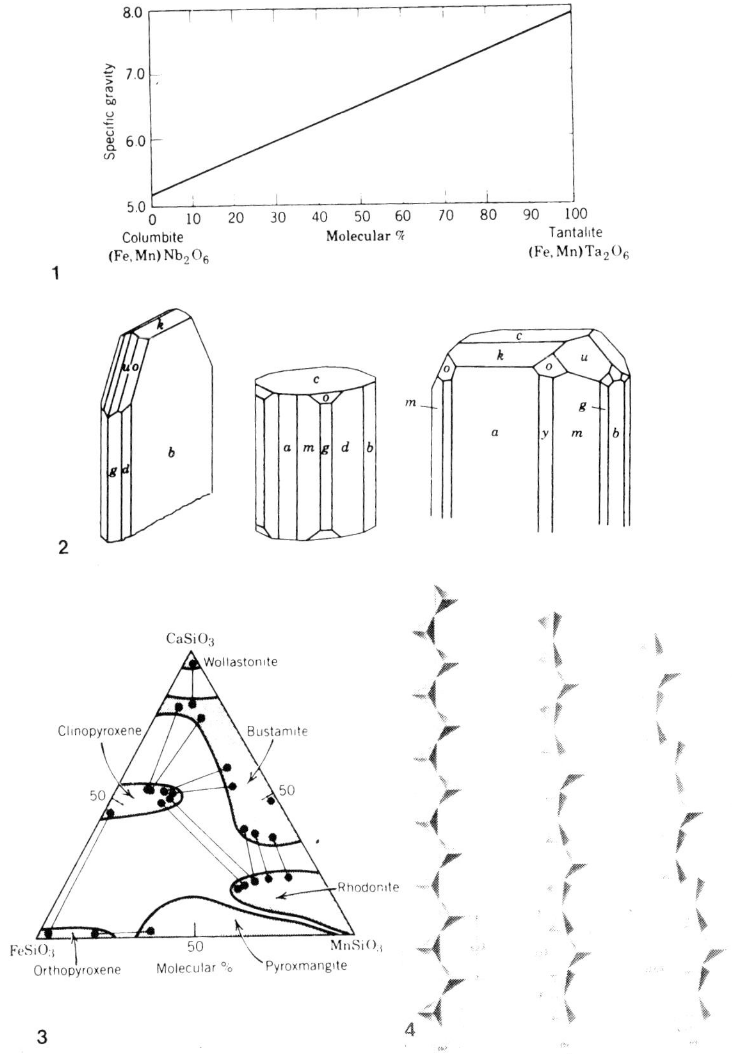

Figure 3. The evolution of the minerals. (1) Variation of specific gravity with composition during the transition between columbite and tantalite. A complete solid solution exists from columbite to tantalite. (2) Crystals formed during the transformation from one to the other of these minerals. (3) Transformation of one mineral species into another as a result of partial changes in chemical composition. Transformations among inosilicate minerals belonging to the pyroxenoid and pyroxene groups. (4) Silicates, chains of SiO_3 showing different types of atomic orientation in pyroxene and pyroxenoid minerals.

formations, and in addition, combinations of these types of crystals. All have the same chemical composition and belong to the same crystal system (the cubic). (2) Intermediate chemical forms are known to exist. Pyrite may contain small amounts of Ni (nickel) and Co (cobalt). A series of stages exist between pure pyrite and bravoite which is $(Fe,Ni)S_2$. (3) Crystal formation can occur in two different systems: e.g., pyrite which is cubic and marcasite which is orthorhombic, yet both have the same chemical composition (FeS_2). Their formation depends on temperature.

Other minerals such as columbite and tantalite (which are also iron compounds) show a continuous transition between the two species and their crystals have intermediate forms (Figure 3).

More complex relationships exist between pyroxene and pyroxenoid groups. These are inosilicates in which the element silicon has a primary role in the formation of long chemical chains. They have properties similar to those of carbon chains involved in the formation of cellular macromolecules (Figure 3).

Calcite displays more than 2000 combinations of crystallographic patterns, all of which are within the hexagonal system. The same is true of the common mineral quartz.

THE RIGID ORDER THAT HAS DIRECTED MINERAL TRANSFORMATION

The formation of crystals in minerals is subject to severe restrictions. This becomes evident from the fact that the about 3,000 minerals recognized as "valid species" are built by only five two-dimensional lattices. And when these lattices are combined perpendicularly, only fourteen three-dimensional space lattices are possible. These are unique. Seven of them are considered primitive, and represent the seven crystal systems to which all minerals belong.

The number of symmetry combinations is limited and small. The total number of non-identical symmetry elements and their combinations results in 32 crystal classes. When the 14 space lattice types are combined with the symmetry inherent in the 32 crystal classes, only 230 space groups are derived. Other arrangements based on the magnetic moment and other properties can arise, but the number of periodic arrangements is still finite. These are the ways in which atoms in crystals are arranged in space as polymers, i.e., as homogeneous periodic arrays (Klein and Hurlbut, 1985).

The smallness of the numbers allowed is impressive: 5, 7, 14, 32, 230.

PRINCIPLES OF MINERAL EVOLUTION

Mineral evolution has been conditioned by that of the chemical elements. The evidence may be summarized as follows:

1. The species of minerals that were formed were controlled by the electrons of the elements involved.
2. Mineral crystallization is determined by bonds in atoms that involve different types of electron behavior.
3. Several minerals show a continuous transition between two mineral species and their crystals occur in intermediate forms.
4. The number of symmetry combinations in crystals is limited and small.
5. Minerals with different chemical compositions may have the same pattern.
6. The same chemical composition may give rise to different patterns, depending on the physico-chemical conditions of the environment.
7. All mineral transformations occur by self-assembly and self-disassembly of atoms and molecules.
8. The energy state of the constitutive atoms governs the crystal pattern that is formed and maintained.

Crystals are the structures that suddenly reveal the order that is concealed in nature. They display, in a startling way, the rigid channels that atoms are obliged to follow.

SIMILAR FEATURES OF THE THREE EVOLUTIONS THAT PRECEDED THE BIOLOGICAL ONE

In summary, elementary particles originated from primeval matter and energy. The particles led to the organization of the chemical elements which in turn guided the formation of the minerals. Several features are common to the three evolutions.

1. Few primary forms were used: e.g., two in elementary particles (quarks and leptons), only one in chemical elements (hydrogen), and by and large 18 elements in minerals.
2. All secondary forms arose by combination of the primary ones.
3. All combinations occurred solely by self-assembly and self-disassembly.
4. The number of combinations was finite and small: 20 elementary particles, 106 chemical elements and about 3000 minerals.
5. Severe restrictions were involved at each stage, only certain forms being allowed to emerge.
6. Symmetries participated at every level.
7. Every evolution led to an increase in complexity.
8. The main factor governing the type of form that prevailed in evolution was the energy state of the structure.

These considerations make it possible to gain an understanding of the emergence of biological evolution as an immediate and obligatory consequence of these earlier physico-chemical processes.

PART IV

THE PRODUCTION OF "CARBON COPIES" AT THE DIFFERENT ORGANIZATION LEVELS

Chapter 12

ATOMIC, MOLECULAR, AND BIOLOGICAL MIMICRY AND ITS SIGNIFICANCE FOR PERIODICITY

ATOMS MIMIC OTHERS IN THEIR PROPERTIES

The term *mimicry* has been applied rather frequently in zoology to the situation in which animals belonging to different species have characteristics resembling one another; or to the situation in which animals have characteristics similar to those of plants, and plants to those of animals (Holmes, 1979). This is not surprising, particularly since mimicry already exists at the atom level. Atoms having different numbers of protons and neutrons in their nucleus, and different numbers of electrons in the surrounding shells can have similar chemical properties. Atomic mimicry made it possible to establish the Periodic Table of the elements.

Knowledge of the organization of the electrons around the nucleus revealed that the atoms mimicking one another are those that have the same number of electrons in their outer shell. An example is provided by group IIA in the Periodic Table where one finds beryllium, magnesium, calcium, strontium, barium, and radium having similar properties mainly due to the presence of two electrons in the outermost shell (Chapter 1, Figure 1 and Chapter 11, Figure 2).

MOLECULAR MIMICRY IN MINERALS

The minerals zincite (ZnO), bromellite (BeO), greenockite (CdS), wurtzite (ZnS), and swedenborgite ($NaBe_4SbO_7$), though having different chemical compositions, produce crystals of similar external symmetry. They fall in the same crystal class (Bloss, 1971). Interestingly, inspection of their formulae discloses that they have certain atomic features in common that may well account for such molecular mimicry. The five minerals are compounds of oxygen or sulfur both of which are located side by side in the same group of the Periodic Table. This clearly means that they have similar chemical properties.

Another but quite different group of minerals are chalcanthite ($CuSO_4{\cdot}5H_2O$), albite ($NaAlSi_3O_8$) and anorthite ($CaAl_2Si_2O_8$). They belong to the same triclinic crystal class. What is important here is to realize that copper sulfate, though strikingly different from an aluminum silicate, produces a crystal which is of the same type as that of an aluminum silicate.

THE ELECTRONIC MECHANISM RESPONSIBLE FOR MOLECULAR MIMICRY

The question now to be asked is: What is the atomic origin of molecular mimicry? This is perhaps best answered by discussing fluorite (calcium fluoride, CaF_2). Fluorite is isostructural with barium chloride, $BaCl_2$. The atoms are arranged in the same way in both crystals. Barium occupies the same position in the crystal lattice as calcium, and the chlorine atom the same position as the fluorine atom (Bloss, 1971). This substitution ability derives from the fact that calcium and barium are members of group IIA in the Periodic Table which means that they have 2 electrons in their outer shell, and as a consequence, similar chemical properties. This is also the case with fluorine (F) and chlorine (Cl) which are members of group VIIA, and have 7 electrons in their outer shell. They have similar chemical properties.

Not less than 15 compounds are known to be isostructural with fluorite. Clarification of the electronic causes of similarities seems to be forthcoming (Pauling 1949, 1960; Dickerson and Geis, 1979; Sharp 1988).

MINERALS CONTAIN ATOMIC PROCESSES THAT ALLOW CHEMICAL VARIATION WITHOUT CHANGING THE FINAL PATTERN AND ALLOW VARIATION OF PATTERN WITHOUT CHANGING THE BASIC MOLECULAR CONSTRUCTION

The carbonates of calcium, magnesium, iron, and manganese all produce the same crystal form. It turns out that the atoms of calcium can be replaced by the same number of atoms of magnesium without changing the final pattern. Chemical analysis reveals that: (1) The same form can be produced by different chemical compositions that have carbonate as part of the molecule. (2) Atoms as different as iron and magnesium, when combined into carbonate, are able to produce exactly the same pattern. (3) The carbonate component of the molecule by itself could lead to the production of a different crystal growth because, in the presence of heavier atoms (e.g., strontium or barium), it forms a different atomic edifice.

It is thus clear that the minerals already contain the basic atomic processes that allow variation by substitution of atoms without the occurrence of a consequent change in the final pattern. They also contain the processes that allow variation of the pattern (by introducing new types of atoms) without there being a change in the basic construction (which in this case is the carbonate part of the molecule).

IODOFORM CRYSTALS ARE "CARBON COPIES" OF WATER CRYSTALS

Water teaches us the same lesson that has been learned from minerals since the same structural pattern can be produced by substances with very different chemical compositions. Iodoform which is triiodomethane consists of three atoms: hydrogen, carbon and iodine (CHI_3). When a saturated solution of this compound is allowed to crystallize, it forms the same type of crystals that are formed by water (H_2O), and crystallization follows the same sequence of stages (Nakaya, 1954). The final result is a flat six branched crystal with secondary ramifications with a size gradient similar to that of water crystals. The fact that the only atom shared by iodoform and water is hydrogen raises the possibility that hydrogen bonds are responsible for the production of the hexagonal and ramified pattern obtained in both situations. However, it is not excluded that oxygen and carbon may be involved in the determination of this pattern.

MIMICRY IN PLANTS AND ANIMALS MAY NOT ONLY DEPEND ON SIMILAR DNA SEQUENCES, BUT ALSO ON THE ELECTRONIC CONFIGURATION OF KEY ATOMS, PRESENT IN PROTEINS AND OTHER MOLECULES

Proteins are built from only 5 atoms: carbon (50%), oxygen (25%), nitrogen (15%), hydrogen (7%) and sulfur (in small quantities). These atoms have the same electronic properties as those that occur in minerals, and although their assembly results in more complex molecules, their electron and proton effects remain the same. Such information provides insight into the emergence of structural and functional mimicry at higher levels of organization.

At the organism level four situations are known to exist. There are plants that mimic other plants; animals that mimic other animals; animals that mimic plants, and plants that mimic animals. A prime example is the flower of the orchid *Ophrys* that resembles so closely a wasp that this insect copulates with the flower (Curry-Lindahl and Tinggaard, 1965). This situation is illustrated in Figure 1. Crystals of ammonium sulfate and potassium sulfate, though consisting of different atoms have essentially the same form. They are orthorhombic crystals. Consider now the flower of *Rosa pendulina*: it has 5 petals distributed radially that are of the same form and size. This pattern is mimicked by the flower of *Oxalis adenophylla* that belongs to

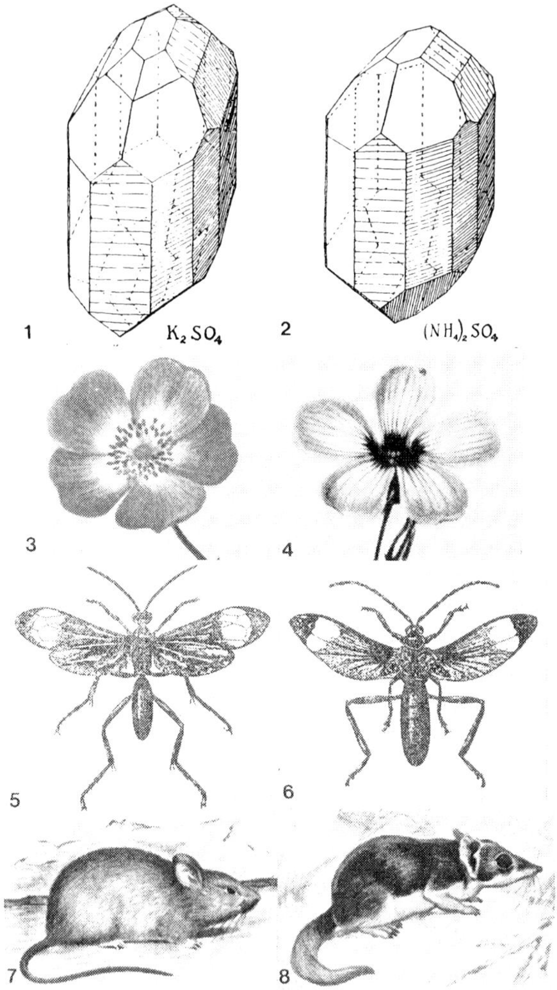

Figure 1. Mimicry in crystals, plants, invertebrates, and vertebrates. (1) Crystal of potassium sulfate, K_2SO_4. (2) Crystal of ammonium sulfate $(NH_4)_2SO_4$. Both crystals have essentially the same form and are orthorhombic. (3) The flower of *Rosa pendulina* with 5 petals distributed radially (Family Rosaceae). (4) Flower of *Oxalis adenophylla* with 5 petals also distributed radially, but belonging to a different family (Oxalidaceae). (5) The wasp *Mygnimia aviculus* of the order Hymenoptera. (6) A most similar species, *Coloborhombus fasciatipennis*, belonging to the order Coleoptera. (7) The house mouse *Mus musculus* which is a placental mammal. (8) The marsupial mouse *Dasycercus* which belongs to the non-placental mammals but is practically identical to the house mouse.

a different family. Next, in animals the classical example of this phenomenon is provided by the wasp *Mygnimia aviculus* which is nearly indistinguishable from another insect *Coloborhombus fasciatipennis* belonging to a different order (Coleoptera). Last, is the house mouse *Mus musculus*, which is a placental mammal. Its equivalent among the marsupials is the pouched mouse *Dasycercus*, which is called also a mouse because of their striking similarity both in shape and habit.

It is well established that different orders and classes have different genetic constellations, but they have retained many similar structural genes. This represents conservation for millions of years. The existing differences are found to lie mainly in regulatory DNA (King and Wilson, 1975; Wilson, 1976). Thus, an animal or plant, though not closely related phylogenetically, may be expected to produce the same type of proteins (translated from structural genes) that lead to a similar configuration or function. This is for instance the case with hemoglobin. Not only animals but plants, such as clover, contain genes for this protein. The amino acid sequence analysis reveals close similarities between plant and human hemoglobin (Ellfolk, 1972; Dilworth and Coventry, 1977). Moreover, studies of these genes in plants and animals, involving DNA sequencing and DNA-hybridization, have disclosed that they have a common origin (Landsmann et al., 1986).

However, as it turns out, genes do not even need to be the same for a similar structural or functional pattern to emerge. Animals or plants may possess a series of different DNA sequences, as found in unrelated organisms. Such different genes can still produce the same structure or function if the proteins they give rise to (or other molecules), have comparable effects. An example is found among the four proteins: staphylococcal nuclease, yeast asp-tRNA synthetase, cholera like cytotoxin and Shiga-like cytotoxin. The amino acid sequences show no significant similarity. However, all four proteins perform the same function; they bind oligonucleotides or oligosaccharides (Murzin, 1993).

Another example is provided by the 15 proteins, with completely different amino acid sequences that all form the eight-stranded alpha-beta-barrel structure, a most regular protein feature. This is an outstanding case of structural mimicry (Branden and Tooze, 1991).

In essence, then, if the molecules contain pertinent atoms that have similar effects, one can expect "carbon copies" to emerge. As pointed out by Schwabe and Travers (1993) in proteins, very similar structures can be formed from widely differing amino acid sequences and the same biological function has more than one structural solution.

Chapter 13

THE COLLABORATION BETWEEN MINERAL AND GENE PRODUCT

METALS AND OTHER ELEMENTS THAT ARE PART OF MINERALS CANNOT BE CREATED BY CELLS AND ARE NOT PART OF THE GENETIC CODE

It is not widely recognized that the cell is unable to put together protons, neutrons and electrons, nor build an iron atom. Simply put, the iron of hemoglobin does not derive from any information in the genetic code. As remarked by Weinberg (1989) "Iron can be neither created nor destroyed by living cells." This also applies to sulfur, copper, zinc and other elements. These elements are present in many proteins whose primary function depends on them. For example, zinc-containing proteins number more than 200. They constitute a family called zinc-finger proteins that function as transcription factors. The zinc domains bind specifically to the DNA sequences of structural genes activating or repressing them (Sluyser et al., 1993). Yet, one seldom asks: What is the origin of these elements and how have they been coupled and coordinated with the amino acids of the polypeptide? The metals and other elements occurring in cells have their origin in the mineral world.

THE MINERAL IS THE KEY COMPONENT OF MANY MACROMOLECULAR CATALYSTS — GENES DO NOT NECESSARILY CREATE BASIC FUNCTIONS: THEY MAINLY ACCELERATE THEM

Calvin (1983) has studied the problem of transition from naturally occurring inorganic catalysts to biologically generated highly specific catalysts. The evolution of a catalyst, exemplified by the iron atom, can be traced from the inorganic aqueous ferric ion to the haem molecule, and finally to the enzyme catalase (Figure 1).

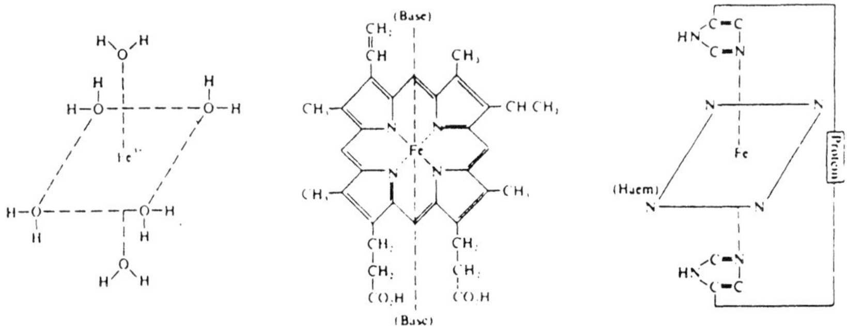

Aqueous ferric ion — Haem — Catalase

10^{-5} — 10^{-2} — 10^{5}

Catalytic activity $ml^{-1}s^{-1}$ at 0° C

Figure 1. Catalytic activity occurred in ferric ions before iron appeared in enzymes. Catalytic activity ($ml^{-1}s^{-1}$ at 0 °C) in an aqueous ferric ion is 10^{-5}; in haem (a combination of the iron atom with a protoporphyrin), it increases to 10^{-2}, and in the enzyme catalase it attains the value of 10^{5}.

Source: From Calvin, 1983.

Briefly, the iron atom alone rapidly oxidizes in the presence of water molecules and shows a variety of oxidation states. Haem is the nonamino acid portion of hemoglobin, a respiratory protein of the red blood cells, in which the iron atom combines with oxygen. Catalase occurs in blood and contains haematin. Its activity involves the oxidation and reduction of the iron in the molecule. The noteworthy aspect is that the aqueous ferric ion (Fe^{3+}) has a catalytic activity of 10^{-5}, whilst that of haem is 10^{-2} and of catalase 10^{5}. The haem group per se does not create any new function, but rather it exerts a more rapid potentiating action than pure iron. This also holds true for catalase since it produces a more rapid and greater action than that of haem. The combination of iron with amino acids results in a much more canalized event. The key property of the macromolecule does not lie in the peptide sequences but rather resides in the mineral from where function originates.

Similar considerations apply to numerous metals that are an integral part of macromolecules. These include: zinc, cobalt, nickel, copper, molybdenum, vanadium, magnesium, sodium, and potassium. These minerals, or mineral components, are known to confer upon the proteins and other molecules their main functions.

THE CALCIUM ATOM DETERMINES THE TYPE OF MINERAL AND THE PROTEIN DECIDES THE CRYSTAL SYSTEM IN BIOLOGICAL STRUCTURES

Calcium is the key atom that determines the construction of the minerals aragonite and calcite. Both minerals have the same chemical composition ($CaCO_3$) but they crystallize in different systems i.e., orthorhombic and hexagonal, respectively. Calcium forms the center of polyhedra which it builds together with oxygen. These polyhedra are formed by 9 oxygens in aragonite but only by 6 in calcite. Hence, the polyhedral configuration determines the type of mineral that will be formed.

In nature, the amount of oxygen available in the environment, as well as the temperature and pressure determine whether aragonite or calcite are produced. In living organisms, however, it is the type of amino acids in the protein that governs the crystallographic arrangement of the calcium with oxygen. This is the situation in the earstones of fishes. They consist of the mineral aragonite associated with a protein (similar to keratin) that is only 0.2 to 10% of the total content. One finds in these hearing organs $Ca^{2+}O_9$ polyhedra rather than $Ca^{2+}O_6$. Apparently, the governing factor is the amount of oxygen made available by the amino acids. An abundance of acidic amino acids in this protein makes it possible for many oxygen atoms to be available, thereby producing the $Ca^{2+}O_9$ configuration, and hence, aragonite mineralization. It is noteworthy that in the shells of molluscs, depending on the protein present, calcite or aragonite is formed.

MINERAL, PROTEIN, AND SUGAR COMBINE IN THE BUILDING OF MOLLUSC SHELLS IN ACCORDANCE WITH CRYSTAL CHEMISTRY MECHANISMS

The mantle of molluscs secretes the shell consisting of: (1) calcium carbonate crystals ($CaCO_3$), (2) a protein (of the keratin group), and (3) a mucopolysaccharide. These molecular components inter-react in a fluid region before assembling as a solid shell structure (Figure 2). First, the mineral combines with the protein as the calcium ions associate with the COO-group of the amino acid, aspartic acid. Secondly, the other part of the mineral molecule, the carbonate (CO_3) is linked via hydrogen bridges to the amino acids. Thirdly, the oxygen of the bicarbonate becomes associated with the polyhedral structures. Hence, a tight link is established between the amino acids and the mineral atoms via the oxygen and hydrogen atoms of the protein. The calcium-aspartic acid configuration, as well as the bonding of the bicarbonate on the peptide chain are shown in Figure 2. The specific protein involved varies from species to species, and this is the source of the variation in shell form (Matheja and Degens, 1968; Degens et al., 1969; Härkönen, 1986).

A similar mechanism occurs in vertebrate bones, scales, and teeth. These are composed of collagen (a protein containing mainly the amino acids glycine, hydroxyproline and proline) associated with apatite crystals. This particular min-

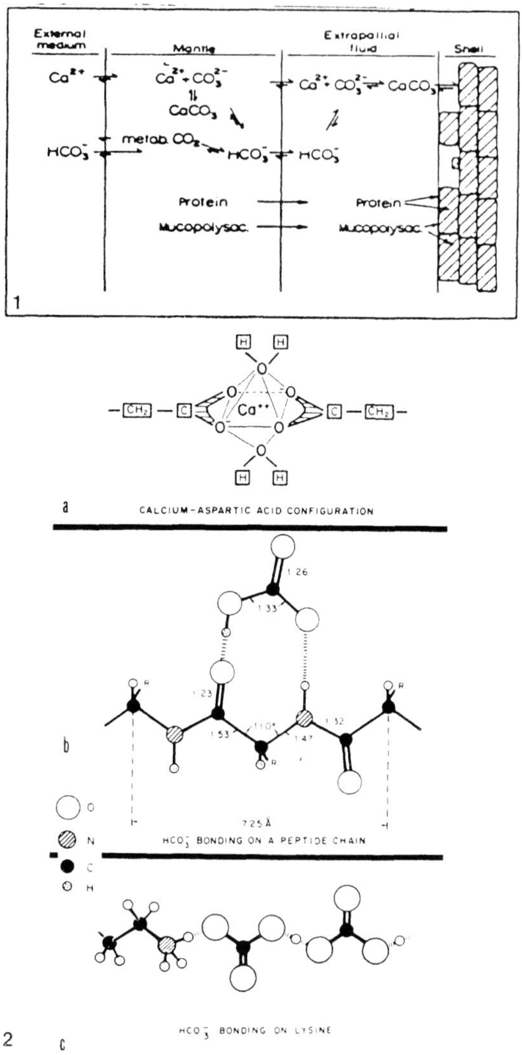

Figure 2. Atomic interactions between calcium ions and amino acids in the building of living structures. (1) Shell secretion in molluscs involves a sequence of chemical reactions in which calcium atoms coordinate with those of proteins and mucopolysaccharides. (2) Visualization of the atomic interaction between calcium atoms and the amino acid aspartic acid, as well as the way in which the bicarbonate part of calcium carbonate establishes bonds in a peptide chain and in the amino acid lysine.

Source: (1) After Wada and Grenway, from Wilbur (1972). (2) From Matheja and Degens (1968).

eral consists of calcium phosphate and fluorine i.e., $Ca_5(PO_4)_3F$. The association of the mineral with these amino acids is known to influence the crystal formation of the apatite (Degens et al., 1969; Härkönen, 1986).

The picture which emerges is that the calcium atom is the primary agent of mineral formation and that the protein is the secondary factor determining the different crystallographic forms. In other words, the protein is not responsible for creating a new situation; rather, it is only controlling the choice between certain alternatives which are governed by the chemistry of the calcium atom.

COOPERATION BETWEEN THE MINERAL AND THE GENE

A similar situation has been described in detail in sea urchins. There is a close interaction between minerals and genes in the formation of their pattern. Biophysical results show that the gene products function only as secondary components, thus leading to the fixation of a given alternative. The spicules present in these animals are the result of calcium carbonate and magnesium deposition. The gene product intervenes by being incorporated into the mineral structure in the form of an organic matrix which constitutes only 1% of the skeleton of sea urchins. The atoms of calcium carbonate assemble in accordance with strictly physical principles without the intervention of gene products, and thus build the bone's basic structure. The gene product incorporates only into the mineral scaffolding, giving it the shape that becomes characteristic of a species. Its secondary intervention decides whether the shell is going to be spherical or ovoid, short or long. The mineral together with the gene govern the characteristics of the organism (Inoue and Okazaki, 1977).

The earstones, the shells, the bones, and the spicules contain the same calcium atom that is now known to function as a second messenger in cell communication (Catterall et al., 1989). In all situations, it is the electronic properties of the calcium atom that direct the structures and functions of those systems in which it participates.

Chapter 14

CELLULAR PROCESSES INHERITED FROM THE ELEMENTARY PARTICLES AND THE CHEMICAL ELEMENTS

RIGHT-HANDED AND LEFT-HANDED FORMS COULD NOT BE ERASED—THEY OCCUR FROM ELEMENTARY PARTICLES TO HUMANS

The discovery of many new elementary particles has led to the finding that neutrinos are either right-handed or left-handed and that their spin of rotation can be clockwise or counter-clockwise. Since atoms result from the combination of elementary particles, it is perhaps not surprising that the carbon atom is also either right-handed or left-handed (Figure 1). This phenomenon is evident among the minerals: quartz crystals, for example, occur as left- and right-handed structures (Figure 1).

When performing a chemical synthesis in the laboratory a 50:50 mixture of right-handed and left-handed forms is usually produced. Though sugars and amino acids in solution occur in these two proportionate forms, this is not so in cells. One of the two tends to prevail (Figure 1). An interesting case is provided by DNA. Most DNAs are right-handed but at very high salt concentrations (e.g., Na^+), they form left-handed helices (Cantor, 1981; Wang et al., 1982). The different configuration of the purine nucleotides is considered to be responsible for the left-handed structure. Its backbone assumes a zigzag look, hence the term Z-DNA (Figure 2) (Hill and Stollar, 1983).

These macromolecules are the main components of living organisms which in turn build organs and structures which are also left-handed and right-handed. This is already the case in plants and invertebrates. Mollusc shells have these two asymmetric forms, whose origin can be traced to the first egg division of the animal

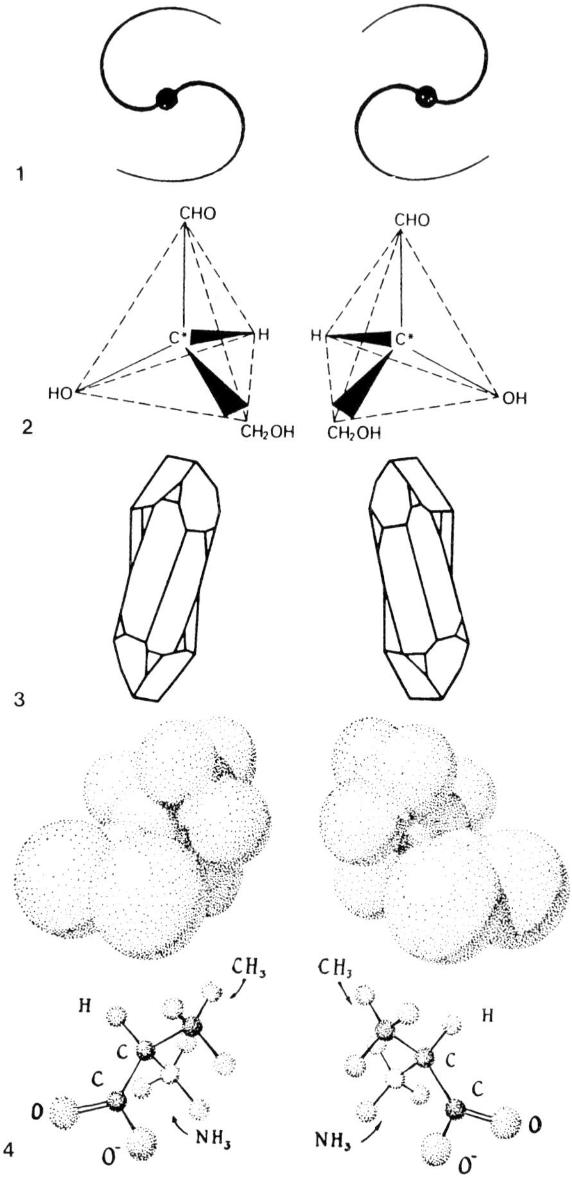

Figure 1. Occurrence of left-handed and right-handed structures from galaxies to humans. (1) Spiral galaxies of the S-type (left-handed) and Z-type (right-handed). (2) Carbon atom bonds in left-handed and right-handed glyceraldehyde. (3) The mineral quartz in the left and right forms. (4) The amino acid alanine in the left and right form. (See Figure 2).

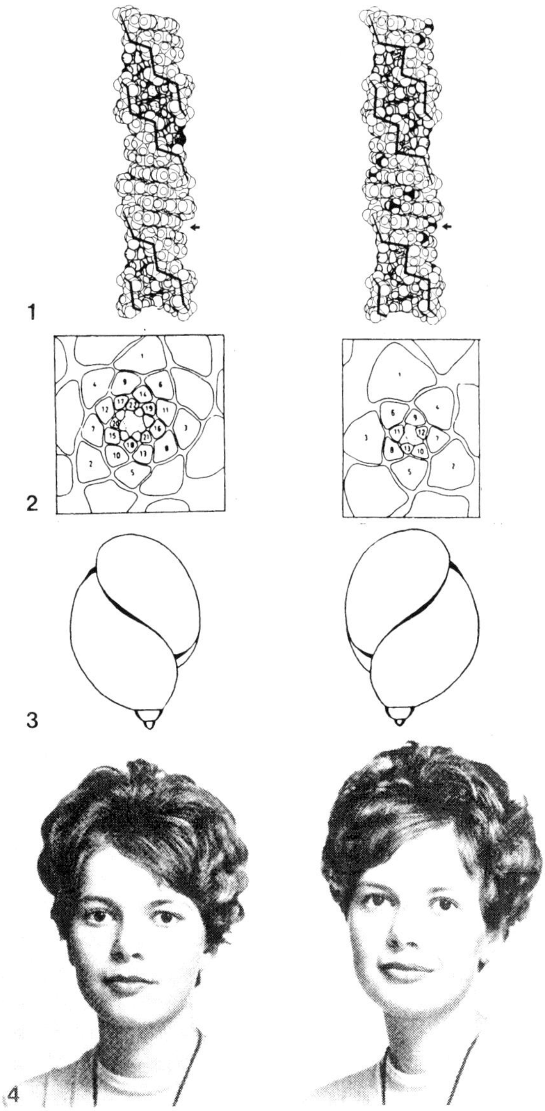

Figure 2. Occurrence of left-handed and right-handed structures from galaxies to humans. (1) DNA (Z configuration) in its left (unmethylated) and right (methylated) forms. (2) Sections through plant shoots (conifer *Araucaria excelsa*) showing spiral divergence left- and right-handed. (3) Shells of the snail *Limnaea* showing left and right orientation. (4) The identical twins Monica and Gerd. Monica is left-handed and has the forelock to the right. Gerd is right-handed and has the forelock to the left. They are mirror images of each other.

(Figure 2). Moreover, humans manifest this universal asymmetry. One egg twins are usually extremely similar in most morphological and functional characters but they may differ in the symmetry of their hair distribution. One of them may have a left-handed form, while the other a right-handed one (Figure 2).

The existence of right-handed and left-handed forms in higher organisms is taken to mean that they do not represent an accidental event. There seems to be a connection with earlier levels of organization. The continuity of the phenomenon from elementary particles, to atoms, to minerals, to molecules and to macromolecules implies the existence of an autoevolutionary process that could not be erased and is sustained as far as the human species.

THE OCCURRENCE OF FIELDS AT EVERY LEVEL OF ORGANIZATION

A field that was discovered at the biological level was originally met with suspicion for several reasons. First, evidence of a molecular component that could be considered responsible for the field was difficult to prove before the advent of molecular biology. Secondly, a field in intercellular communication had no defined material basis until chemical messengers were discovered and shown to be involved in this process. And thirdly, the well-established fields in physical phenomena seemed to have no links with those found in biological structures, since there were no intermediate means by which to establish close links between elementary particles and living organisms. The situation has lately changed rather radically since these three gaps in our knowledge have been bridged. Fields occur at every level of organization, and in addition to having a well-defined material basis, they show the same essential type of organization.

MATTER CONSISTS OF QUANTUM FIELDS

Electric, magnetic, and gravitational fields express themselves by producing forces in space, which act on material objects lying within a specific area. Every field has a definable strength and a direction at a given point in the region surrounding its origin. Whereas gravitational fields have to do with actions at enormous distances, such as the attraction between the sun and planets, quantum fields deal with submicroscopic actions at the level of elementary particles. The whole material universe is considered to consist of fields created by the elementary particles themselves. At one of its extremes, for instance, there is a field which concerns the electron. As Dyson (1953) aptly remarks, "it is astonishing, even to a physicist, that our solid world of animals, trees and stones is built by quantum fields and nothing else."

THE PROPERTIES OF MAGNETIC FIELDS FACILITATE THE UNDERSTANDING OF FIELDS AT OTHER LEVELS

The natural magnetism found in magnetite (Fe_3O_4) is attributed to the mineral cooling from the molten native state in the presence of the earth's magnetic field. This mineral attracts iron.

1. When a magnet is placed in a vial containing small iron particles, the particles are noticed to attach mainly to its extremities, the so-called poles. They are designated south and north because they orient to these regions of the earth.
2. When two magnets are placed close to each other, poles of the same type repel each other, whereas those of opposite type attract one another.
3. A magnet cut into two smaller pieces would be expected to show isolation of the north from the south pole. However, this is not achieved. What one obtains are two complete magnets, each with a south and a north pole. If division continues the result is always the same. New complete magnets arise; this pattern is maintained irrespective of size.
4. The reverse experiment can be carried out. If small magnets are put together and aligned in such a way that the north pole of one coincides with the south pole of the next, then a large magnet is built, whose power equals the sum of the small magnets (Figure 3).
5. The intensity of the magnetic field decreases in a gradient with increasing distance from the origin (Boutaric, 1938).
6. The magnetic field is easily visualized by spreading small particles of iron over a sheet of paper and placing under the paper a magnetized iron bar. The small iron particles can then be seen to dispose themselves around the poles with an orientation that is in accordance with the lines of force that extend from one pole to the other. The field thus becomes self-evident (Figure 3).
7. Magnetism is the result of electrical forces at the atomic level, where it is mainly caused by unpaired electrons in certain orbitals, as well as by changes in atomic spin (Bloss, 1971).

FIELDS IN CRYSTALS AND MINERALS

In crystal chemistry the occurrence of fields is a well-established phenomenon.

The crystal field theory is based on the particular ordering of atoms in minerals. This may be illustrated by the metal ion of chromium which is located within a crystal formed by six oxygen atoms. The result is a regular octahedral configuration. The addition of an extra electron to the crystal alters this configuration but maintains the octahedral form. This is marked by a decrease in crystal energy, and an increase in stability. The field is expressed by the net change in crystal energy accompanied by the maintenance of the general pattern. The field action is mainly electrostatic and it changes with distance. This results from the orientation of specific orbitals in atom groups. The atomic relationships that build crystal fields are mostly evident in the minerals of the spinel group (Ballhausen, 1962; Burns, 1970).

FIELDS IN DIVIDING EGGS, ANIMAL EMBRYOS, AND BODY ORGANS

Embryologists coined the term *morphogenetic field* when they found that the position of the different parts within the dividing egg bore a close relationship to the position of other regions. They noticed that the egg is able to maintain its pattern when its mass was either reduced or increased. This seemed analogous to a magnetic field. Furthermore, experiments carried out with dividing sea urchin eggs, and eggs

Figure 3. Fields present at different levels of organization: Elementary particles, minerals and living organisms. (1) Rings depicting the magnetic field built around a stream of electrons, travelling in an electric current, along a wire perpendicular to the surface of the page. (2) Visualization of the magnetic field created by a mineral such as magnetite which is known to attract iron. A paper is first covered with small particles of iron and then a rod of magnetite is located in their vicinity. The small particles are found to orient themselves according to a field pattern directed by the two poles. (3) Organ specific fields in the developing embryo of the chick. Fragments, when raised on a host embryo, containing cells from these areas, differentiate only into the type of tissue that characterizes each field: (e) eye field and (h) heart field. The maximum and minimum intensities of these fields are marked by the concentration of the shaded areas. (4) A magnet with its South (S) and North (N) poles. When the magnet is divided into two pieces, this process does not result in the isolation of one of its poles, but each piece acquires its own South and North pole, i.e., the magnetic field is reshaped and maintained, irrespective of change in size. If many small magnets are aligned, and put together with all their North poles facing in the same direction, a large magnet is reconstituted in which the magnetic moment is equal to the sum of the moments of the small magnets (the magnetic moment is the force multiplied by the distance between the poles). That is, if there is an increase in size, then the field adjusts accordingly. (5) Egg of a triton (Amphibia) at the two cell stage in which the cells were slightly separated by means of a thread (*left*). Each cell does not give rise to half an embryo, but produces a perfectly normal organism. The embryonal field, like the magnetic one, is maintained when the size is decreased to half.

1

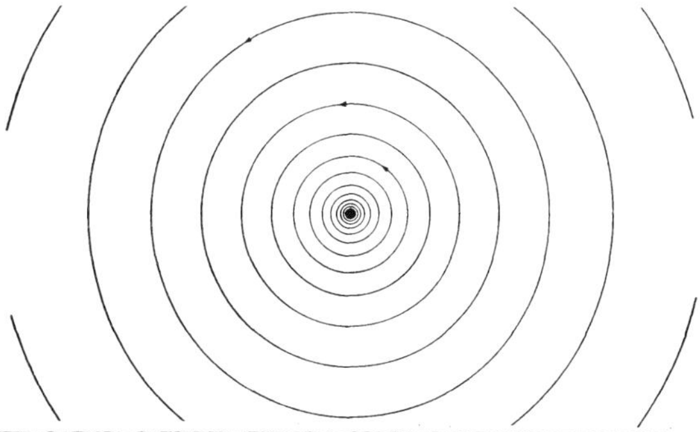

2

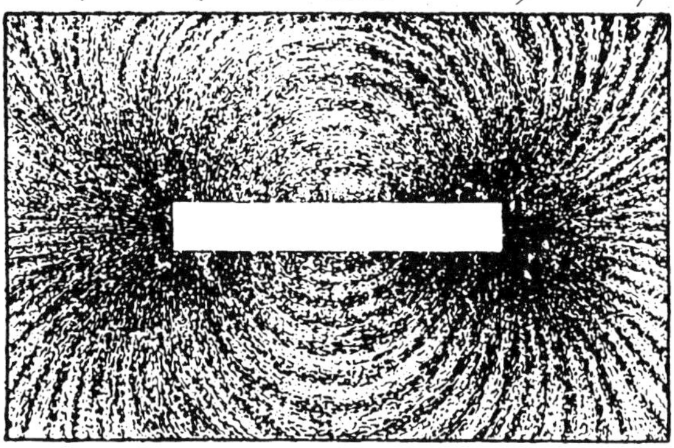

3

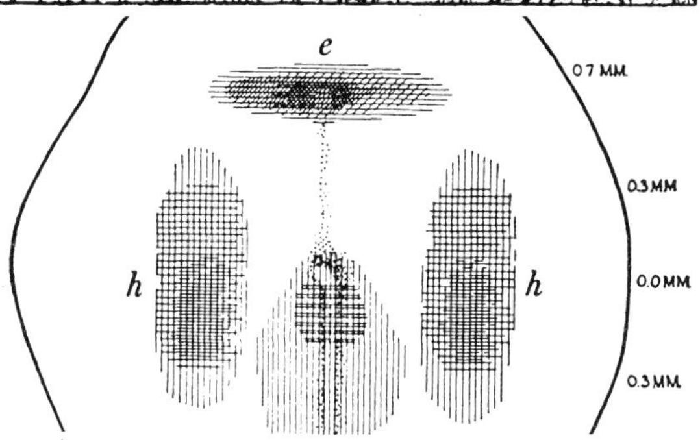

4

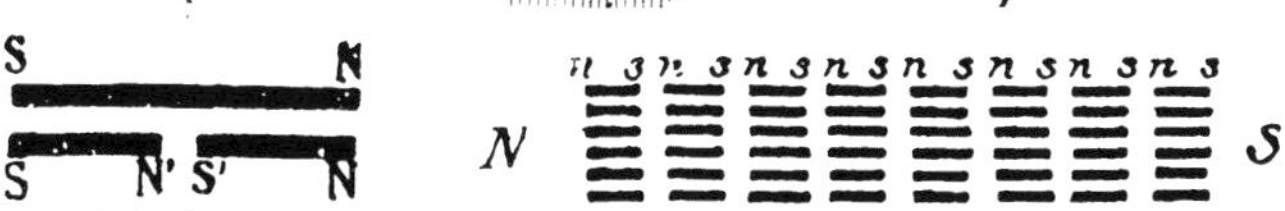

5

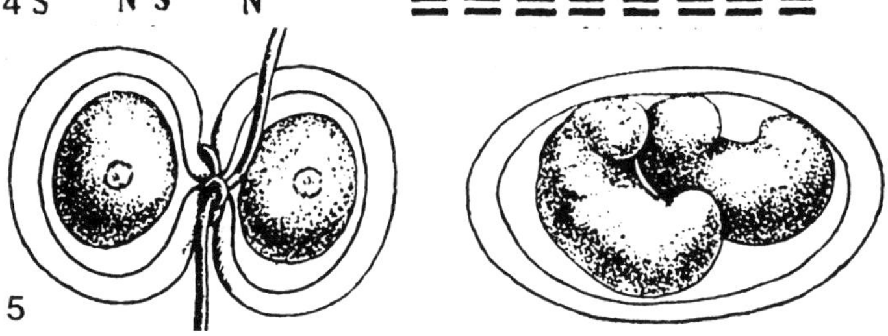

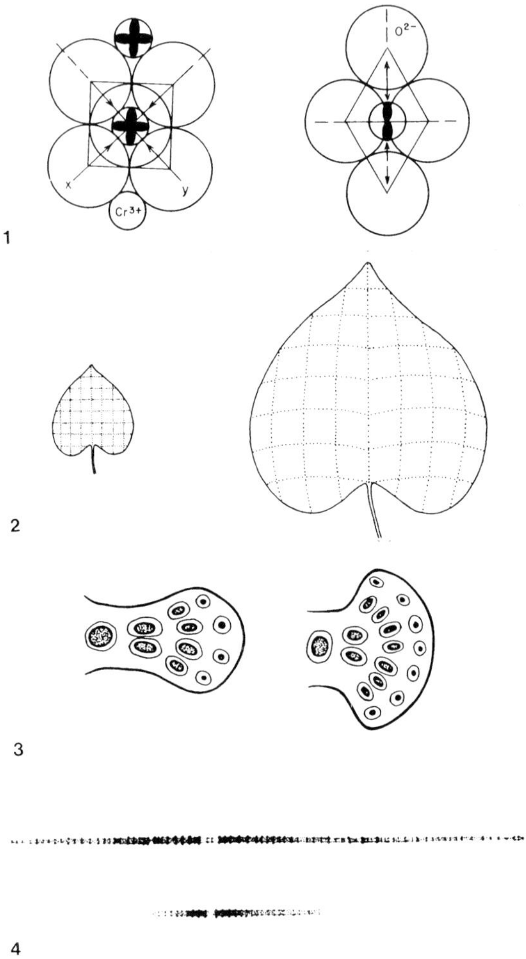

Figure 4. Field formation. (1) Crystal field. This is maintained when a change in crystal energy results from a reorientation of orbitals of a transition metal, such as chromium (Cr), surrounded by oxygen atoms; the regular octahedron (*left*) changes into another pattern that continues to be octahedral (*right*). (2) Field in plants. The growth of a leaf is uniform, thereby maintaining the basic pattern, as shown by inking a grid on a young leaf and noting the spacing of the lines when the leaf reaches its final size. (3) Animal field during cartilage formation in the chicken; normal limb (*left*) and mutated limb (*right*). The increase in the number of formation centers does not change the geometry of the organ. (4) Chromosome field in *Agapanthus umbellatus*. The same pattern is maintained in the chromosome, irrespective of the change in length and shape that it undergoes between pachytene (*upper figure*), and the second prophase of meiosis (*lower figure*).

of other species, have led to the following main conclusions concerning the properties of cellular fields:

1. Each animal field has a focal point of maximum intensity, which declines in strength with the distance from the center. This decrease follows a gradient.
2. When the total mass of the field is reduced by removing groups of cells, this does not affect the pattern of the field as a whole. A perfect larva is produced.
3. Splitting the field into two halves leads to the formation of two complete fields that are equivalent in structure to the original single field. This was demonstrated in the sea urchin by cutting early embryos into halves. Each half produced the same type of sea urchin. This type of experiment was repeated in amphibians and other vertebrates; the results were the same (Figure 3).
4. Fusion of two fields into one results in a single pattern that is identical to each original unit. If, however, the axis of orientation is different at the moment of fusion, then parts of the pattern are found to belong to one field, while some parts belong to the other field. Intermediate forms are absent.
5. The fate of an individual cell during early development depends upon its particular location within the field (Weiss, 1939; Brachet, 1957; Driever and Nüsslein-Volhard, 1988 a, b; Müller, 1990).

Body organs, such as chicken limbs, also reveal the presence of organizing fields. In the case of mutations, e.g., polydactyly, where the number of fingers is increased, a readjustment of structure occurs. The same organ geometry is maintained irrespective of the change in the number of fingers, and of their size (Figure 4).

THE NUCLEUS CAN DETECT ITS POSITION WITHIN THE FIELD OF DIFFERENTIATION DUE TO THE PRESENCE OF MOLECULAR GRADIENTS

Molecular analysis has shown that the locations of maternal messenger RNAs in ascidian embryos correlates with egg polarity and are distributed according to gradients (Jeffery, 1985). In *Drosophila*, the distribution of the *bicoid* protein follows an exponential concentration gradient, with a maximum occurring in the anterior end of the body and reaching background levels in the posterior end of the embryo (Driever and Nüsslein-Volhard, 1988 a, b). A similar antero-posterior gradient distribution for both RNA and protein has been found in this species (Mlodzik and Gehring, 1987). A nucleus seems able to detect its position in the cytoplasm by means of the level of the gradient. That is, the protein behaves like a morphogen that determines the antero-posterior position in the embryo on account of its gradual decrease in concentration. Microinjection of the cytoplasm material obtained from the wild type embryo into a recipient mutant embryo has made it possible to define the capacity of the field of cell differentiation (Anderson et al., 1985; Berleth et al., 1988; Singer, 1993).

GRADIENT AND FIELD IN PLANT GROWTH CAN BE VISUALIZED

Measurements of the growth of the stem apex have shown that a gradient of increasing growth rate extends from the top to the bottom of the apical part of the plant. This creates a field that determines the organization of plant organs (Thornley, 1976; Dale, 1982).

The plant field can be visualized by covering a young leaf with traces of ink. When the leaf reaches maximum size, its growth is found to be essentially uniform throughout. The same pattern, and the proportions within the pattern, are maintained, regardless of any drastic change in size (Figure 4).

THE CHROMOSOME FIELD AND ITS UNDERLYING MOLECULAR BASIS

Chromosomes of higher organisms consist mostly of DNA and proteins. The analysis of their fine structure in rye, *Agapanthus*, and other species reveals that their constituent chromomeres are large on both sides of every centromere, and decrease in size successively to the telomeres. The decrease in chromomere size along the arms builds a gradient (Figure 4). In chromosomes with short arms the decrease is rapid, whereas in long arms it is slow. This is an indication that the telomere (or chromosome end) interferes with this process. Moreover, this chromomere pattern is maintained when the chromosomes change, both in length and shape, during reorganization at different cell stages. These observations formed the basis of the concept of chromosome field (Lima-de-Faria, 1954).

Since then, it has been found that genes, and other types of DNA sequences tend to occupy specific territories along the chromosome. For instance, the 28S and 18S ribosomal RNA genes, in 189 species, occur near the telomeres in 85.2% of the cases (Lima-de-Faria, 1973, 1980 and 1983). Molecular analysis, based on DNA cloning and hybridization of specific DNA sequences e.g., in deer species, has provided evidence in support of the chromosome field concept. Gene territory is found to be maintained when the DNA sequences are moved to chromosomes of different sizes (Lima-de-Faria et al., 1986; Scherthan et al., 1987, 1990; Lima-de-Faria, 1991a).

Order within the field is maintained by well-defined proteins that determine gene expression by binding to specific DNA sequences. These proteins function as a vehicle of both information and order, among DNA sequences (Wolffe and Brown, 1988; Harvey and Melton, 1988).

FEATURES COMMON TO THE FIELDS FOUND AT THE DIFFERENT LEVELS

Fields ranging from those of elementary particles to those of animal organs have the following properties:

1. Every field has a point of origin, a given strength and direction.
2. Fields are accompanied by gradual changes in energy (elementary particles and crystals), or by the gradual decline in concentration of specific molecules that produce discernible gradients in chromosomes, cells and organs.
3. Every field is able to maintain its own pattern when its mass is reduced or increased.
4. The function performed by an individual elementary particle, an atom, a molecule, a nucleus or a cell, is dependent on its position within the specific structure that contains it.

THE FIELDS ARE CAUSALLY RELATED: A FIELD AT ONE LEVEL CAUSES THE EMERGENCE OF ANOTHER AT THE NEXT LEVEL

Electrical fields are present in cells and organisms, and have been described in detail in hydroids such as *Obelia* and *Tubularia*, which belong to the same phylum as sea-anemones.

Regeneration of body parts follows the organism's polarity. A small piece of the animal will differentiate into a complete organism with anterior and posterior ends. In *Obelia* this polarity in cell differentiation is directly caused by an electric field which occurs between the two ends of the regenerating piece. This field can be neutralized or even reversed by sending an electric current in the opposite direction. Similar experiments carried out with *Tubularia* have shown that the electronegative pole favors regeneration, whilst the electro-positive one retards this process (Lund, 1921; Rose, 1970a,b; Berrill and Karp, 1976).

The change observed in the field of differentiation by neutralizing or reversing the electrical field serves as evidence of causality between the fields at the level of elementary particles (electrons) and at that of the cell.

THE ELECTRICITY GENERATED BY MINERALS, CELLS, AND ORGANISMS

Crystals generate electricity under two main conditions: one is when pressure is exerted on them and the other, when the ambient temperature changes. The first is called piezoelectricity, and the second pyroelectricity (Figure 5). When pressure is exerted at the ends of the polar axis of a crystal, a flow of electrons takes place that

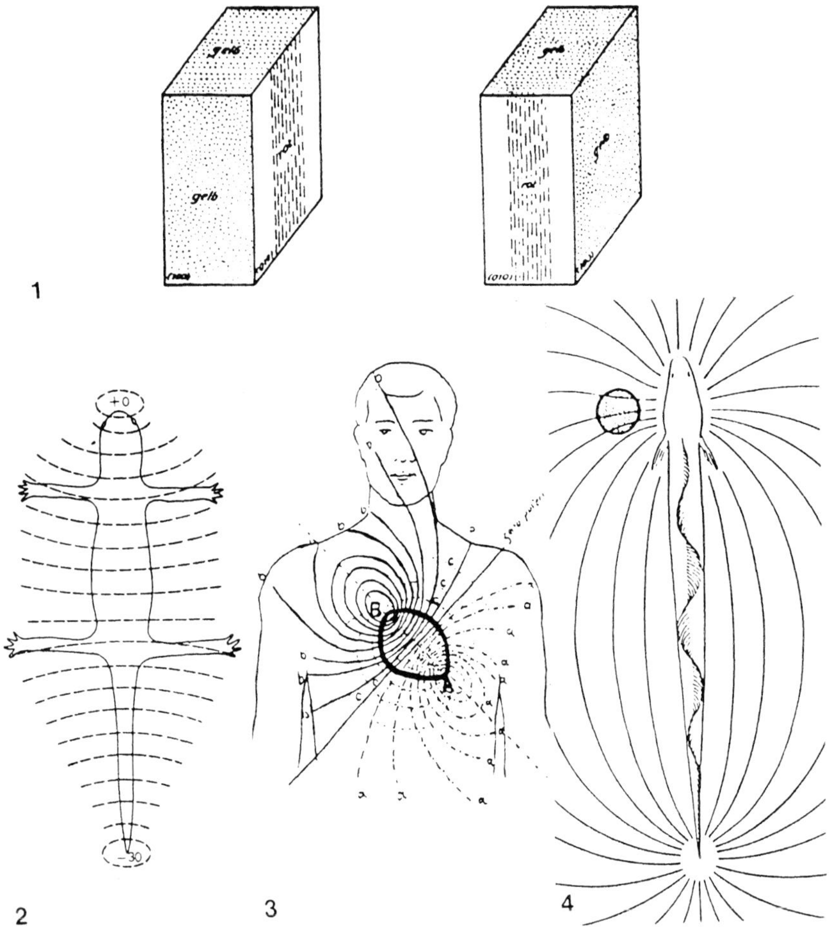

Figure 5. Production of electric current. (1) Electricity produced by a change in temperature in scolecite ($CaAl_2Si_3O_{10} \cdot 3H_2O$) and metascolecite (a form with a different water content). (2) Electric field produced by adult newt determined by two recording electrodes at the tip of the nose and tail. (3) Electricity produced by the human heart during contraction as recorded in an electrocardiogram. (4) Current flowing between the electric organ in the tail of a fish and receptor pores in the head.

charges the crystal with negative electricity at one end and positive electricity at the other.

The phenomenon was discovered in quartz by Pierre Curie in 1881, but it was only after 1921 that it became of general application in radios and watches. As pressure creates electricity, conversely an electric current leads to a deformation of the crystal. The result is that a slice of quartz subjected to an alternating current vibrates, and as such, becomes the essential component of a quartz watch (Pitt, 1988).

Every cell generates electrical currents and many animal organs produce powerful electrical fields. These may extend outside the organ, as in the case of the human heart, its current being recorded routinely in hospitals in the form of electrocardiograms. The electricity may reach outside the organism's body as in the electric eel and the electric ray, with strengths of about 200 volts (Figure 5) (Romer and Parsons, 1978).

Chapter 15

THE TRANSFER OF MINERAL ORDER TO THE CELL

CELLULAR PROCESSES INHERITED FROM THE MINERALS

CELL WATER GOVERNS THE PROPERTIES OF PROTEINS AND DNA

Water in the solid state is a mineral; in the liquid form it is called a mineraloid (Klein and Hurlbut, 1985). The hydrogen bonds of water are able to join to peptide linkages, this being a key property, since it determines the conformation of proteins. As a result, water determines not only structure and catalytic activity but also other properties that depend on the protein's configuration (de Duve, 1984 a, b).

Considering now right-handed DNA, it occurs in two forms, *A* and *B*. These differ in the distance of a complete helical turn, the way the sugar groups are bent, and in the angle of the tilt that the base pairs make with the helical axis (Watson et al., 1987; Saenger, 1988). It is also significant that the amount of water present may determine which form occurs in the cell. In solution DNA usually assumes the *B* form but when cellular water decreases, the *A* form of DNA may appear. To date, however, it is still uncertain whether the *A* form of DNA occurs within cells. In sharp contrast, RNA molecules have a tendency to occur in an *A*-like form. This is because the *B* form cannot exist as the result of the extra hydroxyl group present in the sugar residues. This has led to the view that the helices of DNA-RNA hybrids assume the *A* configuration.

This means that cell water may have played a critical role in the evolution of DNA. Also the preservation of the *B* form must have been a crucial factor governing the type of bases that were allowed to be added to DNA during its evolution, particularly since the angle of the tilt that the base pairs make with the helical axis is different in the two forms.

NEW PROPERTIES ARE FOUND FOR WATER THAT INCREASE ITS IMPORTANCE AS A DETERMINANT OF CELL FUNCTIONS

It is only recently that the properties of water have become better understood. The movement of water molecules becomes restricted when they border on a cell surface, such as a membrane. Most of their hydrogen bonds get fixed to the bordering molecules. This is the phenomenon of solvation and the modified water is termed vicinal.

1. The amount of vicinal water in the cell is relatively large because the cell contains hundreds of surfaces such as those of the nuclear membrane, the endoplasmic reticulum, the Golgi apparatus, the mitochondria and the cytoskeleton. The total area of the internal membranes in the cells of the human body is estimated as being circa 840 acres, i.e. the size of many football fields.
2. Most important is the finding that the properties of vicinal water are quite different from those of ordinary water: (a) It is less dense (about 3%). (b) More energy is necessary to heat it (circa 20%). (c) Solutes have a greater difficulty moving in it (in other words, a greater viscosity). (d) The properties of vicinal water change stepwise and abruptly, instead of gradually as they do in ordinary water. Four distinct temperatures result in dramatic changes, these are: 15, 30, 45 and 60 °C. This indicates that vicinal water occurs in four distinct phases that correspond to these temperature zones.
3. Still difficult to explain is the finding that the solvation capacity of vicinal water seems independent of the chemical properties of the surface to which it adheres.
4. The solvation forces exert a field effect, which encompasses as much as 16 molecular layers. This implies that most of the water in the cell, if vicinal, would drastically change the order of many chemical reactions (Drost-Hansen and Singleton, 1989).

BASIC SIMILARITIES BETWEEN CRYSTAL REPLICATION AND DNA REPLICATION: DNA ONLY IMPROVED ON THE MINERAL CANALIZATION

Living organisms are known to replicate, i.e., to copy themselves. Present knowledge of genetic material shows that DNA replication mainly follows the processes available at the mineral level that existed before DNA emerged.

1. *Water already transmits its pattern to other crystals.* Simpler than most minerals, water is not considered as being able to replicate, but evidence is

accumulating that it can. Secondary crystals are produced at the terminal regions of primary ones. The new crystals follow the configuration of the initial one (Knight and Knight, 1973).

2. *Seeding.* If one cools a saturated solution of sodium chloride (common salt), and seeds the solution by adding a small crystal of this chemical, the crystal starts growing. Subsequently, it breaks up into minor units that also grow independently. The outcome is a large number of long crystals. The phenomenon occurs fairly rapidly and with much precision, since the crystals are similar. Molecular studies reveal that DNA growth requires prior seeding as in the case of the crystal solution. DNA replication is preceded by the formation of an RNA primer, i.e., a small piece of RNA that is formed first and later discarded and which allows DNA growth to start (Dahlberg, 1977; Kornberg, 1980).
3. *Copying ability and size limit.* Once a minute fragment of a crystal is introduced into the solution a cascade of atomic events takes place that obliges all other atoms to organize themselves into a pattern that is limited and identical to the one added. All crystals formed have a given size limitation, otherwise they would not have planes. The same is true of DNA. DNA replication involves the formation of only DNA molecules. The DNA sequences are also limited in size. Newly replicated DNA in *E. coli* is formed in short pieces termed Okazaki fragments (Okazaki et al., 1968a, b). In mammalian cells these replication units may be as many as 30,000 per nucleus (Blumenthal et al., 1973).
4. *Speed of replication.* The formation of crystals in a solution is almost instantaneous but under relatively unfavorable conditions it may take a few hours. The mechanism of DNA replication is equally rapid. It takes 40 minutes to replicate the *E. coli* chromosome, at 10^3 nucleotides per second (Alberts et al., 1977). The period of DNA synthesis in a plant like *Zea mays* (maize) takes 8 hours in the roots, and the same event in human cells occurs in 7.6 hours at 38 °C (Dyer, 1979).
5. *Fidelity of the copying process.* The fidelity of the copying process is evident both at the crystal and DNA levels. If the crystal has the same chemical composition as the saturated solution, but has a different chemical configuration, i.e., left instead of right form, copying fails to occur. Hence, the extreme fidelity of the pattern copied is already present at the mineral level. The fidelity of template formation in DNA replication e.g., in *E. coli* is such that only one mistake is made in 10^9 to 10^{10} base pair replications (Drake, 1969). When DNA polymerases are confronted with a template primer with a terminal mismatch, they clip off unpaired primer residues by using their exonuclease activity. By means of this process they remove their own polymerization errors.

6. *Self-assembly and self-disassembly are necessary for copying.* The forces that hold molecules of salt during the self-assembly of crystals are reversible. Both growth and dissolution occur simultaneously during crystal formation. Molecules are being added and released as the result of physico-chemical changes in the solution. This is one of the sources of the formation of many separate crystals, i.e., their form of replication. Similar properties are found during DNA synthesis and the copying of RNA from DNA (transcription). The hydrogen bonds between the bases are strong enough to keep the two DNA strands together, but weak enough to allow separation of strands at the time of replication and transcription. This phenomenon is also dependent on the physico-chemical conditions existing within the cell.
7. *Dependence on external agents.* Both levels are affected by external physical and chemical agents. Differences in temperature lead to the formation of crystals with various shapes, and even light affects the form the crystal will assume. Different chemical environments oblige a crystal of the same chemical composition to assume different forms, as observed with sodium chloride that produces different patterns in the presence of glycine or formamide (Rinne, 1922). Temperature also affects DNA replication. In human cells the synthesis of DNA takes 22.4 hours at 33 °C, but only 7.4 hours at 36 °C (Dyer, 1979). The chemical composition of the cell also affects DNA replication. The size of the eukaryotic replication unit changes in different tissues, such as in the liver and the spermatocytes of the newt *Triturus* (Edenberg and Huberman, 1975). Formamide that changes the shape of sodium chloride crystals also changes the shape of DNA. This chemical is the key component in the standard procedure used in molecular biology to separate both strands of a DNA molecule when one wants them to hybridize with foreign DNA or RNA.
8. *The primer does not need to be identical to the copied molecule.* Surprisingly, the initiation of replication does not always start with a molecule which is identical to that which is later copied. Instead, in some cases a different molecule is used as a primer. The formation of water crystals takes place in the upper layers of the atmosphere. Experiments have confirmed that sodium chloride molecules (that have their origin in the oceans) and ashes from volcanos can function as primers in the initial building of snow crystals. A type of clay that has a structure similar to that of ice turns out to be also a primary source of condensation of water molecules (Allen, 1984). This is also what happens in DNA replication. The primer is not DNA, but an RNA molecule, without which DNA cannot start its assembly. A similar but not identical molecule is used (Kornberg, 1980). Primers with a different chemical composition occur also in proteins. Mineral structures can promote the formation of a protein crystal. An example is a piece of magnetite

which can serve as the center of nucleation for the crystals of the protein concanavalin B (McPherson, 1989).

9. *Stop signals.*Specific base sequences in DNA, called stop codons, specify RNA chain termination. Crystals are also dependent on molecular processes that govern the emergence of their faces, which is the way by which they stop their growth (Watson et al., 1987).

The picture that emerges is that DNA is mainly using the atomic processes that the minerals employed in their own replication billion of years before macromolecules arose.

THE GROWTH OF RNA MOLECULES DISPLAYS FEATURES THAT REAPPEAR AT OTHER LEVELS

The synthesis of RNA molecules along a DNA double helix and the growth of the branches of a fir tree have a similar configuration that at first sight seems to be accidental (Figure 1). However, close analysis reveals that the molecular processes involved follow similar biological solutions.

1. In both cases a central axis functions as a molecular source. This is provided by the DNA molecule in RNA assembly and by the cells of the apical meristem in the fir. Both are the origin of copying processes. In the former case, a similar but not identical macromolecule is formed, namely RNA. In the latter case, a similar but not identical cell is produced. The macromolecule leads to the formation of a specific protein while the cell leads to the formation of a stem with leaves.
2. Growth is initiated in both processes by a specific molecule. The protein RNA polymerase is needed for the formation of RNA molecules. The growth hormone, indole-3-acetic acid (an auxin) is necessary for the growth of the branches of the tree.
3. The molecules (in RNA synthesis) and the molecular edifices in the form of tissues (of the tree) acquire a form characterized by side branches that grow at right angles from the main axis.
4. Both increase in size successively. The oldest molecules and the oldest branches are the longest, and are located at the base of the main axis. The youngest molecules and branches are the shortest, and are at the apex. This accounts for the triangular shape of both structures.
5. The RNA molecules and the tree branches could assume other configurations such as a globular form or an irregular ramification but both are essentially linear.
6. These structures are not produced continuously but are well separated and spaced at regular intervals. This implies the presence of a process of

inhibition at close range that is overcome after a certain distance, thereby allowing the emergence of a new unit.

7. RNA polymerase initiates the transcription of RNA after binding to a specific DNA sequence termed promoter. The polymerase molecule then moves along the DNA template strand, and hence, extends the growing RNA molecule. This growth continues until the polymerase encounters another specific sequence in the DNA termed the termination signal. At this point the RNA molecule is ready and detaches from the DNA axis (Thangue and Rigby, 1988). Similarly, the auxin initiates the growth of a branch at a specific site in the central stem. The auxin also moves along the longitudinal axis of the plant. The transport is polarized downwards and rapid. The growth continues until the oldest branch at the base of the tree starts to detach as it dies.
8. The RNA polymerase does not function alone; additional proteins are involved in the copying of RNA. The same is true of the auxin which functions together with other plant hormones such as gibberellins (Wareing and Phillips, 1978).
9. Both processes involve not only activation but also repression. RAP1 which is one of the proteins participating in RNA transcription functions as an

---

Figure 1. Molecular and organ growth producing the same pattern as a result of similar structural and functional processes. (1) Electron micrograph of RNA transcription in a ribosomal RNA gene. The axial filament is the DNA of the gene. RNA polymerase molecules are bound to it. The nascent RNA-protein complexes increase in length, with increasing distance from the initiation point of transcription which is at the top. The result is the building of an oriented gradient. (2) Diagram of the same phenomenon; promoter region of the DNA (P), DNA template (double lines), RNA polymerase (circles), RNA transcript (wavy lines). (3) Leaf of the fern *Polypodium vulgare* showing the same type of oriented growth building a gradient. Note the alternate formation of the leaflets as in the RNA transcripts. (4) The fir tree *Picea abies* (conifer): growth starts at the top and the nascent branches increase in length with increasing distance from the initiation point. This is due to the movement of auxin molecules along the stem. The primary branches lead to secondary ones. (5) Growth of ice crystals (windowpane frost). The properties of the water atoms determine the successive increase of the primary branches in a gradient. Formation of secondary branches at right angles with the primary ones occurs as in the fir tree. (6) Gill of the molluscs belonging to the suborder Archeogastropoda with its bipectinate form having alternate side branches which form an oriented gradient.

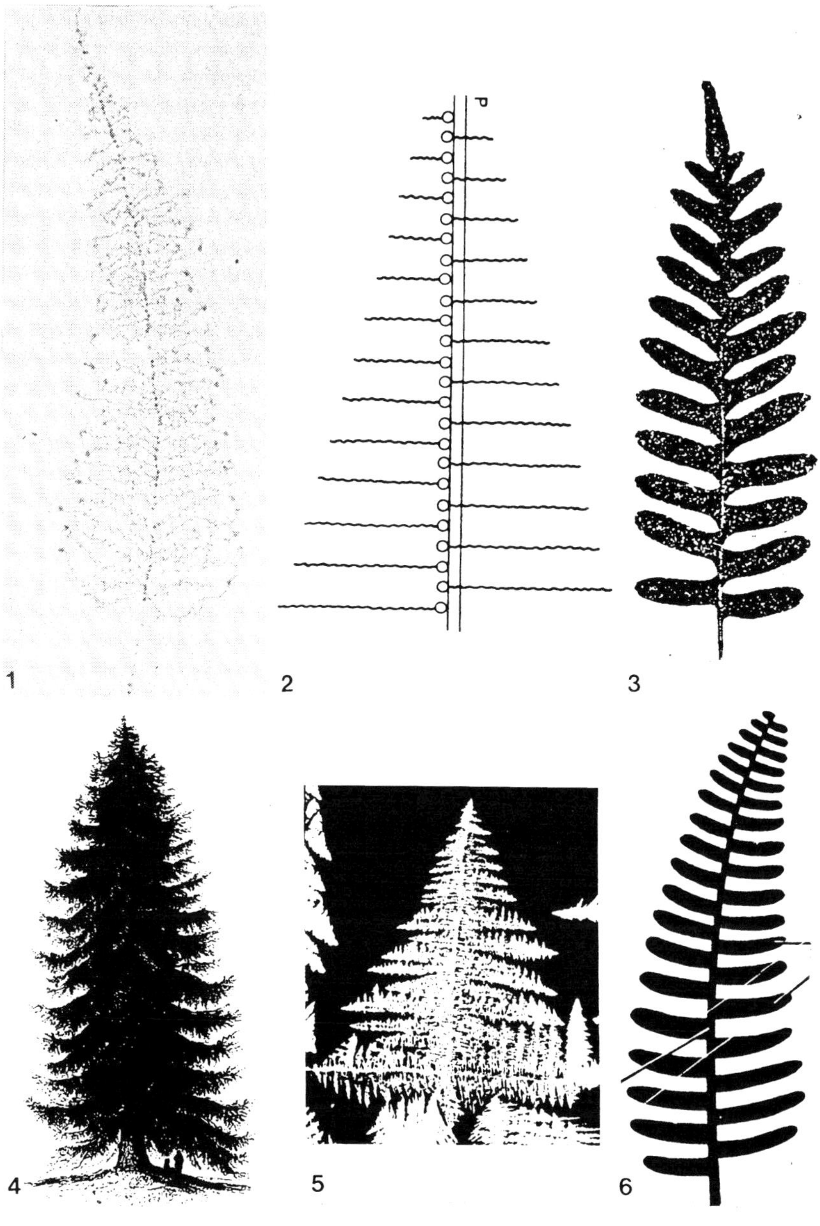
P
1
2
3
4
5
6

activator and repressor (Brand et al., 1987; Nasmyth and Shore, 1987). At low concentrations auxin has an optimal growth effect but as the concentration increases it also acts as an inhibitor (Wareing and Phillips, 1978).

10. Surgical experiments with plants have led to evidence of inhibitory effects associated with neighboring primordia, a phenomenon which explains the occurrence of regularly well spaced tree branches. Usually RNA polymerases do not have nearby neighbors; each molecule completes transcription before the next one begins transcription (Miller, 1981).
11. Growth at an angle of 90°, and the existence of a molecular gradient that leads to the triangular form are events resembling those occurring in water crystals produced by freezing. It is to be noted that the pattern of water crystals is decided by simple atomic interactions (Figure 1). Other plant and animal structures, such as fern leaves and mollusc gills, similarly display growth gradients and alternate formation of the constitutive parts (Figure 1).

It is of course understood that many molecular details concerning these processes remain to be elucidated, and one should not expect the molecular homology to be total, since such homologies represent different levels of organization. However, the significant point is that the molecular construction and function follow essentially the same type of ordered pathways.

CRYSTALLINE STRUCTURE OF MUSCLE

The structure of muscle is so regular that biologists describe it as crystalline. Cross sections of muscle when viewed with the electron microscope show patterns that cannot be easily distinguished from those of a mineral when this is also viewed with the electron microscope. The similarity is striking; in fact, one could easily mistake one for the other (Figure 2). This is the case with the image of tourmaline (a cyclosilicate) which shows a 6-fold pattern that results from the association of silicon and oxygen atoms (Si_6O_{18}). That of insect flight muscle in cross-section is characterized by thick and thin filaments packed with a 6-fold crystalline regularity. These are the result of the association of two proteins: actin and myosin. Six parallel actin filaments bind to one central myosin by means of their atomic configurations (de Duve, 1984 a, b). Though proteins are much larger than silicates, it is solely the atomic processes in protein and mineral that are responsible for the formation of the same hexagonal pattern. Furthermore, the electron microscopic view of rabbit muscle reveals the layered pattern of its fibers. This is practically indistinguishable from the layers of fibrous asbestos in the mineral serpentine. Serpentine, like tourmaline, contains silicon and oxygen atoms in fairly similar ratios ($Mg_6Si_4O_{10}(OH)_8$).

Finally, one cannot overlook the following point. Carbon which is the most abundant atom in proteins (50%), and silicon, the principal atom in silicates, are known to have many properties in common. Both belong to group IVA of the

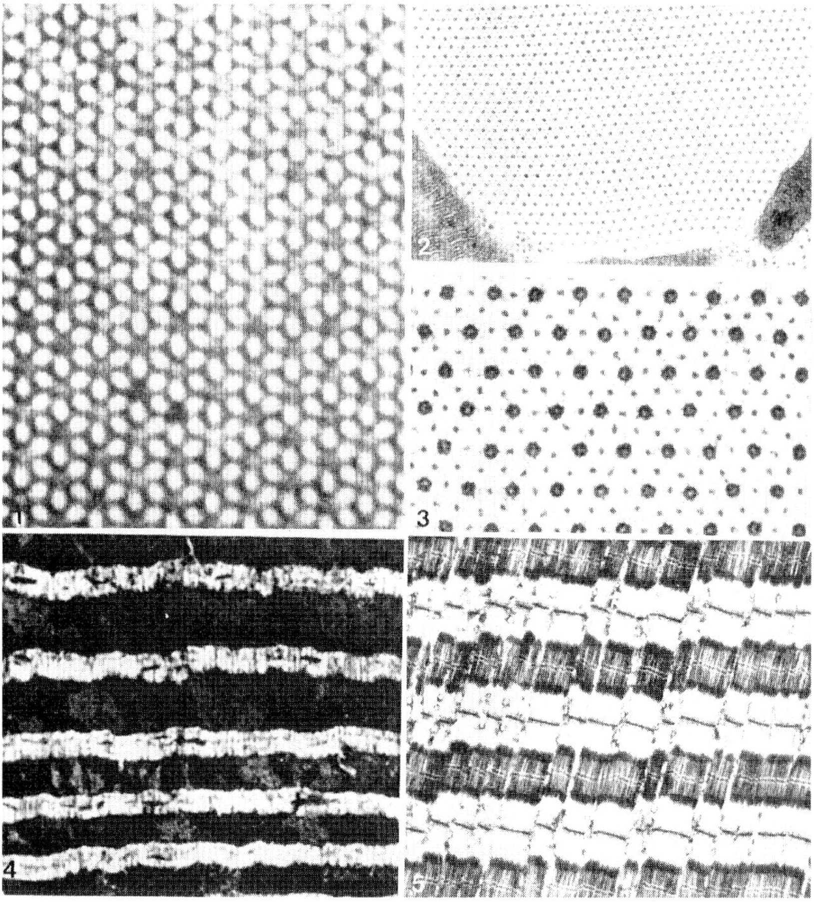

Figure 2. Crystalline structure of minerals and muscle. (1) High resolution electron microscope image of tourmaline (a cyclosilicate). The 6-fold pattern is the image of the 6-fold ring built by silicon and oxygen atoms (Si_6O_{18}). The bar scale represents 15 Ångstroms. (2) and (3) Two electron-micrographs of an insect flight muscle viewed in cross-section, showing the thick and thin filaments packed with 6-fold crystalline regularity. (4) Veins of fibrous asbestos (silicates) in serpentine, a layer lattice mineral ($Mg_6Si_4O_{10}(OH)_8$). (5) Microscopic view of rabbit muscle showing the layered pattern of the muscle fibers.

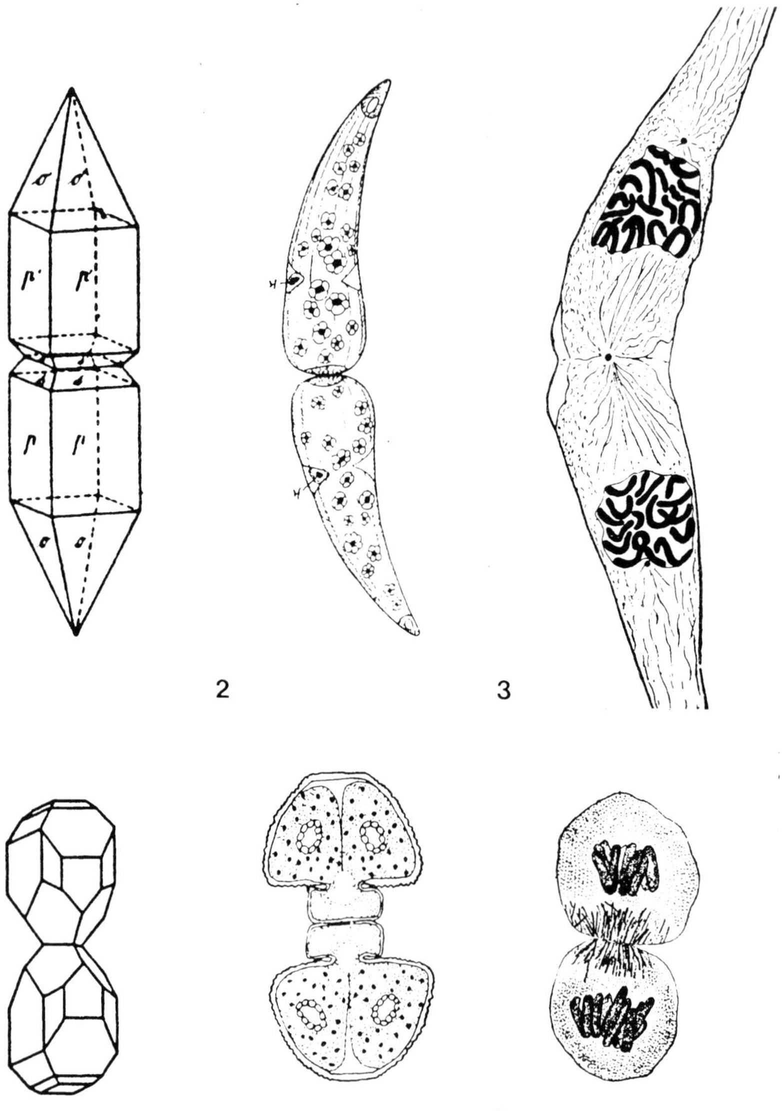

1 2 3

4 5 6

Figure 3. Cell division. (1) Twin crystals of iodosuccinimide $(CH_2CO)_2NI$. (2) Cell division in the green alga *Closterium Ehrenbergii*. (3) Mitosis in salamander cells. (4) Twin crystals of smithsonite ($ZnCO_3$). (5) Cell division in the alga *Cosmarium*. (6) Dividing cell in *Geotriton fuscus* (a newt).

Periodic Table, since they have four electrons in their outer shell. Thus, the similarities observed between muscle and silicates may simply be the result of the electronic configuration of their atoms.

CELL DIVISION

Cells separate into identical but independent units as the result of cell division. Molecular copying and redistribution of organelles are the main events in this process.

Minerals like iodosuccinimide and smithsonite build twin crystals that have some features resembling those of cell division. At least two features are notable. First, the twins are separated by a constricted region in a manner analogous to that occurring in cell separation. Second, the twin crystals are oriented in opposite direction which is just as two newly formed cell halves do. The ordered and oriented pattern that the twin crystals acquired is the sole result of atomic interactions. The atoms found in cells do not differ from those present in minerals. Expressed in another way, the basic features of cell division preexisted in the minerals (Figure 3).

GROWTH WITH A REGULATED CHANGE IN PATTERN

An octahedral crystal (i.e., a solid with 8 plane surfaces) can grow into that of a crystal with 6 equal square sides (i.e., a cube), but it does so by a series of regular events. (1) It passes through intermediate stages in which the faces of the octahedron begin to disappear and those of the cube start to appear. (2) This drastic change takes place without the crystal loosing its coherence. (3) The growth does not involve ramifications or accidental deviations. (4) The transformation is clearly directed, since it involves going from the octahedron to the cube. (5) Growth of the octahedron is limited in that it transforms into the cube, and stops there.

Human growth is not an essentially different matter. A human at birth has a large head and a small body, and at maturity has a small head compared to its large body. This represents a drastic change in proportions like that occurring in the two crystals just mentioned. The remarkable fact is that the growth of the human body follows the same regular events: (1) It passes through intermediate stages in which the relative size of the head diminishes whilst that of the body increases. (2) The change takes place without the occurrence of any disorganization. (3) The growth does not involve ramifications or accidental deviations. (4) The transformation is directed, going only from the infant to the adult form. (5) The growth is limited since it always stops at the adult stage. As shown in Figure 4 the same phenomenon occurs in invertebrates.

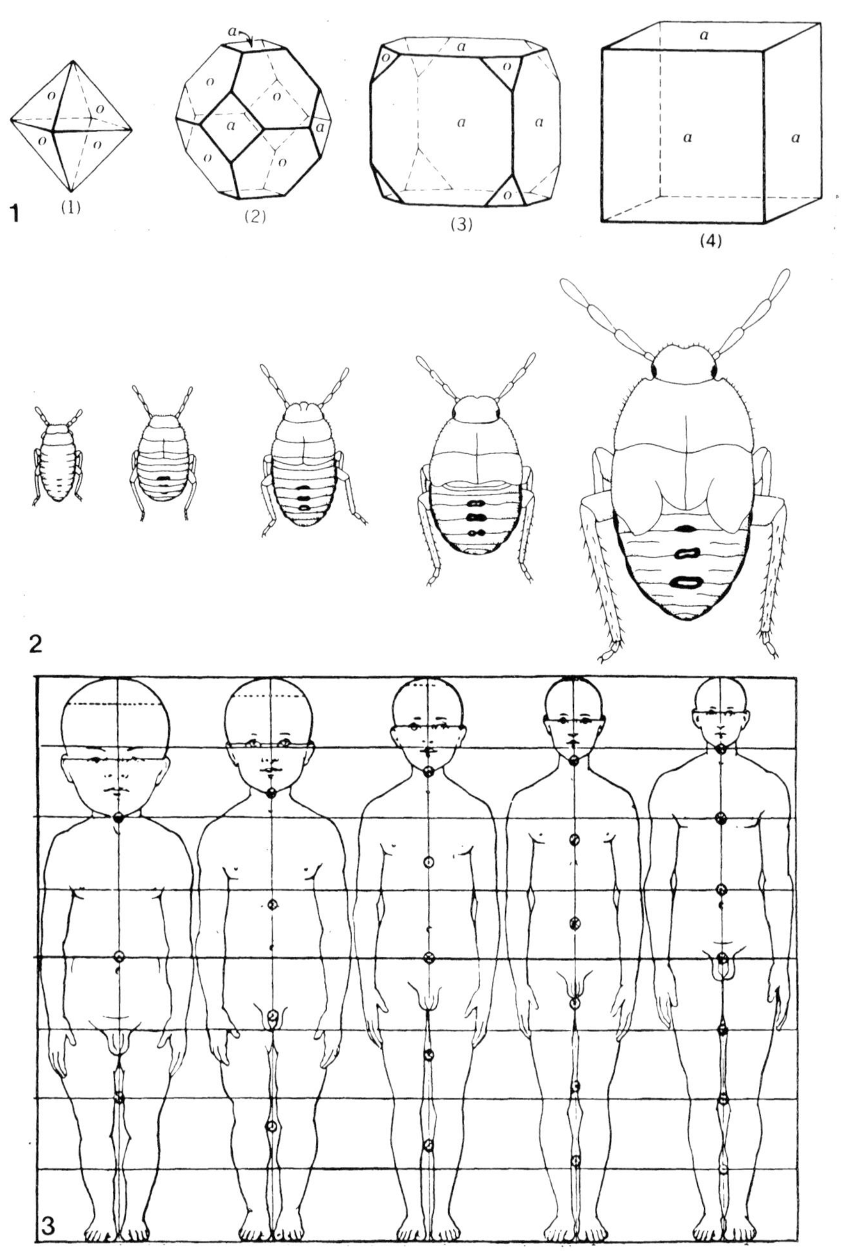

Figure 4. Growth with regulated change of pattern. (1) Growth of a crystal from an octahedron to a cube, showing two intermediate forms which are combinations of these two solids. (2) Larval development in the insect (*Cydnus aterrimus*) showing the formation of the wings. (3) Change in proportions of the human body (at birth, and at the ages of 2, 6, 12, and 25 years).

GROWTH WITH MAINTENANCE OF THE GENERAL PATTERN

Other types of crystals grow by adding new atoms without undergoing an appreciable change in their original pattern. They only become larger and larger. This is the case in quartz. A leaf or lizard, follows this type of growth. (1) All three start from a minor structure. (2) The rapid growth takes place without loss of coherence. (3) The increase in mass is not accidental or distorted. (4) The proportions existing between the different parts are maintained, although the body more than doubles its size. (5) The growth is directed since it only produces a form similar to the original one, and it is limited because it stops at a specific size in all three structures (Figure 5).

GROWTH ACCOMPANIED BY RAMIFICATION

It is important to realize that ramified growth does occur but it failed to appear in the preceding examples. Different atomic combinations lead to different solutions.

Water crystals grow from an original nucleus. They start to develop branches which are equidistant from each other. These branches then acquire side ramifications at regular angles from the main axis. The number of branches is usually six. This ramification pattern is also seen in the leaves of *Cannabis sativa*, as well as the fingers of the limb of a newt. (1) Both grow from a small nucleus. (2) Branches are successively added. (3) As their number increases the average distance between them tends to be maintained. (4) The growth is directed and limited. In the case of the snow crystal, growth stops at the appearance of ramifications on the six main branches. The leaf does not show more than seven leaflets, while the limb does not develop more than five fingers (Figure 6).

TWIN FORMATION

The phenomenon of twin growth observed in minerals parallels that in animal twins in a striking manner. (1) Twinning in minerals arises as a result of a change in the direction of growth of a minor crystal lattice caused by environmental influences. (2) During growth a simple crystal divides along a plane which lies parallel to one of the faces, and one half of the crystal is formed in a way suggestive of a 180° rotation relative to the other half. These so-called rotation twins are mirror images of one another. (3) In some minerals "contact twins" are formed in which the two components are also mirror images but in which they have interpenetrated one another (Whitten and Brooks, 1988).

Twins in amphibians, reptiles and humans follow a similar organizational pattern. (1) They originate from a single egg. (2) At one of the first divisions, changes in the environment induce the formation of two separate individuals. (3) The twins may remain united or be separate. (4) If united, then the interpenetration may occur in the middle of the body, as in quartz (salamanders and humans), or

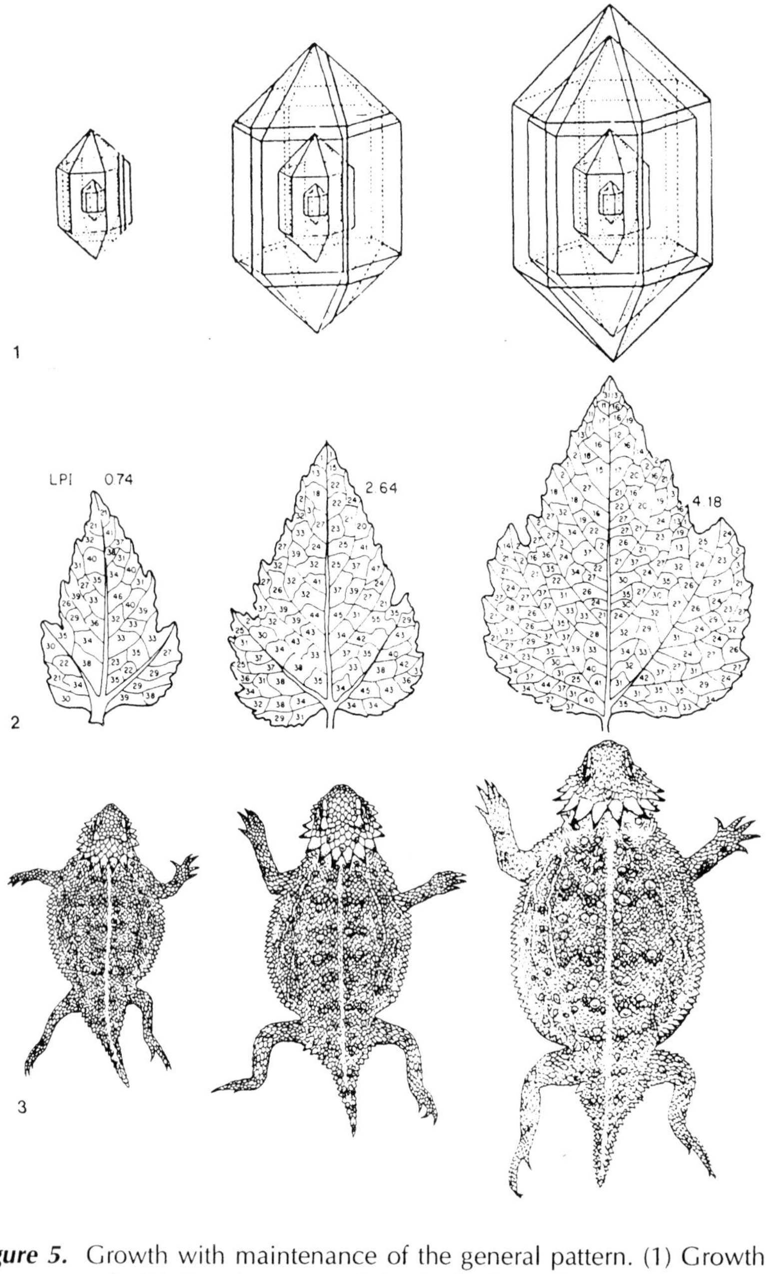

Figure 5. Growth with maintenance of the general pattern. (1) Growth of quartz crystal. (2) Development of the leaf of *Xanthium strumarium*. (3) Different sizes in the lizard *Phrynosoma solare*.

Figure 6. Growth accompanied by ramification. (1) Successive stages in the growth of snow crystals. (2) Sequence in the leaves of *Cannabis sativa* grown in 8 days. (3) Stages in the development of the limb of the newt *Triton taeniatus.*

mainly at the tail, as in calcite (newt and tortoise). (5) In the four animal species cited here, and the two minerals, the twins in each situation are mirror-images of one another (Figure 7).

REGENERATION OF TERMINAL REGIONS

It is generally recognized that the power of regeneration or the capacity to repair in organisms following injury, has declined in the course of evolution. In amphibians and reptiles, for example, regeneration is limited to the limbs or the tail.

The phenomenon of regeneration can be traced to the minerals. Pasteur (1857) described in detail the regeneration of broken ends of crystals of ammonium bimalate which were grown in a saturated solution. The crystal is found to form curved layers that lead to the re-establishment of the original form. Conifer trees, upon decapitation, are able to completely rebuild their shape by inducing a side branch to grow vertically. A single arm of the starfish *Linckia* is able to produce

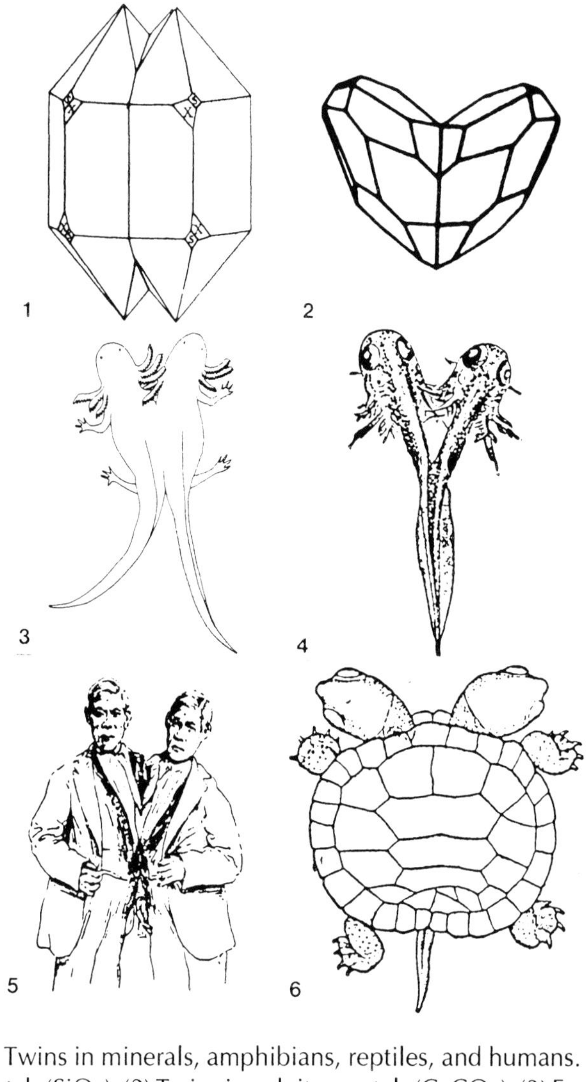

Figure 7. Twins in minerals, amphibians, reptiles, and humans. (1) Twins in quartz crystals (SiO_2). (2) Twins in calcite crystals ($CaCO_3$). (3) Experimentally fused twins in salamanders, which are united in the middle but free in the upper and lower body parts, like the quartz crystals. (4) Experimentally-induced twins in the newt (Triton), which are united by the tail and have the heads free, as in calcite in which the two crystals are united at the base but are free at the upper region. (5) Naturally occurring human twins, united by the middle region of the body but with the heads and legs free as in the salamander and quartz. (6) Naturally occurring turtle twins with separated heads but a single tail showing the same situation as in the newt and calcite.

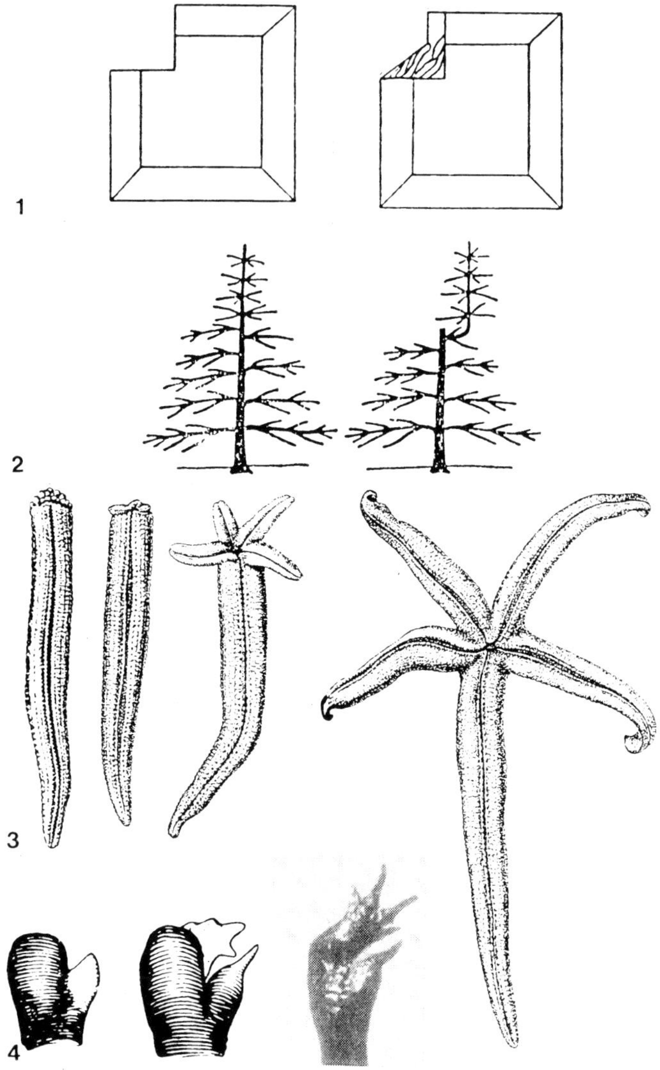

Figure 8. Regeneration of terminal regions. (1) Crystals of ammonium bimalate: crystal with its end broken (*left*) and under regeneration which leads to the restoration of the original pattern (*right*). (2) Conifer tree before decapitation (*left*) and after a new branch, has grown vertically, restoring its complete shape (*right*). (3) Three successive stages of the regeneration of the whole body from the end of a single arm in the starfish *Linckia multiflora*. (4) Regeneration in an amphibian of a whole limb from a limb stump.

the missing four arms, thus restoring the original pattern. A limb stump in an amphibian is equally able to produce the five digits (Figure 8) (Weiss, 1939).

PLANT AND ANIMAL CHIMAERAS

Plant and animal chimaeras are single organisms developing from the union of different individuals, or are individuals composed of tissues of two different genotypes.

In plants the mosaics are usually obtained by grafting. Leaves with green and white tissues, which have different genetic constitutions, are able to build a chimaera. The two different types of cells tend to grow parallel to the margins of the leaf (Dale, 1982).

Animal mosaics are obtained by the fusion of fertilized eggs coming from different strains or species. A sheep-goat chimaera with a fully normal appearance can be produced by combining one 8-cell sheep embryo with three 8-cell goat embryos. Its mosaic character is particularly conspicuous in the horns, hair, and blood cells (Fehilly et al., 1984). Similarly, chimaeric mice can be obtained by fusion of two embryos whose genetic constitutions are not the same. Another procedure involves the injection into an embryo of a light colored mouse, of cancer cells obtained from a black mouse parent. The resulting adult shows a body with stripes of black and light areas that parallel each other (Mintz and Illmensee, 1975; Mintz, 1978) (Figure 9).

No satisfactory genetic or developmental explanation of the parallelism observed in the tissues of plant and mouse chimaeras is yet available. However, this particular phenomenon is seen in the minerals. A well-known case is staurolite which crystallizes with kyanite. Growth of the two minerals is oriented so that the faces of both crystals parallel each other. Furthermore, when plagioclase overgrows on microcline, the same parallel arrangement is found. Clearly, atomic interactions are the decisive factors and one therefore wonders what atomic processes found in proteins (or other compounds) are responsible for the same orientation in living organisms.

HYBRID FORMATION IN ANIMALS AND MINERALS

Geneticists can hardly conceive that anything but genes are responsible for the patterns and functions that arise in plant and animal hybrids following sexual reproduction. Yet close examination of the properties of minerals indicates that the basic rules of combination were present before living organisms appeared. (1) Two different crystal patterns have the ability of combining into hybrids. (2) The intermediate forms are not accidental since they retain the faces of the original crystals. (3) Retention of the original forms is evident even in the most complex crystals when many faces are combined. (4) Consider the mineral garnet (a nesosilicate); a dodecahedron and a trapezohedron produce not just one but a whole

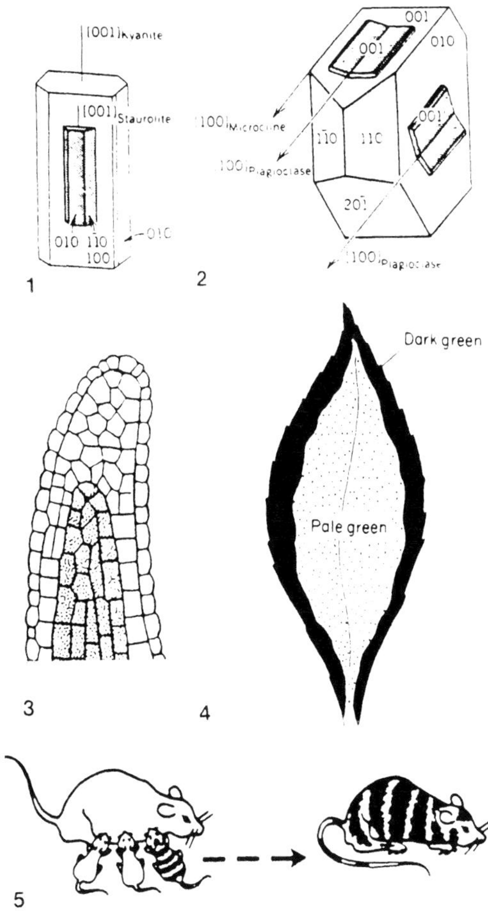

Figure 9. Chimaeras in minerals, plants and animals. (1) Staurolite ($Fe_2Al_9O_6(SiO_4)_4\ (O,OH)_2$) grows together with kyanite (Al_2SiO_5); this growth is oriented, thus, resulting in the faces of both crystals being parallel. (2) Plagioclase ($Na(AlSi_3O_8)$-$Ca\ (Al_2Si_2O_8)$) overgrows on microcline ($KAlSi_3O_8$); the growth of the two crystals is also oriented in relation to each other. (3) Cross-section of the leaf of *Pelargonium* showing the chimaera formed by green and white tissues (having different genetic constitutions). The growth is oriented in relation to the leaf margin. (4) Leaf with different green tissues (with different genetic constitutions) building a chimaera. The internal tissue also grows parallel to the margins of the leaf. (5) Chimaeric mice with striped black and white fur. This is the result of injection of cells into embryos. The embryo was from a white parent and the cells from a black parent. The stripes are vertical and parallel to each other, as in the case of minerals.

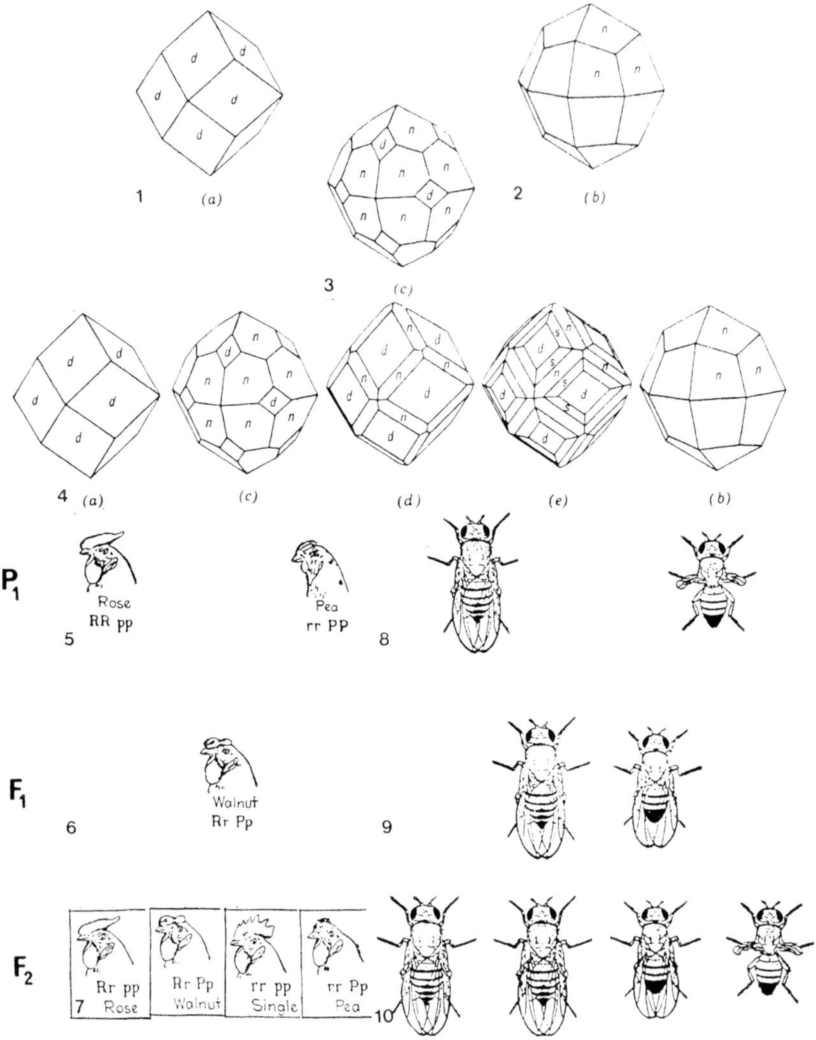

Figure 10. Hybrids in garnet crystals and animals. Crystals of garnet, a nesosilicate with the general formula $A_3B_2(SiO_4)_3$, where the A site houses Ca, Mg, Fe^{2+} or Mn^{2+} atoms, and the B site incorporates Al, Fe^{3+} and Cr^{3+}. The two original forms: (1) the dodecahedron (with faces marked *d*) and (2) the trapezohedron (with faces marked *n*). (3) Combination of the two original forms with *d* and *n* faces. (4) The two original forms and three combinations of them. One of these called a hexoctahedron (*e*) arises less frequently than the others. (5) Cross between fowls (*Gallus gallus*) with different comb forms and colors. (continued)

series of hybrid forms. (5) Some combinations arise less frequently than others, as in the case of the hexoctahedron (Figure 10). Figure 10 also includes a cross made by Morgan (1928) (this is taken from his classical work on *Drosophila*) and a similar cross in birds (Sinnott and Dunn, 1939). In the cross between fowls (*Gallus*) the parents are shown to have different comb forms. The hybrid of the first generation (F_1) shows an intermediate shape consisting of a combination of both original forms in which the features of the parents are retained; that is, just as in the case of the combination of the dodecahedron with the trapezohedron of garnet. In the second generation (F_2), the patterns of the original parents, as well as that of the F_1 hybrid, reappear together with other less frequent combinations (also as in the case of garnet).

The *Drosophila* cross is one between the long-winged (wild-type) and the vestigial-winged fly. The first generation gives rise to two types of hybrids, one of which incorporates the properties of the parents (small body and partly long wings), while the other is like the wild-type parent. The second generation displays the original patterns of the parents, and the combinations of the first generation hybrids (Figure 10). As with the minerals, the original patterns are retained; some of the combinations are more common than others, and although hybrids are formed, the initial structures can be recognized and sorted out.

HYBRIDS IN PLANTS AND MINERALS

Hybrids resulting from crosses in plants follow the same principles of organization. Crosses between two species of *Salix*, one with broad, the other with narrow leaves, gives in the first and the second generation a combination of the shapes of the original forms, showing leaves of intermediate sizes (Heribert-Nilsson, 1918). Two species of *Dianthus* with small and large flowers produce in the first generation an

---

Figure 10. (Continued) The parents P_1 have different comb forms. (6) The first generation hybrid (F_1) shows an intermediate shape consisting of a combination of both original forms. (7) Due to the presence of two dominant genes (R and P) only four types of combs appear in the second generation combinations. These are: the two parent patterns, the F_1 hybrid pattern and a new combination pattern (rrpp). (8) *Drosophila melanogaster*: cross between the wild-type (long-winged) and the vestigial-winged fly. (9) The first generation (F_1) results in two types of hybrids, one of which combines the properties of the parents (small body and partly long wings) while the other is identical to the wild-type parent. (10) The second generation (F_2) displays the original patterns of the parents, as well as combinations of the first generation hybrids (long wings occur in 3/4 and vestigial in 1/4 of all the flies).

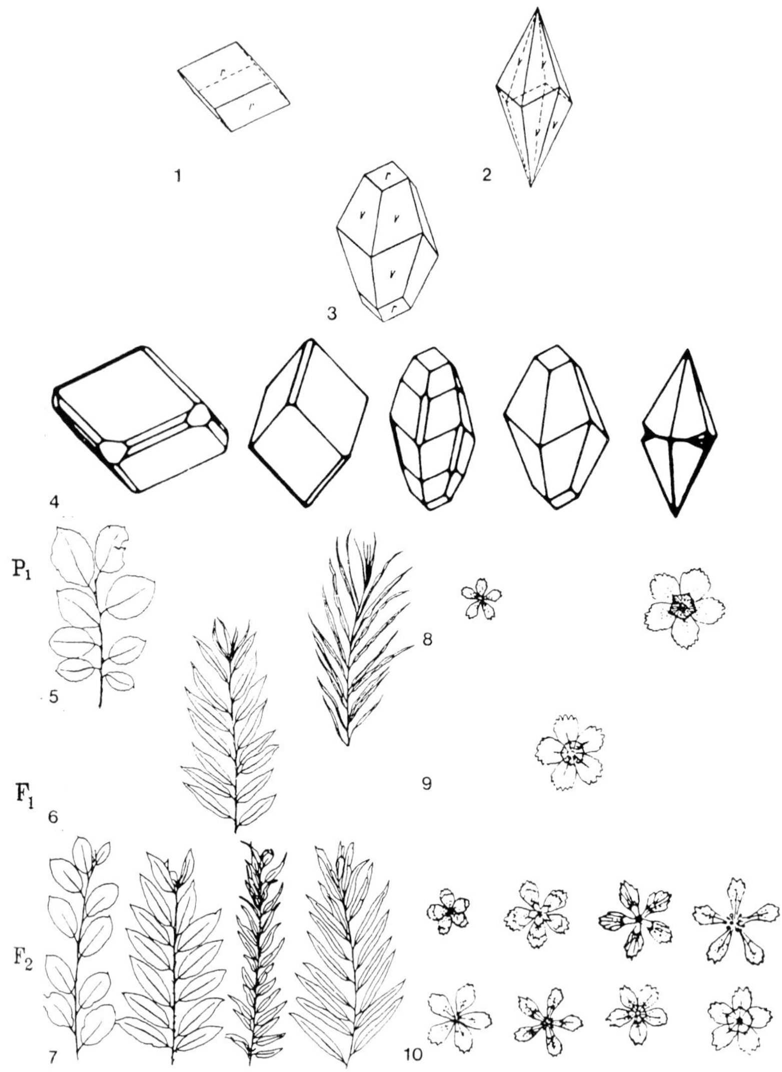

Figure 11. Hybrids in calcite crystals and in plants. Crystals of calcite ($CaCO_3$). The two original forms: (1) the rhombohedron (with faces marked *r*), and (2) the scalenohedron (with faces marked *v*) produce a series of combinations. (3) One of them is the combination of the two original solids (*r* = rhombohedron and *v* = scalenohedron). (4) The solids of the third line are combinations of these two forms and a prism. Calcite crystals may occur in over 600 forms. (5) Cross between the parent species (P_1) *Salix caprea* (broad leaves) and *S. viminalis* (narrow leaves). (6) The first generation hybrid (F_1) is a combination of the shapes of the original forms showing leaves of intermediate size. (continued)

intermediate form, and in the second generation, a series of combinations of the parental and F_1 patterns (Wichler, 1913). The two calcite crystals, the rhombohedron and the scalenohedron, produce a series of hybrids of the two original solids. (1) The faces of the rhombohedron and the scalenohedron can be distinguished in the hybrid. (2) Many intermediate forms arise which in some cases include even a prism. (3) In every case a fully regular crystal is formed. (4) All of them crystallize in the same system. Though the fundamental pattern has been maintained, combinations can occur and some are more common than others, because their energy states are more stable (Figure 11). (Klein and Hurlbut, 1985).

DOMINANCE OCCURS AT BOTH THE CRYSTAL AND BIOLOGICAL LEVELS

There are two other features which are shared by mineral, plant, and animal hybrids. The first is that hybrids only occur between crystals that have the same, or closely related chemical composition. This is also true of living organisms in which the crosses usually occur between varieties within a species, or between closely related species, as a direct result of similarities in the chemical composition of their cells and DNAs.

The second feature concerns the occurrence of dominance in chemical hybrids; this was described by Pasteur as early as 1847. The tartrate of sodium and potassium constitutes a heterozygous salt, which not unlike a hybrid in a living organism shows the dominance of one of the chemical components. The double salt gives a rotation value for the polarized light which is about the same as that of the tartrate of potassium in pure form. Other salts may have an incomplete dominance showing intermediate values. Certain compounds are dominant while others are recessive.

It is thus clear that the dominance in crystal hybrids may vary with the chemical composition of the two salts that are coupled, just as dominance varies with the chemical composition of the two DNA sequences that are coupled in the hybrid cell. The molecular mechanism responsible for dominance remains unknown at both levels of organization.

Figure 11. (Continued) (7) The second generation (F_2) consists of a series of leaf forms in which the patterns of the original parents (P_1), as well as that of the F_1 hybrid, reappear in combinations with intermediate shapes. (8) Cross between *Dianthus armeria* (small flower) and *D. deltoides* (large flower) (P_1). (9) The hybrid of the first generation (F_1) shows an intermediate shape, resulting from the combination of the parental patterns. (10) The second generation (two lowest rows) consists of different combinations of the parental and hybrid forms, of which the most representative 8 are shown.

It has been established for a long time that the production of hybrid combinations in living organisms follows the assortment of chromosomes at meiosis and their distribution in the gametes. It also depends on gene exchanges occurring during crossing over. Although such knowledge represents progress, next to nothing is known of the molecular events that occur between the different gene products when assembling the phenotype of the hybrids.

ATOMIC PRINCIPLES OF HYBRID FORMATION COMMON TO MINERALS, PLANTS, AND ANIMALS

In summary, the combination of molecular assemblies in crystals and living organisms obey the following six principles:

1. Different patterns have the ability to become stable hybrid forms.
2. The intermediate structures do not represent accidental events. Rather, they retain the features of the original components.
3. This retention of the original forms is evident when many traits are combined, even in the most complex hybrids.
4. Of the many forms produced some arise less frequently than others.
5. Hybrids occur only between components that have the same or relatively similar chemical composition.
6. Of the two original components one tends to prevail, thereby establishing a dominant pattern.

These six principles of organization common to the mineral and biological levels carry the important implication that the cell and DNA have inherited the atomic order already laid down in the minerals and show how the events at the level of sexual reproduction have been conditioned by the rules of crystal formation.

PART V

THE CONTRIBUTION OF DIFFERENTIAL REPRODUCTION AND DEATH TO PERIODICITY

Chapter 16

CHROMOSOME BEHAVIOR IS AN INTERNALLY REGULATED PROCESS

THE INDEPENDENT ASSORTMENT OF CHROMOSOMES AT MEIOSIS IS A DIRECTED EVENT

In order to understand the precise evolutionary implications of sexual reproduction, it is necessary to recognize that the events involved are wholly directed by the basic molecular construction of the cell, the chromosomes and the genes. As is known, chromosomes move to the poles in more than one way, and they determine which type of assortment will occur. The assortment of chromosomes at meiosis is not a random event, as usually stated, but rather a directed one. It occurs in such a way that the chromosomes move to the poles *independently* of each other. Maternal and paternal chromosomes are sorted out on an individual basis. However, in organisms such as *Sciara*, the chromosomes influence the behavior of other chromosomes obliging them to move to specific spindle poles. In this species all the chromosomes coming from one parent move together to the same pole (Du Bois, 1933; Crouse, 1961, 1965). This type of segregation occurs as regularly as the independent segregation of the chromosomes. Another example is provided by studies on *Drosophila*: one of the two homologues of chromosome II marked at the early stages of meiosis is eliminated in the sperm. This represents a process directed by the products of three genes (Crow, 1979).

Distortion in the segregation of homologous chromosomes has been found to be a general phenomenon. At the meiotic divisions several genes influence chromosome segregation in such a way that specific chromosomes are transmitted to the progeny in excess of the expected proportion of 50%. Thus far, the phenomenon has been extensively studied in plants, butterflies, mosquitoes, *Drosophila* and the wood lemming (*Myopus schisticolor*). In some species, such as *Drosophila*, the distorted segregation of the chromosomes leads to the production of 100% female progeny (Lyttle, 1991).

Taken together, one can state that chromosomes are able to modify their own distribution at meiosis by means of specific molecular messages. Independent assortment occurring at meiosis is the process chromosomes have chosen to establish in most organisms.

CROSSING-OVER IS A REGULATED MOLECULAR PROCESS

The separation of homologous chromosomes is preceded by their pairing. During this stage an exchange of segments takes place between the homologues (crossing-over). The result is that novel gene combinations are formed which are included in the sex cells. This is a matter of special interest, particularly since crossing-over is an example of how a process thought initially to be random has turned out to be a most ordered event. As Stern and Hotta (1978) remark "The textbook impression that crossovers somehow occur randomly along chromosomes is dimmed by the evidence for a stringent organization of the meiotic process." They go on to provide the evidence which demonstrates the non-random occurrence of crossing-over. This includes: (1) Crossing-over is not a common but an infrequent event at any particular site in a chromosome. (2) This phenomenon is unevenly distributed along a chromosome, as exemplified by the heterochromatic regions of several plant and animal species, which seldom undergo crossing-over. Other chromosome regions are found to have a high frequency of crossovers. (3) The occurrence of two or more crossing-overs in a chromosome region is a directed process. A crossover in one segment inhibits the formation of a second one in its immediate vicinity. This has long since been called positive interference. (4) Chiasmata have been correlated with crossing-over and their occurrence has been found to be limited to very well-defined regions of the chromosomes in many plant and animal species. (5) The non-random appearance of crossing-over is also evident from cases where a decrease in chiasmata in one chromosome arm, or in one pair of chromosomes, results in an increase of crossing-over at other regions. (6) Molecular evidence coming from studies in eukaryotes is now strengthening the view of the existence of ordered events, and makes the phenomenon of crossing-over appear even more regulated as an integral part of the molecular program of meiosis. Thus, three proteins namely, recA, recB, and recC are found to initiate the unwinding and nicking of the DNA at specific sites and additional proteins, participating in DNA synthesis, are also necessary for crossing-over to occur (Kobayashi et al., 1984; Ponticelli et al., 1985; Nicolas et al., 1989).

THE MINOR ROLE OF REPRODUCTION IN PERIODICITY

It is obvious that sexual reproduction results in new chromosome and gene constellations that provide the basis for variations in pattern and function. Differential reproduction not only allows some of these to go farther, but also inhibits the propagation of others. There can be no doubt that reproduction plays some role simply because it contributes to the formation of new genetic combinations. However, the crucial point is that these combinations are in most cases the result of internally directed processes which have contributed toward the maintenance of periodicity. Had the genetic processes not been internally directed, then periodicity could not have been easily maintained.

Admittedly, progress in this field seems to have been rather slow at the molecular level. Among the many key questions that need to be more fully addressed are the following: (1) What are the chemical mechanisms that decide the sorting out of chromosomes at meiosis? (2) What are the molecular pathways that direct crossing-over? (3) What are the molecular processes that create dominance? (4) What are the atomic processes which determine the way in which gene products assemble, and thereby decide the final shape of the structures and functions of organisms?

Chapter 17

MUTATION IS A DIRECTED PROCESS IN BACTERIA AND HIGHER ORGANISMS

MUTATION IN PRO- AND EUKARYOTES

Mutation provides an additional source of variation which may occur in connection with sexual reproduction or independently of it. Various lines of evidence indicate that mutation is a directed event. These include the following:

(1) Not all possible sites in DNA mutate with equal efficiency; some change more often than others. These sites are called "hot spots," a term coined by Benzer (Benzer, 1955; Benzer and Freese, 1958).

(2) Beneficial mutations are not the result of chance as was previously thought. Cairns et al. (1988) working with bacteria have shown that the earlier experiments were poorly designed and that mutations may represent a response to a new requirement. These workers used a strain of *Escherichia coli* which could not metabolize the sugar lactose. The addition of lactose to the bacterial medium led to the occurrence of mutations that enabled them to metabolize the sugar. The mechanism employed by *E. coli* to direct mutation turns out to be quite simple. The bacteria deleted a specific gene that blocked lactose metabolism, and at the same time activated a dormant gene that conferred the ability to metabolize the sugar.

(3) There is still the tendency to think that directed mutation does not occur in higher organisms but recent molecular analysis reveals the contrary. Temperature per se, can actually change the genotype of higher organisms in a directed form. Bernardi and Bernardi (1986) have made an extensive study of the base composition of the DNA of both cold-blooded (fishes, amphibians, and reptiles) and warm-blooded (birds and mammals) vertebrates. They found that the non-coding DNA and the sequences that code for proteins (structural genes) are richer in guanine-cytosine in warm than in cold-blooded animals. To ascertain whether this

canalization of DNA evolution is ascribable to the action of temperature, they compared the base composition of related species of fishes living in cold (20–25 °C) and in warm (37–40 °C) waters. Again, guanine-cytosine-rich DNA was more prevalent in the fishes inhabiting warm waters, specifically, hot springs. The conclusion drawn from these experiments was that temperature plays a key role in the canalization of the evolution of DNA.

Another piece of evidence arguing against random mutation comes from the work of McClintock (1950, 1978, 1980) who discovered that each seedling in maize exhibited a characteristic rate of mutation, which failed to change within the life cycle of a given plant. These studies involved the determination of the number of color patches. This provided a measure of the frequency of mutations occurring in the seedling. Thus, on this basis, mutation was found to have a constant rate, a sort of time chart and hence, not to represent a random process. McClintock also discovered that occasional sectors of variegated maize tissue show a rate of mutation which is not the same as that of the whole plant. This took place in adjacent sister sectors in which the rate of mutation had increased in one sector, and declined in an adjacent one. Thus the picture which emerges is that the rate of mutation was reset in these tissues.

Recently it has been found that in humans there are mutations, called "dynamic," that are responsible for diseases, such as the fragile X syndrome and muscular atrophy. They are due to a change in the number of repeats of a particular DNA sequence. The number of repeats decides the behavior of the gene in future human generations and influences the age of onset of the disease. Moreover, the DNA site is chemically modified by the addition of methyl groups. The behavior of these mutations "defies Mendelian laws of inheritance," since they, moreover, behave differently depending on the sex of the parent transmitting the gene (Sutherland and Richards, 1994)

A CONNECTION BETWEEN FOOD INTAKE AND DIFFERENT GENETIC CONSTITUTION

The situation described in bacteria by Cairns et al. (1988) in which the genome responds to a new need is also found in animals. The fruit fly, *Rhagoletis pomonella*, is a parasite of the hawthorn tree. The females lay their eggs on its fruits and the larvae feed on them. This species, diverged into two races some 150 years ago. One race continued to live on hawthorns while the other started to infest fruits of apple trees. Analysis of their genomes has revealed significant differences in the frequency and type of enzyme genes present in the two races (McPheron et al., 1988; Feder et al., 1988). Moreover, emergence from the pupa, which is an event synchronized with the ripening of fruits of both the hawthorn and the apple, has proven to be an inherited trait (Wright, 1989).

Over 600,000 species of parasitic insects are known to exist. Since new species may emerge as the result of new feeding habits or changes in host-parasite relationships the question arises: How many species have this type of origin?

DNA REPAIR AS A FORM OF DIRECTING MUTATION

Few phenomena could serve as a more striking example of autoevolutionary regulation than DNA repair. The base cytosine in DNA undergoes spontaneous atomic change (deamination) to the base uracil (which is normally present in RNA). The expectation is that during the life-time of a human being, 1 in 1000 cytosines will have changed to uracil, thus leading to the formation of abnormal genes and chromosomes. The repair process reverses this transformation by converting the uracil back to cytosine, and thereby, maintaining order in the DNA. This is carried out in the cell by the uracil-DNA N-glycosylase pathway which prevents deamination of cytosine to uracil.

This built-in process of repair is responsible for the preservation of the original organization of the chromosome. It is also regarded as a mechanism of directing mutation (Brash and Haseltine, 1982; Lindahl, 1982; Haseltine, 1983; Friedberg, 1985; Bohr and Wasserman, 1988; Imlay and Linn, 1988; Lahue et al., 1989).

SPECIFIC PROTEINS COMPEL DNA TO MAINTAIN ITS PATTERN

Not only the base cytosine, but the other bases in DNA are kept under control. Broadly speaking, the mere substitution of a single base in the DNA of human beings leads to lethal hereditary diseases, notably sickle-cell anemia. Assuming that the human genome comprises about three billion bases and that substitution mistakes are as rare as one in a million, then one would expect the occurrence of 3,000 mistakes during a single replication of the genome. Such an amount of error would have been intolerable. To obtain a human being from a fertilized egg, a million billion replications are necessary. These replications result in a normal individual because the actual rate of mistakes is about one in 10 billion. The maintenance of this order is due to three enzymatic processes: (1) The nucleotide added to the nascent strand is the only one that pairs correctly with the complementary base, i.e., it is the nucleotide that produces the most energetically favorable state. (2) "Proofreading" whereby a nucleotide that is non-complementary gets expelled from the DNA chain. And (3) upon completion of DNA synthesis, there is final control of the effectiveness of processes 1 and 2 in which DNA methylation is used to distinguish between the parental and the newly formed strand.

It is noteworthy that it is the enzymes that are mainly responsible for the maintenance of order in DNA, and not DNA itself. This is shown quite convincingly by the fact that when DNA is synthesized in the absence of enzymes, the number of errors occurring increases to one in every 100 bases. But in the presence of enzymes replication is found to be 100 million times more accurate (Radman and Wagner, 1988).

MUTATIONS IN 5S RIBOSOMAL RNA ARE DIRECTED BY ITS BASES

The evolution of 5S ribosomal RNA at the level of its base organization has been studied in detail in *E. coli* and other organisms. The secondary structure of this RNA is characterized by areas of base pairing which form helical stems. The molecule has canalized its own evolution as the result of the following possible ways:

1. Differences in strength between the paired bases: guanine-cytosine pairs form three hydrogen bonds, adenine-uracil only two, and between guanine and uracil there is only one bond.
2. Although bonding is normally only between adenine and uracil, and between guanine and cytosine, bonding between adenine and cytosine is possible. In fact, it occurs in some fungi.
3. Unusual base pairing between two uracils may also occur.
4. Base changes on the opposite side of the RNA stem are not entirely independent events, as previously supposed.
5. Most important is that changes at adjacent positions are not entirely independent.
6. On a random basis guanine can change to the other three bases but in the secondary structure of the 5S RNA there is an ordered substitution. Only two such possibilities are realized.
7. Moreover, a change of a pair of nucleotides from guanine-cytosine to guanine-uracil occurs more often than a change from guanine-cytosine to cytosine-guanine (which is the normal base pairing).
8. Insertions (due to duplications of adjacent nucleotides) and deletions are also known to have occurred (Ghiselin, 1988).

On balance, the above evidence shows that the mutations in the 5S ribosomal RNA are mainly directed by the properties of its bases. These are not changing at random, as was previously supposed, but are being substituted in accordance with rigid schemes dictated by their atomic properties.

AUTONOMOUS EVOLUTION DISPLAYED BY MOLECULES

Patterson (1987) has produced evidence that some molecules evolve to a large extent independently of the organism's body. A prime case is ribosomal RNA, which is an integral part of the protein-producing apparatus. It changes to a large extent independently of the rest of the organism. As will be recalled, long stretches of rRNA molecules have remained unchanged for 3.5 billion years. This autonomous evolution of molecular processes is in agreement with the emergence of periodicity in a way not necessarily connected with the organism's phylogenetic location.

GENETIC VARIATION IS DIRECTED BY THE CELL'S INTERNAL ORGANIZATION

To recapitulate, the following conclusions can be drawn:

1. In sexual reproduction the assortment of chromosomes at meiosis is an event controlled by the chromosomes themselves.
2. Crossing-over is also a regulated molecular process that is under the direction of certain proteins.
3. Mutation, which is the result of internal changes in the DNA molecule, is canalized in response to a new requirement as found both in bacteria and higher organisms (e.g., insect parasites).
4. DNA repair proves to be a most effective mechanism in the control of the mutation rate. Specific enzymes are responsible for maintaining order within DNA.
5. Mutations in certain RNA molecules are directed by the properties of their bases.
6. Some RNAs have evolved in a way independent of the organism.

DIFFERENTIAL DEATH DOES NOT AFFECT APPRECIABLY THE PERIODIC TREND OF EVOLUTION

Living organisms die due to various causes: (1) other living organisms use them as a source of food, (2) adverse conditions in the environment lead to their death, and (3) cellular internal clocks terminate their existence at a specific age. Animals and plants are known to have a maximum living period that is characteristic of each species (Röhme, 1981).

Differential death resulting from poor food intake or from adverse environmental changes, seems to have had little or no impact on the maintenance of periodicity. Since the channels that establish periodicity are mainly internal, the same solution can arise again and again, independently of its erasure by differential death.

Extinction of a given species such as the sabretooth cats, or a large group, such as the dinosaurs, is another matter. This has blocked the perpetuation of a unique cellular and genetic constellation which cannot be readily repeated. However, when evolution is seen in its totality, and periodicity is recognized as a phenomenon pervading all evolutionary processes, it becomes evident that essentially similar solutions will occur time and again. Thus, sabretooth species arose independently not less than 3 times among the placental mammals and at least twice among the marsupials (Halstead, 1978; Savage and Long, 1986).

The disappearance from the earth of the pterosaurs (the flying reptiles) did not stop the birds, and still later the mammals, from acquiring wings, as well as full flying capacity. The extreme example is of course bioluminescence that has arisen independently more than 30 times (Cormier, 1974; Campbell, 1988).

PART VI

THE ROLE OF DEVELOPMENT IN THE ESTABLISHMENT OF PERIODICITY

Chapter 18

DEVELOPMENT AND EVOLUTION ARE TWO ASPECTS OF THE SAME PHENOMENON

THE SAME MOLECULAR MACHINERY IS USED IN EVOLUTION AND DEVELOPMENT

Development involves a whole series of events that: (1) follow a rigid cascade sequence, (2) lead to the repetition of the same basic solution from generation to generation, (3) are a source of both preservation and innovation, since they lead to the repetition of the same structure for millions of years, and at the same time, create novel solutions. An example of the ability to preserve a pattern is body segmentation—a process from which invertebrates, and especially most vertebrates, could not depart. Another instance is the gills, either fully developed or rudimentary that are a constant feature of many vertebrate embryos.

Evidence of innovation is mainly provided by events such as novel genetic combinations and gene activation. Shifts in the timing of gene expression are also considered as a main source of novelty.

A point of particular importance is that development and evolution do not use separate molecular machineries to produce cellular patterns and functions. The cell, with its enormous array of molecules, its collection of distinct organelles and access to so many different RNA species, could have easily created an independent molecular compartment that would have been used, specifically for directing the necessary embryonic processes. But this is not the case. The same genetic material and the same transcriptional and translational systems are used in evolution and development. This has been a significant factor in the establishment of periodicity. Just as electricity and magnetism are two different aspects of the phenomenon of electron function, development and evolution are also two inseparable aspects of the same phenomenon, namely gene function.

WHAT IS CALLED METAMORPHOSIS IS EVOLUTION OCCURRING WITHIN THE INDIVIDUAL

Two points are of first importance to an understanding of the involvement of development in evolution. One is that there is no need to change the genetic constitution to create a totally different individual within the same species. The other is that there is no need to change the genetic constitution to create an organism that is radically different. Although two individuals have the same genetic constitution, the difference between their organization may be so large as to represent a jump of a genus or even of a sub-phylum.

Examples supporting the first argument include:

1. The queen honey-bee is of the same genetic constitution as the workers of the colony. She becomes a queen, i.e. a fertile female, as a result of a specific queen producing chemical, the "royal jelly," that it ingests. A similar situation occurs in termites in which chemical substances, produced by one type of animal, control the form and function of the other individuals in the colony (Wigglesworth, 1970; Matthews and Matthews, 1978). The queen in termites has an enormous abdomen whereas the male is minute. Yet, both have the same genetic constitution and the same diploid number of chromosomes (Figure 1) (Crozier, 1979).
2. The genetic constitutions of male and female fishes are the same, except for the difference in the sex chromosomes. The two sexes nonetheless are so different that no one would imagine that they belong to the same species. The male has the appearance of a simple appendage that attaches to the abdomen of the female (Figure 1). This is also the case in *Bonellia* (Echiurids). The female is over 1 meter in length, whereas the male is only 1.5 millimeters. The male has no mouth and anus, and has a totally different morphology and function (Figure 1) (Pierantoni, 1944).
3. The same genetic constitution is present in the larva and the adult of insects. Yet, the two are totally different in their habitat, form and functions. For instance, the larva of a dragonfly lives in water, and has a different body shape from that of the adult, the latter being aerial with large wings. No resemblance in form or function exists between the larva and the adult of the silk worm *Bombyx mori*. The same is true of *Myrmeleon formicarius*, another insect species (Chapter 21, Figure 1).

Examples supporting the second argument include:

1. The frog in the larval stage lives in water, has gills and a notochord but no skeleton. The adult lives in air, possesses lungs, and has a complete skeleton with associated muscles. The difference is as big as that between an urochordate and a vertebrate, i.e., a jump of a sub-phylum. The transforma-

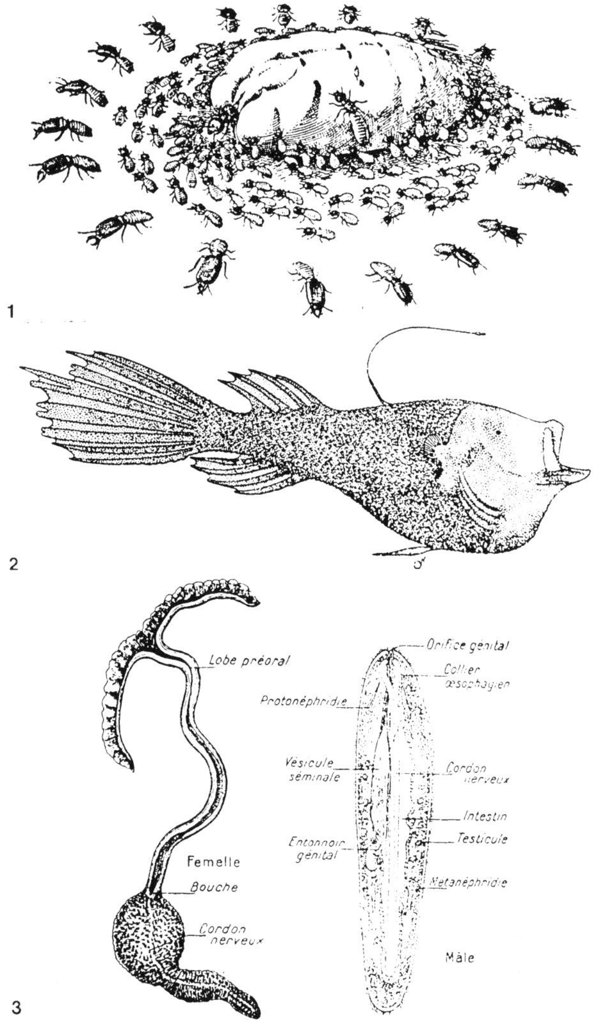

Figure 1. The same genome produces totally different animals without the intervention of reproduction. (1) Body difference in termites (insects). The queen located in the center, has a white and distended abdomen of enormous size compared with that of the male, and that of the "soldiers" encircling her. (2) Female of the fish *Ceratias Holboelli* with a male (of very small dimensions) fixed to its lower abdomen and marked ♂. (3) The female and the male of *Bonellia* (phylum Echiura) are totally different in shape and size. The male (*right*) is enlarged 60 times.

tion occurs by the simple addition of a small molecule, thyroxine. In the presence of this hormone, the transformation occurs within days (Fox, 1981; White and Nicoll, 1981).

2. The larvae of the axolotl (*Amblystoma mexicanum*, Amphibia) are aquatic; they have external gills, and their tails are flat. The adult becomes terrestrial and acquires lungs, while the tail becomes round. The axolotl larvae can reproduce. This led initially to the classification of the two forms not only as distinct species but as different genera: *Siredon pisciformis* and *Amblystoma mexicanum*. The administration of hormones to the *Siredon* form resulted in its conversion into an *Amblystoma* (Perrier, 1936; Aron and Grassé, 1939; Raff and Kaufman, 1983). Another factor participates in this transformation. If the aquatic larva (*Siredon*) is transferred to a dry habitat, it is then transformed into the aerial type with lungs and a round tail. The lack of water has the same effect as an injection of thyroxine (Freeman, 1972). Hence not only an internal, but also an external chemical can produce the transformation, just as in the bees.

These metamorphoses have the following in common:

1. They are achieved without the creation of new genomes.
2. They occur in invertebrates and vertebrates.
3. They are determined by internal agents.
4. They are also determined by external chemicals.

These developmental transformations are as radical and distinct as those occurring in evolution. This is why the same individual can be classified as a different species, genus, class, or even a different sub-phylum. What is regarded as metamorphosis is evolution occurring within the individual.

AN ANIMAL THOUGHT TO BE A LARVA OF A COELENTERATE TURNS OUT TO BE AN ADULT OF ANOTHER PHYLUM

The phylum Placozoa is represented by a single species, *Trichoplax adhaerens*. This invertebrate is the simplest known. It lacks tissues, organs and mouth, head and tail. The animal looks like a large amoeba covered with cilia, and is composed of a few thousand cells. It is found swimming in the sea and has sexual reproduction.

The organization of the body of *Trichoplax* makes it look like the planula larva of coelenterates which belong to another phylum. Its resemblance to planula larva is so striking that before careful studies were undertaken the animal was considered to be the larva of coelenterates. Only later was it recognized as an adult, and as forming a separate phylum (Margulis and Schwartz, 1982).

What was interpreted as a stage of metamorphosis became suddenly an evolutionary event.

LARVAE DO NOT REPRODUCE BUT HAVE EVOLVED NEARLY AS MUCH AS ADULTS

The problem of the transformation of larvae into adults has been studied in detail in many species but it has been considered solely from the point of view of the post-embryonal development of the organism, as if the subject had little or no bearing on evolution.

As a rule larvae do not have sex organs and as a consequence cannot reproduce. As such, differential survival, connected with death but not with reproduction, has occurred. Surprisingly, larvae have diverged from each other to the same extent as the adults. The representatives from different phyla are totally unlike. As mentioned earlier, the planula larva of a coelenterate with its ciliated and amoeboid form hardly resembles the caterpillar of a butterfly.

LARVAL EVOLUTION HAS FOLLOWED ITS OWN PATHS

Garstang (1922) stated that "Ontogeny does not recapitulate phylogeny: it creates it." He stressed that there were adaptations at the larval stage and that these had evolutionary significance. His main contribution lies in expanding Haeckel's idea (1875) that embryonic and larval development is one of the main causes of biological evolution.

Garstang's (1928a and b, 1929) studies carried out on larvae led to the following conclusions:

1. Larval evolution occurs in a way that more or less parallels that of adult evolution. This is evident from a comparison of the larval forms of marine invertebrates.
2. The secondary characters that appear in the larva of Mollusca are essentially adult characters in the making; they are mainly incipient adult features.
3. These secondary characters have no relation to the larva's form of life. The shell-plates of a *Chiton* larva may encumber its movements, and the digging foot of a *Dentalium* larva is useless at this stage of development.
4. The emergence in the first Gastropods of the reversal of the mantle-chamber did introduce some serious functional difficulties into their adult life. "The torsion was imposed upon the adult stage and not primarily developed in its interest."
5. There is a gap in the adult succession between normal symmetrical Mollusca and Gastropoda. The cause cannot be attributed to natural extinction but is a consequence of a sudden jump in the evolution of the larva in earlier times.

LARVAL AND JUVENILE TRANSFORMATIONS AS THE SOURCE OF EVOLUTION IN INSECTS AND HUMANS

De Beer (1958) showed that the larval forms of a Myriapod at the time of hatching greatly resemble an insect. Hardy (1985) concludes from this and other similarities between the two groups that insects, like vertebrates, have arisen by a process involving larval and juvenile transformations.

Bolk (1926) thought that the human species was the result of a larval-juvenile transformation occurring in the anthropoid stock. He termed the process fetalization. By that he meant that embryonic characters of past forms had become incorporated into the adults of the present. Support for this interpretation is provided by recent molecular studies on the great apes.

The chimpanzee, the gorilla and, the orangutan share a common ancestor. The upright posture and large brain size characterize humans. However, these characteristics are not present in the newborn infant but arise during post-natal development as a result of physiological changes conditioned primarily by hormones. The upright posture is the result of the neck and lumbar curvatures of the vertebral column that arise during post-natal development at 3 and 9 months (Napier and Napier, 1985). As for the volume of the human brain, it continues to increase long after birth, but in the chimpanzee it augments only slightly (Changeux, 1985).

The results of DNA-DNA hybridization have confirmed the homology between human and chimpanzee DNA. The difference amounts to only 1.1% and lies mainly in regulatory DNA sequences (King and Wilson, 1975). The similarity of the proteins in both species is equally impressive; it exceeds 99% (Wilson, 1976).

Chapter 19

THE ONSET OF REPRODUCTION DETERMINES WHETHER A DEVELOPMENTAL STAGE WILL BE CONSIDERED AN EVOLUTIONARY OR AN EMBRYOLOGICAL EVENT

DEVELOPMENTAL STAGES CANNOT BE EASILY DISTINGUISHED FROM EVOLUTIONARY STAGES

If reproduction is suddenly incorporated into a developmental stage, this may be regarded as an evolutionary event. No clear line of demarcation can be readily drawn between the two situations. The distinction is an artificial one based on the presence or absence of reproduction.

The class of coelenterates, called Anthozoans, which includes the corals and sea anemones, are never found as medusae. They have remained at a larval stage, the polyp, that has acquired sex organs. For instance, the sea anemones are solitary polyps that are larger and heavier than the polyps of hydrozoans and have a well-developed gastrovascular cavity but otherwise have retained the essential features of the larva.

The stages of the larval sequence in the Scyphozoa change in many ways. As some zoologists have pointed out, they represent an "unbelievable plasticity of pathways" (Russell-Hunter, 1979). What is of special interest is that some larvae can suddenly reproduce asexually (by budding). Not only the adults but the larvae can reproduce (without sexual organs).

The transformation of larvae into adults by the sudden addition of reproduction occurs also in vertebrates. In amphibians, the larvae can also acquire reproductive organs, as is the case in *Amblystoma mexicanum*.

VERTEBRATE ANCESTORS TURNED THEIR LARVAL STAGES INTO ADULTS

The vertebrates are chordates that are considered to be derived from early forms such as the Ascidians. However, they are not supposed to have evolved from adults of this class but from their larval forms. This is because the majority of Ascidians are completely sessile and lack locomotion when they reach the adult stage. The vertebrates became the evolutionary branch that retained the active and free swimming form of the early chordate larva. This is characteristic of the first vertebrates such as the fishes. As Russell-Hunter (1979) points out, vertebrates are aberrant members of the phylum Chordata, whose ancestors turned their larval stages into adults.

Vertebrates could also have evolved from the free-swimming larva of an invertebrate (that looked like a modern tunicate), and not from an adult reproductive form of an invertebrate. Preexisting morphological and functional features were incorporated into later stages in the life cycle (McNamara, 1989).

Chapter 20

THE SAME LARVAL TYPE GIVES RISE TO DIFFERENT PHYLA

THE SAME LARVA TYPE DEVELOPS INTO COMPLETELY DIFFERENT ADULTS THAT BELONG TO DIFFERENT ORDERS, CLASSES, OR PHYLA

The trochophore is a free-swimming larva possessing a girdle of ciliated cells and a tuft of cilia at the apical region. This larva gives rise to completely different phyla: Annelida, Echiura, Sipuncula, and Mollusca (Figure 1).

A comparable situation occurs in insects. The campodeiform larva gives rise to three very different orders: Coleoptera, Neuroptera, and Trichoptera (Figure 2). Another type of insect larva, the vermiform, develops also into three different orders: Diptera, Hymenoptera, and Siphonaptera (Figure 3).

Among the echinoderms, the starfishes, holoturoids, and sea urchins, represent three different classes, which are considered to be derived from a single larval type (Figure 4).

Hence, depending on the animal group, the same larva type is able to produce different orders, classes or phyla.

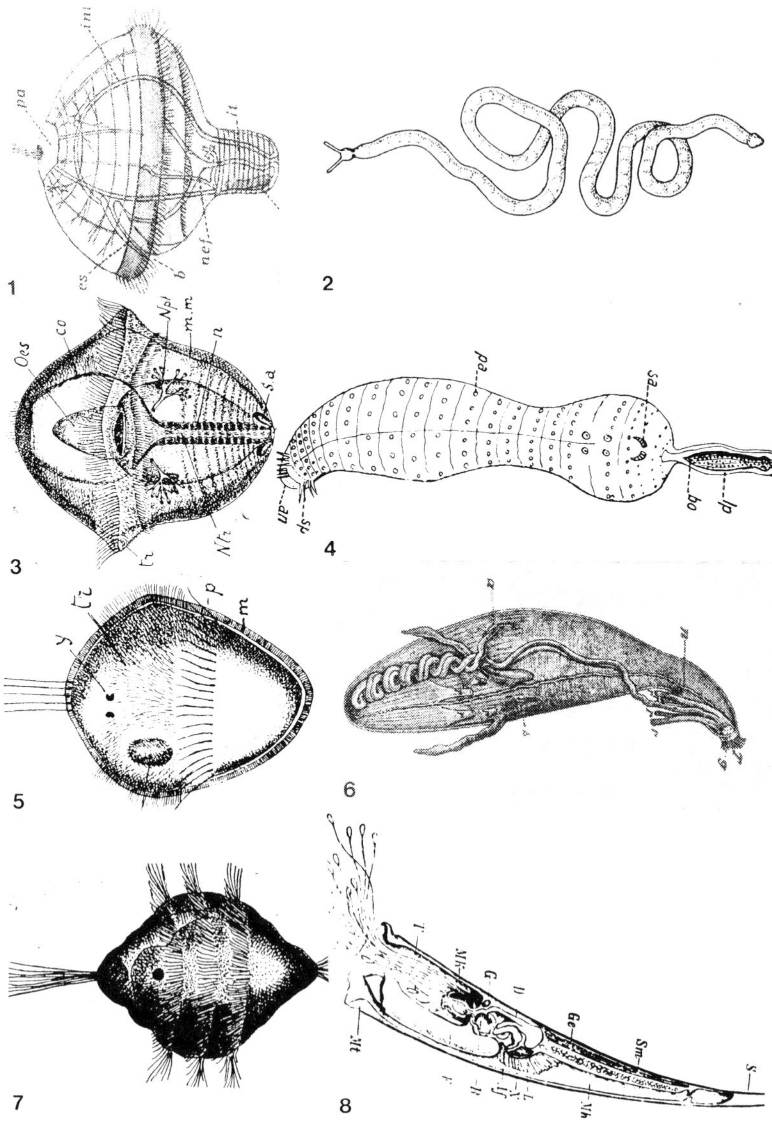

Figure 1. The same larva type, called the Trochophore, gives rise to four different phyla: Annelida, Echiura, Sipuncula and Mollusca. (1) Trochophore larva of the annelid *Polygordius* and (2) the adult *Polygordius neapolitanus*. (3) Trochophore larva of *Echiurus* and (4) the adult *Echiurus Pallasii*. (5) Trochophore larva of the sipunculan *Phascolosoma Punta-arenae* and (6) its adult. (7) Trochophore larva of the mollusc *Dentalium* and (8) adult of the same genus showing internal anatomy.

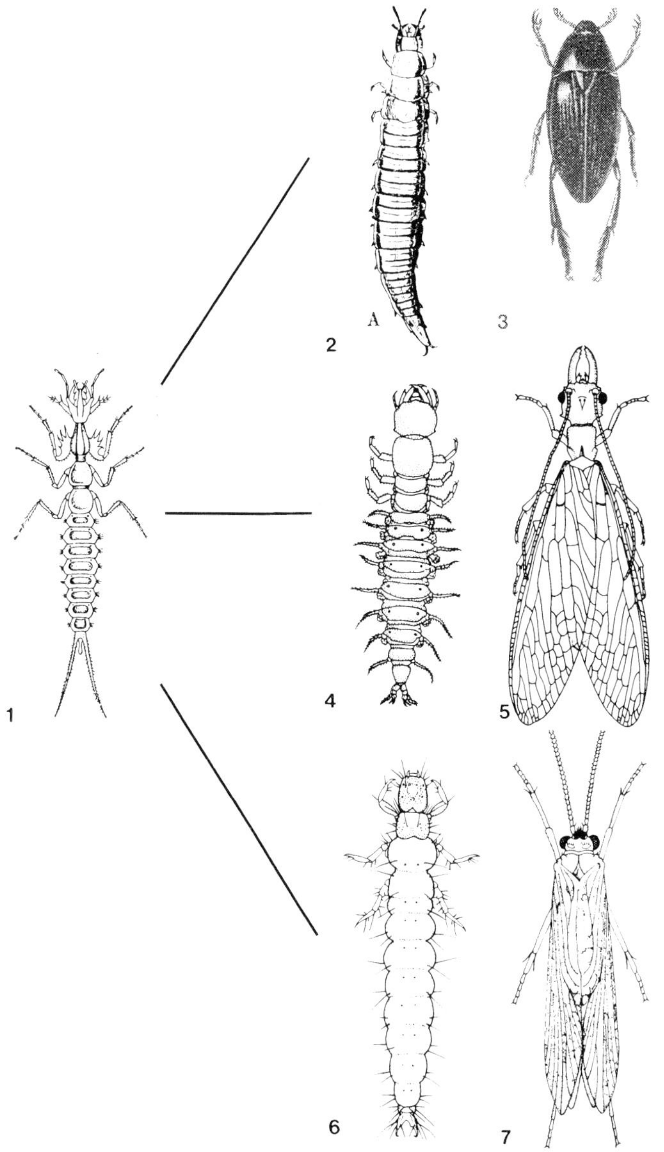

Figure 2. The campodeiform larva of insects gives rise to three different orders: Coleoptera, Neuroptera and Trichoptera. (1) Larva campodeiform of the insect *Galerita*. (2) and (3) Coleoptera: (2) larva campodeiform and (3) adult of *Hydrophilus piceus*. (4) and (5) Neuroptera: (4) campodeiform larva and (5) adult of *Corydalis cornutus*. (6) and (7) Trichoptera: (6) campodeiform larva and (7) adult of *Rhyacophila fenestra*.

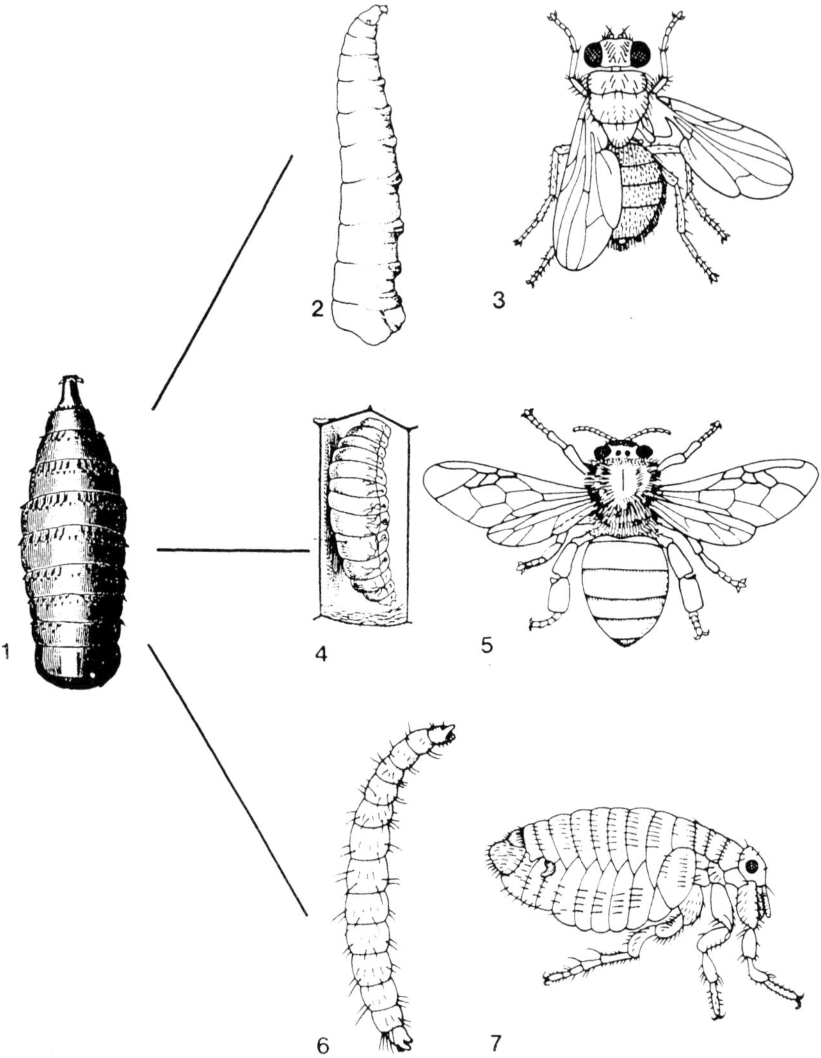

Figure 3. The vermiform larva of insects gives rise to three different orders: Diptera, Hymenoptera and Siphonaptera. (1) Vermiform larva of a fly. (2) and (3) Diptera. The house fly *Musca domestica* as adult (3) and its vermiform larva (2). (4) and (5) Hymenoptera. Adult honey bee *Apis mellifera* (5) and its vermiform larva (4). (6) and (7) Siphonaptera. An adult flea *Pulex irritans* (7) and its vermiform larva (6).

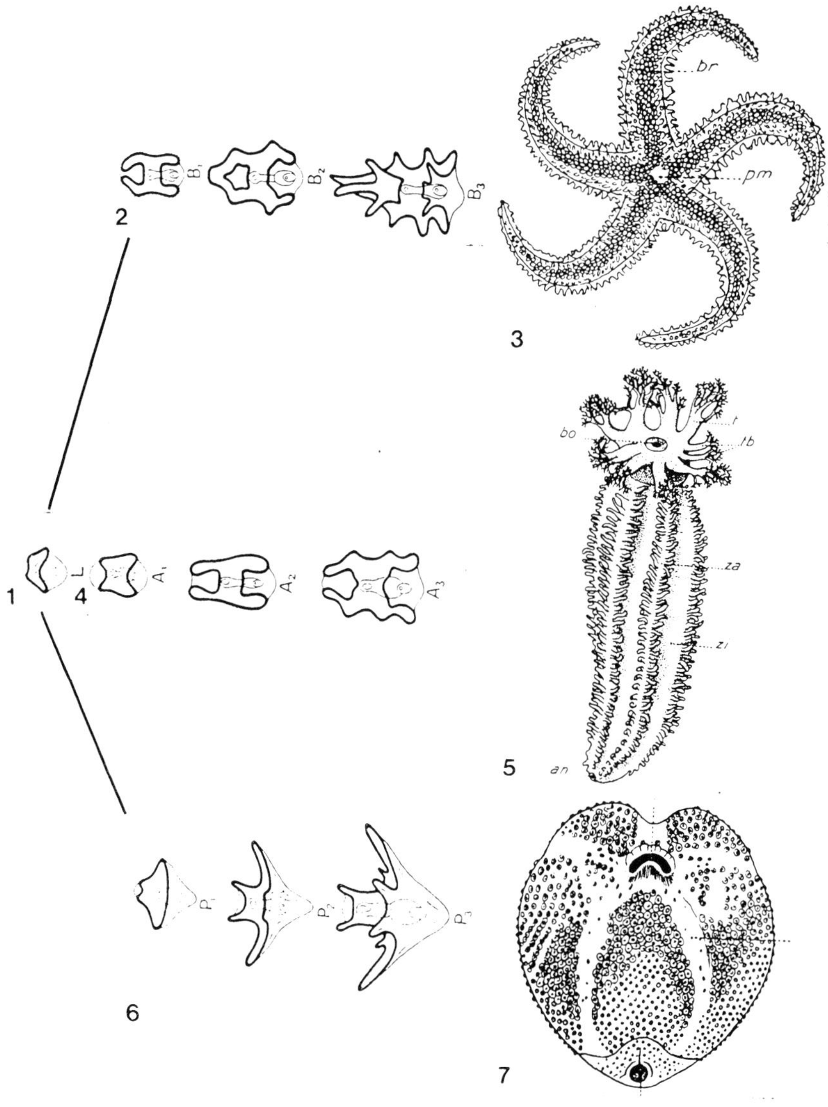

Figure 4. Three similar larva types of echinodermata give rise to different adults that constitute three classes: Asteroidea, Holothuroidea and Echinoidea. (1) Common type from which the three types of larvae are considered to be derived. (2) Three stages in the development of the larva of starfish and an adult starfish *Asteria glacialis* (3). (4) Three stages of development of larvae in holothuroids and adult holothuroid *Cucumeria Planci* (5). (6) Stages in the development of the larva of sea urchins and adult animal of the genus *Schizaster* (7).

Chapter 21

LARVAE AND ADULTS OF THE SAME SPECIES ARE AS DIFFERENT AS ANIMALS BELONGING TO DISTANT GROUPS

IT WAS EASIER TO ACCEPT SPONTANEOUS GENERATION THAN PASTEUR'S EXPERIMENTS

One of Pasteur's well-known contributions to science was the demonstration that spontaneous generation does not occur. He was actually not the first to prove this. A century earlier, Spallanzani carried out experiments in which he showed that animals had their origin in eggs deposited by their parents. Why, then was there such a reluctance to accept the new view, even in Pasteur's time? The reason is simple. It was not easy to conceive that the larva of a fly could have any developmental relationship to an adult fly. The two were totally different in terms of structure and function. The white creeping and nearly amorphous larva, and the black flying adult, with its large eyes, wings, and legs, could not be easily accepted as being the same individual at two different ages. Although spontaneous generation got disproved, the relationship larva-adult, continued until quite recently to be obscure in other groups of organisms. Difficulties in elucidating this process had arisen from the extreme changes in morphology and function that occur regularly during development. This resulted in the accumulation of a vast and confused terminology, as well as a large number of antiquated explanations (Gould, 1977).

LARVAE AND ADULTS OF THE SAME SPECIES ARE USUALLY TOTALLY DIFFERENT BOTH STRUCTURALLY AND FUNCTIONALLY

Larvae have structures and functions which are usually different from those of their adults. The caterpillar has a form unrelated to that of a butterfly. Their main functions are also different. The butterfly has a full flying capacity resulting from its body shape, musculature and wings, whereas the caterpillar crawls over the leaves or soil. Their feeding habits are also unrelated. A main function in the larva is the production by the salivary glands of a protein that is used in the formation of

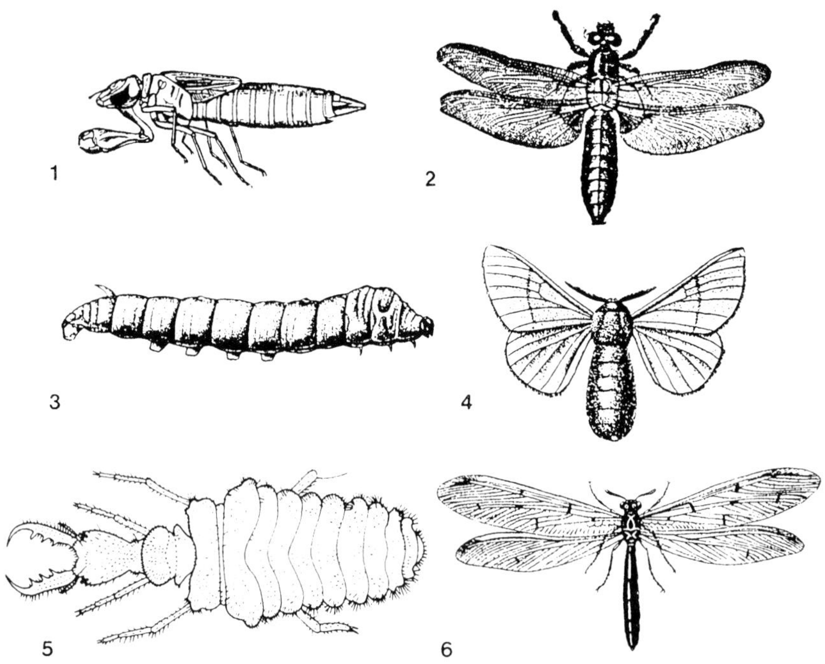

Figure 1. The existing differences in form and function between the nymphs and the adults, as well as between the larvae and adults of insects. Order Odonata. (1) The nymph of a dragon fly has an aquatic life. (2) An adult of a dragon fly lives in the air. Order Lepidoptera. (3) The larva of the silk worm *Sericaria mori* has no wings and produces silk. (4) The adult of the same species does not produce a cocoon and flies. Order Neuroptera. (5) Larva of *Myrmeleon formicarius* with its strong and large mandibles. (6) Adult of the same species in which the large mandibles are absent and four wings allow it to fly.

Table 1. Differences Between Larvae and Adults in Lepidoptera

		Lepidoptera *Butterflies, Moths, and Skippers*	
		Stage 1 Larva	*Stage 2 Adult*
Structures	Mouth parts	Strong mandibles, chewing	Long sucking proboscis, Vestigial in some species
	Eyes	Clusters of lateral ocelli	Large compound eyes and 0 or 2 ocelli
	Wings	No wings	Two pairs covered with scales
	Legs	3 pairs plus usually 5 pairs of prolegs	3 pairs, (5-segmented tarsi)
	Glands	Large labial silk glands	No large silk glands
Functions	Habitat	Most terrestrial	Most terrestrial-aerial
	Production of chemicals	Many have repugnatorial glands	No such glands
	Construction	Production of silk used in building cocoon	No production of cocoon
	Food	Voracious, masticate Plant material	Feeding on flowers, Mainly fluid meals
	Toxin production	Many possess urticating hairs	No urticating hairs

a cocoon, and it feeds on leaves that it cuts with its sharp mandibles, whereas the adult does not produce a silk cocoon and uses its mouth to suck nectar.

This is not only the case in the Lepidoptera but it is evident as well in the Odonata and the Neuroptera. The nymph of a dragon fly has an aquatic life, whereas the adult is aerial. The larva of Myrmeleon with its strong and large mandibles and broad abdomen bears no resemblance to the adult (Figure 1) (Romoser, 1973).

Other insect orders show the same phenomenon (Tables 1 and 2).

Table 2. Differences Between Nymphs and Adults in Odonata

		Odonata *Dragonflies and Damselflies*	
		Stage 1 Nymph	*Stage 2 Adult*
Structures	Mouth parts	Chewing, prehensile labium	Reduced mouth parts
	Wings	One pair of wing buds	Two pairs of wings
	Abdomen	No penis	Complex penis
Functions	Habitat	Aquatic	Terrestrial-aerial
	Respiration	Closed tracheal system, Gill ventilators	Open tracheal system
	Food	Prey mainly upon vertebrates: e.g., fish, amphibians	Prey mainly upon invertebrates: e.g., mosquitoes, bees

Note: The immature instars that occur in insects during metamorphosis are called larvae or nymphs, depending on whether the metamorphosis is complete or partial.

Biologists would have difficulties to think that the larva and the adult had any direct relationship, or could be transformed into the other, if they did not know in advance that this was so, as a result of the careful studies that led zoologists to establish their life cycles.

Chapter 22

SIMILARITIES BETWEEN TUNICATE AND VERTEBRATE LARVAE

TUNICATE AND AMPHIBIAN LARVAE ARE SIMILAR AND GO THROUGH EQUALLY DRASTIC CHANGES

Within the phylum chordata the tunicates are considered urochordates because of their most primitive features. The ascidians are one of their classes whose larvae are what zoologists call a tadpole, on account of their close resemblance to the larvae of amphibians. Paradoxically, amphibians represent a different class belonging to the advanced vertebrates. In addition to having the general form of a tadpole, the larvae have other structural and physiological features in common. They swim with a long posterior tail and they have an anterior mouth, eyes, a pharynx, and an intestine. The larvae have a notochord, which is the characteristic feature of all chordates. And the pharynx has gill slits.

At the end of the free swimming stage the ascidian larvae settle at the bottom of the sea, where they undergo structural and physiological changes (metamorphosis) which are as radical as those that occur in amphibians, and which also parallel to some extent their transformations. The tail is reabsorbed. The internal organs rotate by 180°. The mouth now opens at the upper end. The notochord and the muscles of the tail are pulled into the body, and reorganized. Thus, the adult acquires a totally different appearance and becomes sedentary and barrel shaped (Figure 1 and Table 1).

Frogs during the larval stage are also called tadpoles; they live an aquatic life similar to that of the tadpole larvae of ascidians. They also have a flat tail, round head, external gills and a notochord. These free swimming larvae undergo drastic structural and physiological changes as they become adult frogs. These changes involve the adoption of a terrestrial mode of life which includes the reabsorption of the tail, pulmonary respiration, formation of a skeleton with accompanying musculature, reshaping of the intestines, and a change to a carnivorous diet (Figures 1 and 2, Table 2).

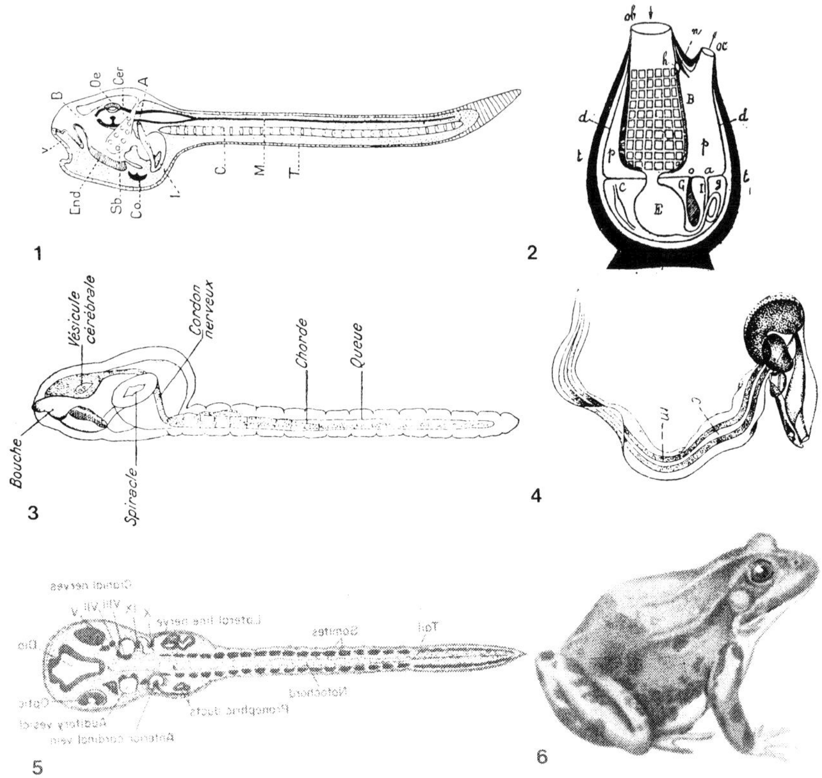

Figure 1. Ascidiacea, Larvacea, and Amphibia are three classes that belong to the phylum Chordata. The larvae of these classes are so similar that they are called by zoologists tadpole larvae, but the adults are totally different. Ascidiacea: (1) Tadpole larva of the ascidian *Polycitor*; the body contains a notochord, a neural tube and a pharynx perforated by clefts. (2) Adult of an ascidian which has only a cerebral ganglion instead of a brain. Larvacea: (3) Tadpole larva *Oikopleura* which has a notochord and is most similar to an ascidian larva. (4) Adult of *Folia aethiopica* which also looks like a typical ascidian tadpole larva. The adult has retained the larval characteristics but has become sexually mature. Amphibia: (5) Tadpole larva of frog with its notochord. (6) Adult of the frog *Rana temporaria* with its well developed skeleton and brain. The larvae are shown in cross-sections so that the anatomical details can be compared.

Table 1. Transformations of Structures and Functions in Tunicates (Ascidiacea)

		Solitary Tunicates Urochordata	
		Stage 1 Larva	*Stage 2 Adult*
Structures	Notochord	Notochord well developed	Notochord degenerates, Absent notochord
	Gills	Gills with slits	Internal gills forming pharynx
	Internal organs	Distributed along body	Internal organs rotate
	Tail	Long tail	Tail reabsorbed, no tail
	Body shape	Like amphibian tadpole	Barrel-shaped
Functions	Movement	Free swimming	Inactive, attached to substrate
	Food	Seek food	Filter-feeders

Although the chemical processes responsible for metamorphosis in ascidians have not yet been studied, those in frogs are well elucidated. There is a tenfold increase in thyroid hormone levels in the blood of the tadpole, which wholly accounts for this cascade of structural and functional events (Alberts et al., 1989).

A matter of special significance is that these tadpole larvae are found to function as adults in other animal groups. Consider the class of tunicates called the Larvacea:

Table 2. Transformation of Structures and Functions in Frogs (Amphibia)

		Frogs (Rana Temporaria)		
		Stage 1 Larva	*Stage 2 Larva*	*Stage 3 Adult*
Structures	Skeleton	Notochord	Change from notochord to vertebrae	Vertebral column
	Tail	Flattened tail in sagittal plan	Tail present	Tail is eliminated, only a rudiment is left
	Gills	External gills	Internal gills	No gills
	Digestive apparatus	Long intestine with spiral form	Intestine shortens and changes shape	Liver and pancreas also change shape
	Members	No members	Development of posterior and anterior members	Posterior and anterior members bones accompanied by muscles
	Circulatory apparatus	Only 4 aortic vessels	Branchial vessels disappear	Aortic vessels change
Functions	Habitat	Aquatic	Aquatic	Terrestrial
	Respiration	Branchial	Branchial	Respiration through lungs and skin
	Nutrition	Mainly herbivorous	Mainly herbivorous	Mostly carnivorous
	Blood circulation	Single branchial circulation, heart receives only venous blood	Drastic change from one circulation type to the other	Double blood circulation; heart receives arterial and venous blood

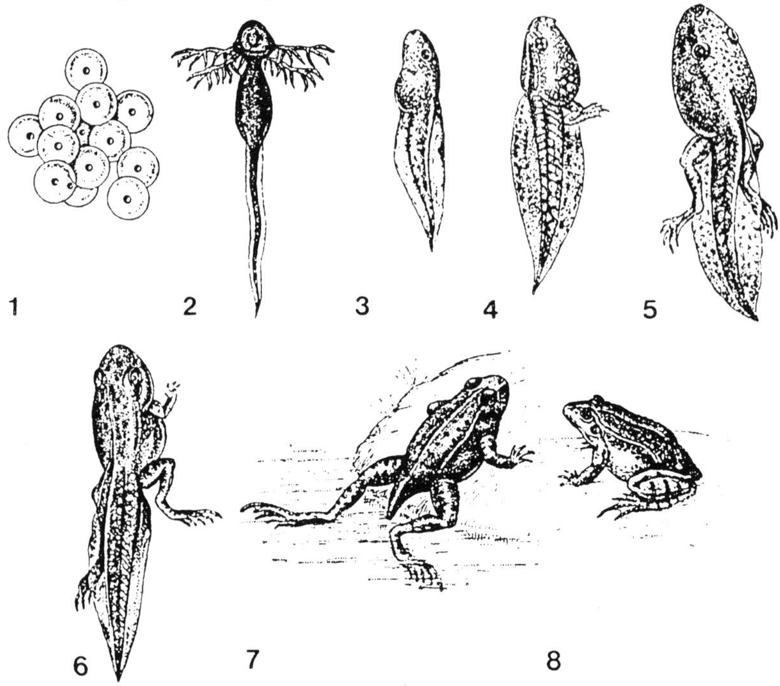

Figure 2. Transformations occurring during frog development. (1) Eggs, (2 to 7) six intermediate larval stages, and (8) the adult in the frog *Rana esculenta*. The structures of the first larval stage are characterized by: the presence of a notochord, a flattened tail, external gills, no members, a long intestine, and only four aortic vessels. Within days the frog modifies its structures in the following ways: it acquires a complete skeleton, eliminates the tail, eliminates the gills, the intestine changes shape, posterior and anterior members are formed, and the aortic vessels are modified. The functions of the first larval stage include: an aquatic life, branchial respiration, herbivorous nutrition and a blood circulation in which the heart receives only venous blood. The adult emerges after the following functional transformations: acquirement of a terrestrial habitat, respiration through lungs and skin, the heart being supplied by both arterial and venous blood and nutrition becoming mainly carnivorous (see Table 2)

Source: (From Curry-Lindahl and Tinggaard, 1965).

they are so designated because not only the larvae, but also the adults closely resemble the larvae of ascidians. The adult looks like the typical tadpole larva with its fusiform body shape. The mouth also lies anterior. There is a pharynx, spiracles, stomach, intestine and a long undulated tail. As adults they have acquired ovaries and testes (Figure 1).

Table 3. Stages in Humans Following Birth

	Stage 1 New Born	
	Environment	*Changes from Liquid to Gas Environment (Amniotic Fluid to Air)*
Structures	Vertebral column	At birth vertebral column has single posterior curve like that of a gorilla
	Brain	Volume small at birth
	Eyes	Eyes closed in darkness of uterus; open on contact with air
	Gills	Gill rudiments at early embryonic stage are reabsorbed before birth
	Intestines	Not used in uterus
Functions	Respiration	Oxygen acquired through blood of mother; it changes to absorption by respiration through lungs
	Vision	Eyes open at birth but vision not fully developed
	Nutrition	Food available via placenta, changes to intake by intestines that become active
	Immune system	Immune defense mainly supplied by mother via placenta

		Stage 2 Infant	*Stage 3 Pre-Adult and Adult*
Structures	Vertebral column	Neck curvature appears at 3 months, and lumbar curvature at 9 months, resulting in upright posture	Upright posture
	Brain	Volume continues to increase long after birth	Large brain
	Thymus gland	Large	Becomes smaller
	Teeth	Teeth appear and are shed, except first molar which is added at 6 years of age	Second set of teeth replaces first set and molars are added. third molar appears at 18 to 26 years of age
Functions	Locomotion	Stationary and creeping to walking in first years	Walking
	Memory coordination	Little memory and partial coordination of movements in early years	Memory and coordination established
	Immune defense	Under transition and development	Immune defense well developed
	Oral digestive function	Initially, ingestion of liquid food (no teeth), followed by deciduous dentition that allows partial cutting of food	Food being cut, nipped, grinded and chewed

Thus, the picture which emerges is that among the chordates an animal may function either as a larva or as an adult. Clearly, two distinct evolutionary stages have been established due to the intervention of reproduction.

HUMANS GO THROUGH THREE STAGES FOLLOWING BIRTH

While in some fish and amphibians, development is divided into larva and adult stages, in reptiles, birds, and mammals this division does not exist, since at birth the animals are minor copies of the adults. However, closer inspection of their characteristics indicates that they go through a post-natal development that can also be divided into several stages. This is especially true of humans. Indeed, some of the structural and functional changes are so dramatic that they include the shedding of teeth and the formation of new ones (Table 3).

Chapter 23

JUVENILE STAGES AND THE ONSET OF REPRODUCTION IN PLANTS

IN PLANTS THE HAPLOID AND THE DIPLOID GENERATIONS BECOME DIFFERENT STRUCTURALLY AND FUNCTIONALLY AS IS THE CASE IN INSECTS

In plants there are no larval stages; instead, they alternate between a haploid and a diploid generation in an orderly fashion.

In the bryophytes, such as the mosses, the more conspicuous and green plant stage is the leafy haploid phase containing the sex organs that produce eggs and sperm. The diploid generation is small and brown in color and different in appearance; it usually grows on the haploid plant. In the flowering plants (Angiosperms), the opposite is the case: the diploid generation is green and large. The haploid generation is reduced to a point where it is not even a separate part. It consists of a small group of cells hiding inside the flower, which are entirely dependent on the diploid phase.

Just as the larvae of animals may assume directly adult forms and in this way produce other classes of organisms, the plants have also the ability to transform their haploid phase into a conspicuous green plant or to reduce it into a microscopic structure.

The use of diploidy and haploidy as a means of producing different structural and functional solutions is also found in insects. In honey bees (*Apis* species) the fertilization of the eggs leads to a diploid organism that develops into a female worker, whereas the unfertilized eggs give rise to haploid individuals that are males. The two sexes have very different functions in the bee society. The same phenomenon occurs in ants and social wasps (Hermann, 1979).

THE ONSET OF REPRODUCTION IN PLANTS IS DETERMINED BY PHYSICAL AND CHEMICAL AGENTS

Botanists established long ago that there is a relationship between the change in organ shape and the onset of reproduction in many plant species. The stage preceding reproduction is known as the juvenile form, whereas the emergence of reproductive organs marks the onset of the adult plant.

The similarity between the plant and animal situation is much greater than this. Juvenile plants, like the larvae of animals, also possess a characteristic shape and symmetry that differs rather remarkably from that of the adult. In *Campanula rotundifolia* (a blue bell), the first leaves are rounded and orbiculate, and when the plant starts flowering the leaves become ovate and linear. *Cannabis sativa* (hemp) leaves change successively and become more lobed in shape and serrated as the plant develops. Once flowers appear, the leaves become less lobed and serrated, and smaller. In these two species, it is the amount of sunlight which reaches the plant that causes these changes. When the days are short, flowers are formed and the leaf pattern is changed. Both morphogenetic processes are dependent on a simple physical agent.

Juvenile plants of *Helix hedera* (ivy) have: (1) pubescent stems, (2) adventitious roots, (3) palmate leaves, and (4) an alternate leaf arrangement. The flowering adult part of the same individual has: (1) smooth stems, (2) no adventitious roots, (3) entire and ovate leaves, and (4) a spiral leaf arrangement. Experiments show that cuttings from the adult and juvenile parts only grow as juveniles and that adult stems grafted on to juvenile stock also grow as juveniles. In the latter case the underlying mechanism is hormonal in nature. When the hormone, gibberellic acid, is applied to adult plants, it causes the growth of the juvenile form. It is thus the high levels of gibberellin in the roots that determine the formation of the juvenile pattern. Once the concentration of the hormone along the stem decreases below a critical threshold, the adult flowering condition emerges (Dale, 1982).

An increase in the hormone thyroxine starts a cascade of transformations that leads the juvenile frog to becoming a reproductive adult, while in this plant it is a decrease in gibberellin that produces a parallel effect. The crucial factor governing transformation in these plants and vertebrates is hormonal in both cases.

HORMONE CONTROL OF METAMORPHOSIS IN CRUSTACEANS AND INSECTS

Crustaceans are known to have certain hormones which were detected in experiments where both eye stalks were removed. This procedure led to precocious growth, as well as to rapid and repeated molting. Further investigation disclosed the involvement of three structures, namely: (1) the Y-glands that produce a hormone leading to the initiation of molting, (2) the X-organ, within each eye stalk, which produces a molt-inhibiting chemical, and also stops ovary formation, and (3) the sinus gland which functions as an endocrine reservoir.

Recent work has confirmed that in all arthropods (i.e., insects, crustaceans, spiders, and other groups) the transformation of body structures and functions (molting and metamorphosis) is mediated by the same types of hormones. The hormone produced by the Y-gland in crustaceans is the steroid beta-ecdysone. It differs only by one additional hydroxyl group from alpha-ecdysone which is the steroid in insects that performs the same function. The two compounds have been chemically isolated, purified and synthesized. Either will induce the structural and functional changes connected with molting in arthropods of all groups (Russell-Hunter, 1979).

HOW DEVELOPMENT PARTICIPATED IN THE ESTABLISHMENT OF PERIODICITY

The emergence and establishment of periodicity must have demanded above all certain ordered phenomena; development seems to have been one of the main agents in this process. Several lines of evidence include:

1. The molecular machinery used in evolution and development is the same.
2. No change in genetic constitution is necessary to create an organism that is so different as to represent a jump of a genus or even a sub-phylum.
3. The same larva type develops into adults that belong to widely different phyla.
4. Many larval characteristics are anticipations of adult features.
5. The distinction between developmental and evolutionary stages is not sharp. In several cases it is the onset of reproduction that determines whether a given set of structures and functions will be classified as one or the other. Both in invertebrates and vertebrates, larvae can acquire reproductive capacities and, as such, can start to function as separate organisms. The distinction is mainly based on the presence or absence of reproduction.
6. The onset of reproductive organs both in plants and animals depends to a great extent on physical, as well as chemical agents, e.g., hormones.
7. It is widely recognized that vertebrate ancestors were able to turn larval stages into adults by incorporating preexisting structural and functional processes into the later stages of their life-cycle.
8. Changes in gene splicing occur during development. This has been well documented at the molecular level in the case of the immunoglobulin genes, which change their position and function during embryonic life in both mice and humans (Brack et al., 1978; Alt et al., 1987).

Hence, periodicity could be established with relative easiness, thanks to the participation of development in evolution.

PART VII

RELATIONSHIP BETWEEN THE ENVIRONMENT AND PERIODICITY

Chapter 24

THE ABILITY OF ORGANISMS TO CHANGE WITH THE ENVIRONMENT IS ALREADY PRESENT IN MINERALS

THE CHANGE IN PROPERTIES WITH TEMPERATURE IS CONDITIONED BY THE ATOMIC CONSTITUTION

When the environmental temperature is increased above 265 °C the crystals of boracite change their shape and color. Below this temperature the crystals are cubes but above it they are orthorhombic (Figure 1). If one considers plants e.g., oat species, they are able to attain different sizes and shapes if grown at 8 °C or 15 °C. In invertebrates, the butterfly *Araschnia* has a color and size in spring that differ from those in the summer. Vertebrates follow the same pattern. In the winter, the stoat *Mustela erminea* is mainly white, whereas in the summer it is dark. The effect of temperature on the pigmentation of higher mammals has been studied in rabbits. If black hair is cut and the temperature lowered, white hair is found to grow in the same spots, and if it is cut again and the temperature raised, hair then grows dark again. In all cases, a higher temperature results in darker color (Figure 1) (Schmalhausen, 1949).

What seems quite apparent is that the atoms in the chemical pathways that lead to pigment formation are, for all practical purposes, behaving in a way comparable to the atoms in the mineral. The main pigment of animal hair is melanin. A change in hair color results from the combination of melanin with oxygen atoms (Lerner, 1967; Atkins, 1987). Change in temperature in both cases leads to a different chemical configuration (a different pigment molecule in the mammal, and a different crystallographic form in the mineral). At the same time, in both cases, the

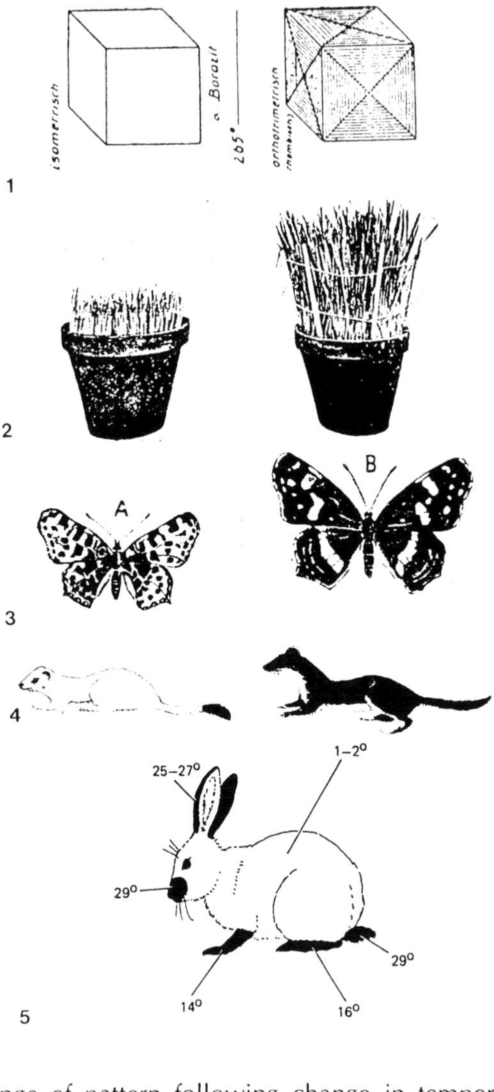

Figure 1. Change of pattern following change in temperature. (1) Two boracite crystals ($Mg_3ClB_7O_{13}$); one crystallizing below (*left*) and the other above 265 °C (*right*); note difference in form and color. (2) Oat plants (*Avena sativa*) grown at 8 °C (*left*) and at 15 °C (*right*). (3) The butterfly *Araschnia levana* in spring (*left*), and in summer (*right*). (4) The stoat, *Mustela erminea*, in winter (*left*), and in summer (*right*). (5) Distribution of temperature thresholds for pigment synthesis in the skin of the Himalayan rabbit (in degrees centigrade). In all cases higher temperature results in a darker color.

changes are well delimited by the properties of the atoms. The rabbit can not produce green, blue, or yellow hairs; it can only change from white to black and from black to white. The crystal of boracite cannot become hexagonal, triclinic, or monoclinic but only cubic and orthorhombic. The constituent atoms in both cases allow a change but only within well-defined pathways.

PRESSURE AND THE ATOMIC CONSTITUTION DETERMINE SHAPE

Pressure is considered as the main factor (sometimes combined with temperature) determining different shapes in both minerals and living organisms. Diamonds crystallize differently when these conditions differ. Pressure changes the pattern of egg cleavage in the worm *Nereis*, and the development of plants in *Taraxacum* (growing at different altitudes). Fishes that live in the oceans at depths of 1900 to 2500 meters suffer enormous pressures, and change their body properties accordingly. Altered pressures, at every level of organization, do not produce all types of forms but only those that the constituent atoms allow (Figure 2).

SALT CONCENTRATION MODIFIES FORM AT ALL LEVELS

A rock salt crystal immersed in a solution of urea and sodium chloride changes its form successively. The crustacean *Artemia salina* gives origin to three races resulting from increased water salinity. Eye formation may be suppressed in fishes as the result of the presence of magnesium chloride in water. *Salicornia* plants develop differently, depending on the amount of sodium chloride found in the soil (Figure 3).

COLOR CHANGE IN MINERALS AND ORGANISMS

The color of a molecule or of an organism depends on the wave-length of light that it receives and that it emits. In all molecules light absorption causes shifts in electrons. A mineral, such as fluorite, occurs in several colors that result from electron substitutions in its molecule. Another example is retinal which plays a key role in the vision of animals. This is attributable to its ability to shift electrons inside its molecule (Atkins, 1987). Among living organisms the pupae of the insect *Pieris*, change their color in accordance with the background in which they grow. The same holds for flat fish. Blinded fish are not able to modify their color when moved to a different environment. The wave-length of light received by the eyes results in the release of adrenalin and other hormones that modify the color of the animal by altering the shape of the pigmented cells of the body (Figure 4) (Veil, 1938; Eckert and Randall, 1978).

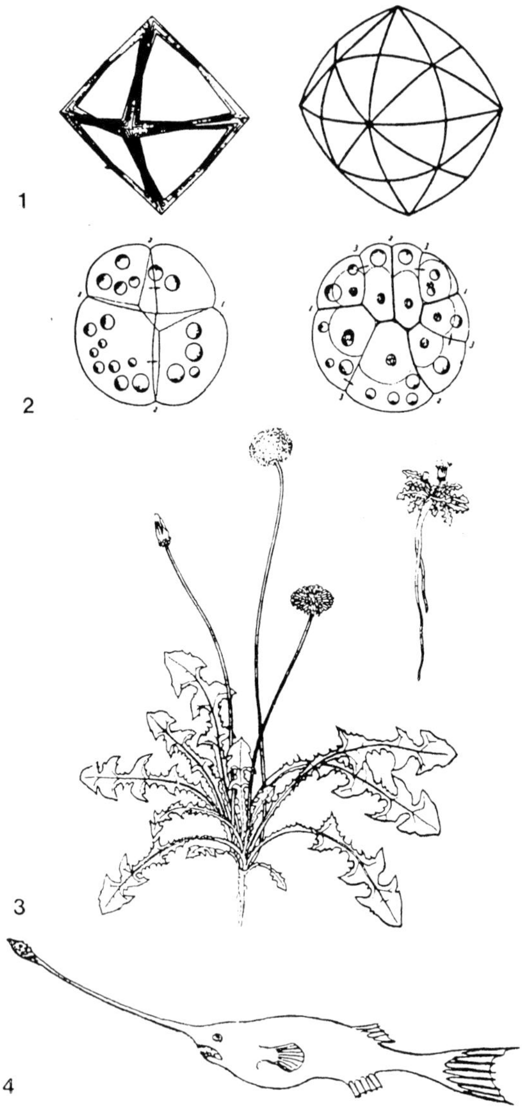

Figure 2. Change in pressure produces different forms. (1) Two diamond crystals formed under different pressures and temperatures. (2) Modifications in the pattern of egg cleavage by pressure in the worm *Nereis.* Normal development (*left*) and an eight-cell stage from an egg subjected to pressure (*right*). (3) Development of *Taraxacum officinale* plants at sea level (*left*) and at high altitude (*right*). (4) The fish *Gigantactis Vanhöffeni* which lives in the Indian Ocean at a depth of 1900 to 2500 meters, and which has a luminous organ at the end of a tentacle.

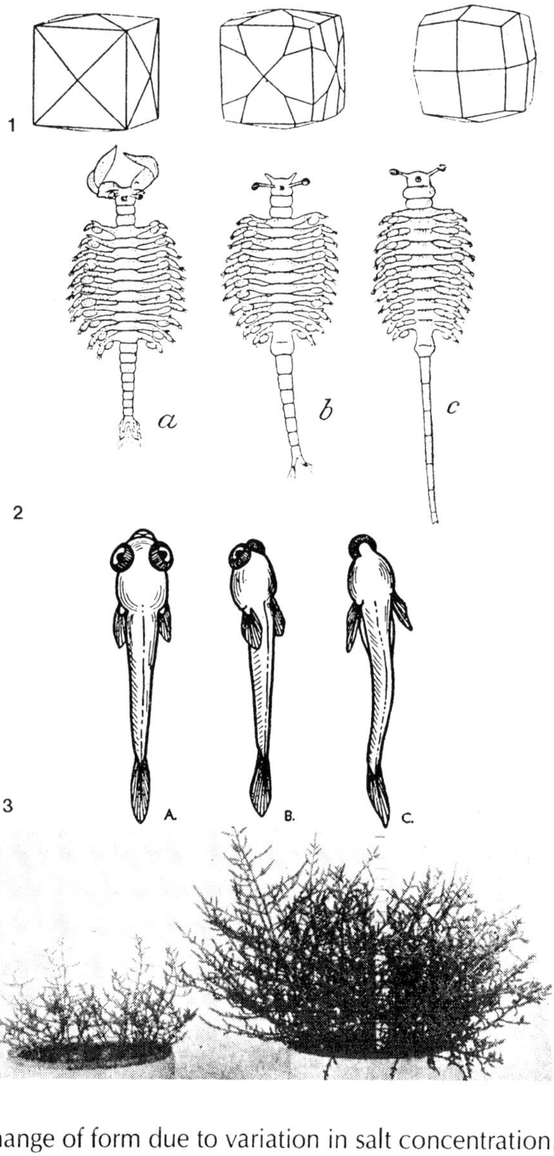

Figure 3. Change of form due to variation in salt concentration. (1) Successive change in crystal form, in a rock salt crystal (NaCl), immersed in urea (diluted in sodium chloride). (2) Three races of *Artemia salina* (crustacean) resulting from increased water salinity. (3) Suppression of eye formation in fish, as a result of the presence of magnesium chloride in the water: normal individual (*left*), fish with eyes growing together (*middle*), fish with single median eye (*right*). (4) *Salicornia* plants growing without sodium chloride (*left*) and growing in its presence (*right*).

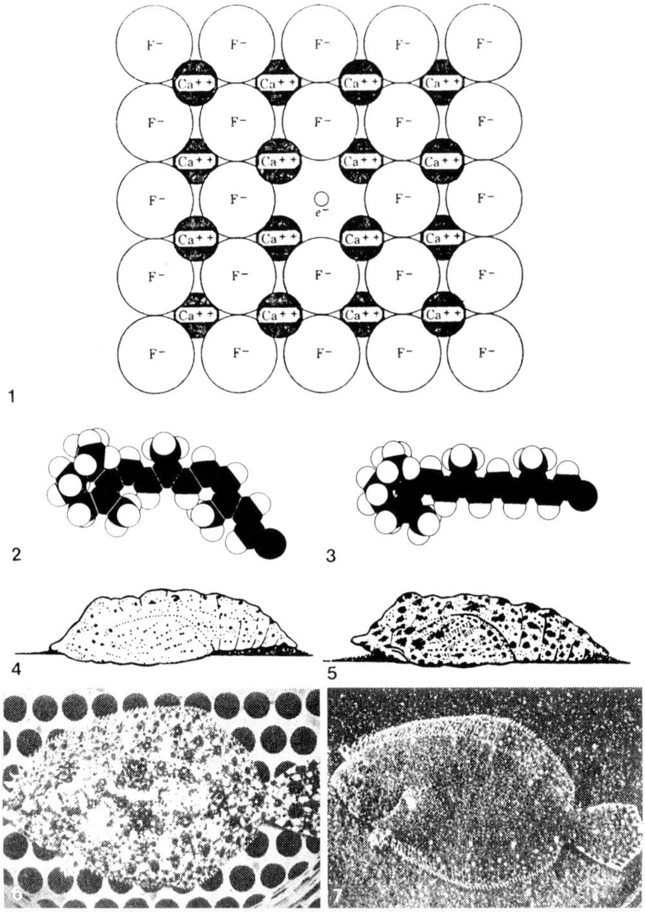

Figure 4. Change of color. (1) Fluorite (CaF_2) is a mineral that occurs in several colors. These may result from substitutions of atoms of fluorine (F) by electrons. In this molecule a color center was created by substituting a fluorine ion (in the center of the molecule) by an electron e^-. (2) The retinal molecule in its *cis* form, and (3) in its *trans* configuration. The *cis* form absorbs and stores energy from light by modifying the arrangement of its electrons, the result being the change of the retinal to the *trans* configuration and its anchoring to a protein molecule. All image-resolving eyes in invertebrates and vertebrates contain retinal as the light-sensitive component. (4) and (5) Pupae of the insect *Pieris.* Green pupa formed on a green background (4) and darkly pigmented pupa formed on a black background (5). (6) and (7) A color change in a fish when placed on a light or dark background. The change in color is the result of information transmitted through the eyes by means of hormones. Blinded fish do not change their color.

The changes described above, that are due to temperature, pressure, salt concentration, or light, are not necessarily transmitted to the progeny. In other words, they do not result in genetically stable situations. However, in other cases, molecular signals from the environment lead to permanent genetic changes, as described in the next chapter.

Chapter 25

GENETIC CHANGES AND ENVIRONMENT

MOLECULAR SIGNALS FROM THE ENVIRONMENT CHANGE THE EXPRESSION OF YEAST GENES

The yeast *Saccharomyces cerevisiae* has two mating types determined by genes, but its chromosomes also contain two other genes which are silent, representing each mating type. The recognition of cells of opposing mating type is in itself a molecular process, which is accompanied by the emission of pheromones that attach to surface receptors. These, in turn, transmit this information to the interior of the cell by means of a protein called transducin. The interaction of a series of proteins leads finally to DNA binding. This provokes both activation and repression of gene sequences located at a distance up to 2.5 kilobases away from the promoter. Moreover, an endonuclease makes double-strand breaks at a specific gene, and the resulting free DNA sequence invades other genes, thereby transforming their expression (Nasmyth and Shore, 1987; Lewin, 1990).

Several features of this process represent new genetic mechanisms. (1) It is a molecular signal from the environment, a pheromone, that initiates the genetic change. (2) This is transmitted by means of a specific molecule, the protein transducin, to the interior of the cell. (3) As a result, protein-DNA interactions occur. (4) Repression takes place at a long distance from the initial site. (5) Segments of DNA are excised and transferred to other regions of the chromosomes creating a new gene expression.

IN FLAX PLANTS THE ENVIRONMENT IS ABLE TO CHANGE THE GENOME IN A PERMANENT WAY

We are not accustomed to the use of the term cataclysm either in genetics or evolution. Most gene mutations are considered to represent minor changes in the genome and evolution is equally considered to have taken place by successive discrete adaptations. Yet the evidence accumulating in the last few years prompted Alberts et al. (1983) to write on the cataclysmic changes that occur in the genome of eukaryotes.

Flax plants grown in soil with too much phosphate or too little nitrogen produce seeds that generate daughter plants that are different in stem, leaf and seed properties. These plants breed true in future generations even if the soil conditions

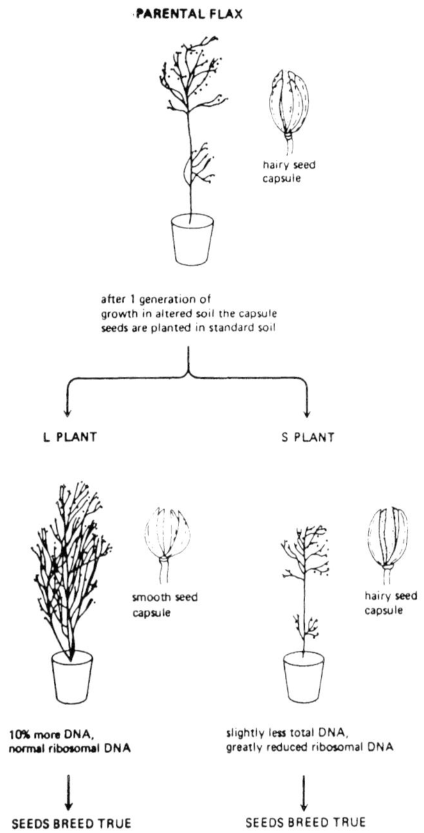

Figure 1. Induction of permanent change in the genetic constitution of flax (*Linum usitatissimum*) by alteration of the chemical composition of the soil. In *S* plants the number of ribosomal RNA genes is greatly reduced. The seeds from these plants breed true, indicating that the genetic change is permanent.
Source: (From Alberts et al. (1983), based on the work of Cullis, 1977).

are changed. Two types have been studied in detail called *L* and *S* (Figure 1). The transformation from the parent plants to the *S* and *L* types has involved changes in their DNA. There has been differential amplification of several DNA sequences in the *L* plant. On the contrary, in the *S* plant the number of ribosomal RNA genes has been halved (Cullis, 1977). Similar stable changes have been described in *Nicotiana rustica* (Hill, 1965). In both plant species, not just one, but several characters were changed simultaneously.

Recently it has been found that heritable effects were also obtained in flax plants (*Linum usitatissimum*) by treatment of seeds with 5-azacytidine. Changes in flowering time and plant length were transmitted to the second generation progenies (Fieldes, 1994).

ORGANISMS POSSESS INTERNAL MECHANISMS THAT ALLOW THEM TO COPE WITH ENVIRONMENTS TO WHICH THEY HAVE NEVER BEEN EXPOSED

An organism does not need to be initially exposed to a given environment to produce a form or a function that fits into it. An animal or plant may display an "adaptation" to a type of environment to which neither have been subjected.

Absolute zero, that is – 273.15 °C (Pitt, 1988), is a temperature only theoretically possible and does not exist under natural conditions. It is only obtained experimentally. Consider now the tardigrades, a phylum of invertebrates that are resistant to extreme environments. In the desiccated stage, they may survive for as many as 7 years, and stand exposure to laboratory temperatures of –272 °C. Tardigrades in diapause can also be submerged in absolute alcohol and survive this treatment (Russell-Hunter, 1979).

The temperature that the tardigrades are able to survive differs only by one degree from absolute zero. Despite the fact that they were never exposed to this temperature, they have the functional means of standing it.

HEAT SHOCK PROTEINS APPEARED BEFORE THERMO-REGULATION

It is now well established that animals exposed to higher temperatures produce heat shock proteins. These proteins are present in most eukaryotes. In animals that have internal thermo-regulation mechanisms, such as birds and mammals (body temperature around 37–38 °C), the heat shock proteins provide protection against deviations in cell temperature. It is not without significance for the understanding of periodicity that heat shock proteins are not only found in animals but are also present in simpler living organisms such as protozoa, fungi and plants (Nover et al., 1984; Nelson et al., 1992; Goldfarb, 1992).

Hence, the molecular solution existed before the emergence of thermo-regulation of body temperature in higher vertebrates. This implies that the protective mechanism was available before it was needed.

ENVIRONMENT AND PERIODICITY

The relationship between the environment and periodicity is not a simple one. In some situations it results in a coordinated interaction, whilst in others it does not.

Coordinated Processes

1. Minerals, plants, and animals change their pattern in the presence of physical and chemical agents.
2. Temperature, pressure and salt concentration provoke structural modifications that are characteristic of the mineral and the living organism, as well as of the agent. At the different levels of organization, the type of modification that arises is limited and specific.
3. In yeast, chemical agents alter gene expression responsible for specific functions.
4. In flax, modification of phosphate and nitrogen concentrations in the soil changes the number of ribosomal RNA genes and creates a permanent genetic change that is transmitted to the plant progeny.
5. Both in plants and animals water causes transformations in gene activity. This occurs when the plant crown is immersed in water, eliciting a change in leaf pattern to the elongated form. Contact with water or dried conditions (desiccation) changes radically the body shape and functions of amphibians. Cell water content affects the release of internal chemical signals (hormones) that in turn activate genes (see Chapter 8).
6. The return to water in different orders of mammals led to similar biological solutions. Properties that are part of the mammalian construction, such as the ability to withstand asphyxia, and hormonal signals, are involved in the mammalian transformations (Lima-de-Faria, 1988).

Uncoordinated or Partly Coordinated Events

1. Vision is a phenomenon that in certain situations is closely related to the amount of light in the environment while in others it is not. This is exemplified by animals living in caves or deep sea waters that are blind, while other animals inhabiting the same locale have well developed eyes. Moreover, animals receiving strong illumination at the surface of the sea may be completely blind (Chapter 4). The same applies to bioluminescence.

Thus, deep sea fishes have luminous organs but mammals inhabiting similar regions lack them completely (Chapter 6).

2. Neither the emergence of the placenta, nor that of flight, is directly connected to the general environment. The placenta developed in marine fishes, amphibians, and in terrestrial, as well as marine mammals. Flight is found to have emerged rather suddenly in fish (teleosts), thereby obliging them to move into a medium (air) that is so adverse to their functions that they can only stand it for a few minutes. Other fish such as rays actually fly in water. They use their large fins, beating them like the wings of the birds of prey, to fly in a different medium.
3. Many plant families have flowers with 3 or 5 petals radially distributed, irrespective of the environment to which they are subjected. The same is true of insects which inhabit the most diverse environments, yet all have six legs.
4. Organisms show physiological solutions that allow them to withstand extreme environments to which they have not been exposed previously.

The conclusion that can be drawn from the above data is that living organisms have properties that are independent of the environment, but at the same time possess other properties that are dependent on its action. This is due to the fact that both independent and dependent properties are the result of ordered molecular processes. This problem leads us to the study of structural periodicity which is treated in the next section.

It is additionally important to realize that some of the changes caused by the environment are not transmitted to the progeny, whereas others lead to permanent genetic modifications.

PART VIII

STRUCTURAL PERIODICITY : THE ORDERED ADDITION OF COMPONENTS IN ATOMS, MOLECULES, AND LIVING ORGANISMS

Chapter 26

ORDERED ADDITION OF COMPONENTS IN ATOMS AND MOLECULES

STRUCTURE AND FUNCTION ARE DUAL ASPECTS OF THE SAME PHENOMENON

The first part of this book largely dealt with the periodicity of function, but in all the periods described, the coherent "packages" consisted of both structures and functions. For instance, flight only arises when a well defined "package" of structures and functions is available.

This part of the book deals wholly with the periodicity of structure. Yet any attempt to treat solely structure is limited by the fact that there is a functional component in every pattern. The five stamens of a flower are sexual organs, and the five arms of a starfish are used in locomotion.

Just as energy is converted into matter, and matter into energy, as is well known from Einstein's equation, form and function are dual aspects of the same phenomenon. Form cannot be separated from function, nor can function be separated from form.

WATER DOES NOT NEED GENES TO PRODUCE A CONSTANT PATTERN AND A SPECIFIC VARIATION

Humans can only give rise to humans. This is also true of other living organisms. With the copying process, called reproduction, they are only able to produce the same basic pattern. Only minor variations seem to occur, such as brown or gray eyes, long or short legs.

This is also the case with water: it has no genes, and is even simpler than most minerals, consisting of only hydrogen and oxygen. Every time a snow crystal is built the same basic pattern consisting of six rays is repeated. Minor variations occur, every crystal being slightly different, but the differences are not accidental and are typical of the structure. Water cannot depart from the rigid construction, namely that of the 6-rayed crystal; no DNA or RNA is involved in the transmission of this extreme order. A snow crystal with a diameter of 8 to 9 mm and a thickness of 0.01 mm, consists of the association of some 10^{18} molecules of water. This is a gigantic structure in atomic terms.

FEATURES OF CONSTANCY AND VARIATION IN THE PATTERN OF WATER CRYSTALS

Most of the studies carried out on snow crystals have centered on the classification of their patterns according to atmospheric conditions, or on their crystallographic properties (Nakaya, 1954; Bentley and Humphreys, 1962; Knight and Knight, 1973; Allen, 1984). However, the available data elucidate biological problems in several ways.

1. Under appropriate atmospheric conditions water only produces highly symmetric six branched crystals. The rays irradiate from a center and form constant angles of 60° with one another (Figure 1).
2. The branches tend to be all of the same length. In a few cases, three are shorter than the other three. This transforms the dominating radial into bilateral symmetry (Chapter 28, Figure 4). The length is not determined by their weight since in some crystals the rays have large plates at their ends.
3. The six branches do not radiate in every possible direction but lie on a single plane. The crystal is formed as a flat surface, being essentially two-dimensional.
4. The primary branches may produce secondary ones. These are also formed at 60° to the primary branches (Chapter 15, Figure 6).
5. The secondary branches lie parallel to each other.
6. They are not distributed in all directions but tend to lie on the same plane as the rest of the crystal.
7. The secondary branches are built opposite to one another. Each has the same length and shape as the one located opposite to it (Chapter 15, Figure 6).
8. The branches tend to grow at regular intervals from one another.
9. A size gradient is formed by the secondary branches. They are usually short near the center of the crystal but become largest in the median region, decreasing in length as they approach the apex of the primary branches. The shape of this gradient and its location are the main sources of variation of crystal shape.

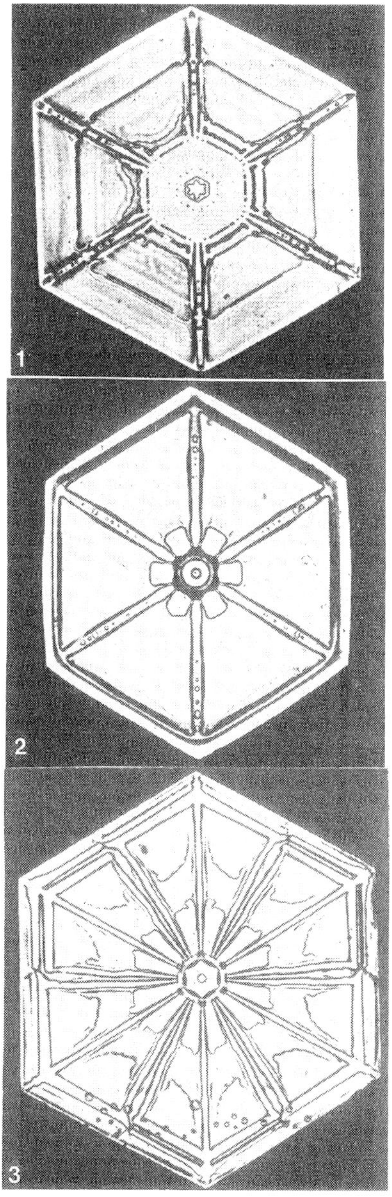

Figure 1. The distribution of six and twelve rays in water crystals. (1) Snow crystal with six rays. (2) Transition form in which a second set of six rays is forming between the previous one. (3) Crystal with 12 rays which are distributed at equal distances from one another as are the six rays.

10. The water molecules are able to ramify for a third time, thus creating tertiary branches. These continue to grow on the same plane as the rest of the crystal and continue to be formed at 60° angles. They are also built opposite to each other, and tend to have the same length and shape. They also occur at regular intervals (Chapter 15, Figure 6).
11. X-ray analysis of snow crystals has shown that they belong to the hexagonal crystallographic system. Forms belonging to the rhombohedral and cubic systems have also been described but the differences may well depend on pressure conditions.
12. Secondary centers may arise in a crystal where branches produce side growths with a three-dimensional orientation. These are not disposed at random, but follow the pattern of the primary crystal center.
13. Two crystals (with their 6 branches) may get superimposed at their centers. This results in a 12-armed structure. However, the fusion is a non-random process. The branches, instead of having angles of varying values, form only 30° angles; all 12 arms are distributed at the same average distance (Figure 1).
14. Forms other than the radial star configuration are known to occur. Pyramids, bullets, columns and plates are found in snow, depending on the temperature and pressure of the atmosphere, but hexagonal symmetry is maintained.

The number of ordered processes in the construction of water forms are not less than 14; these being the product of the properties of the two simple atoms: hydrogen and oxygen.

Photographs of over 5,000 snow crystals show none that are completely similar. Though variation is enormous, it is orderly,and takes place within a rigid atomic frame. Approximately 98% of these crystals are found to be in a six branched condition.

CRYSTAL CHEMISTRY IS NOT ENOUGH ADVANCED TO EXPLAIN THE ORGANIZATION OF SNOW CRYSTALS

Present knowledge of the symmetry of snow crystals though limited shows that in ice two hydrogens must be near each oxygen atom before the water molecule assumes hexagonal configurations (Pauling and Hayward, 1964). However, the theory of crystal lattices cannot adequately explain the following features of water crystals (Knight and Knight, 1973; Hill, 1990):

1. The equal length of the 6 primary branches.
2. The equality of shape and length of the secondary and tertiary branches.
3. The opposite formation of secondary and tertiary branches and their occurrence at regular intervals.

4. The formation of size gradients in the secondary branches.

As Nakaya (1954) remarks, the initial observations gave no indication of constancy in crystal size. However, an accurate analysis carried out during four winters reveals that each crystal type has a characteristic size. For instance, the most ramified crystals among the common radiating type, are about 2.5 mm in size. One suspects that the molecular field existing in crystals governs the length of the primary and secondary radiating arms of the snow crystal. This is possible because

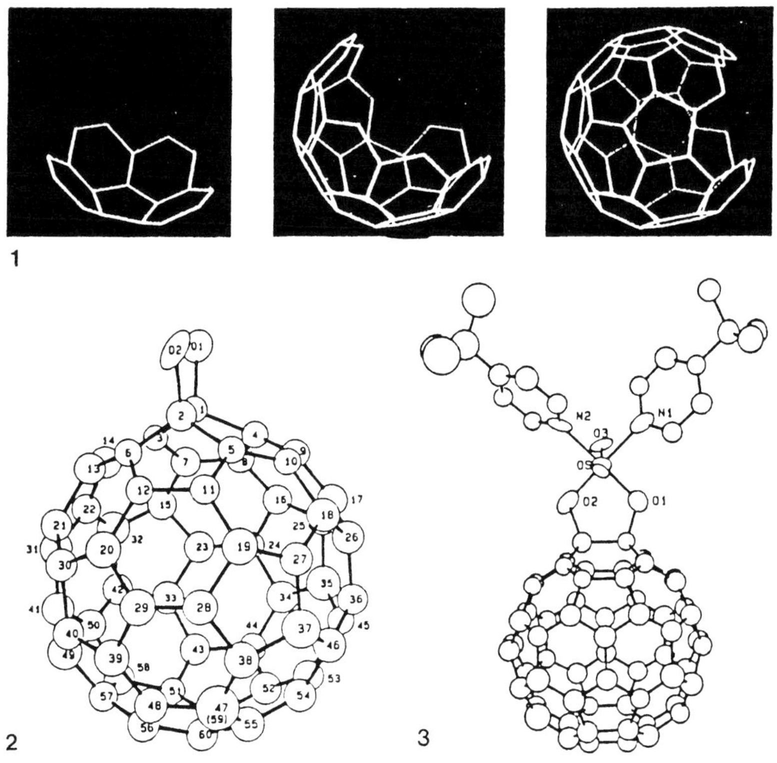

Figure 2. Carbon atoms build balls with icosahedral symmetry. (1) Sixty carbon atoms form a particularly stable structure which assumes an icosahedral form. The figure shows their successive association into a ball. This is a novel carbon framework—the so-called Fullerene C_{60}. (2) and (3) Two molecules resulting from the combination of C_{60} with osmium tetroxide, showing the geometry and relationship between these atoms.

Source: (1) From Smalley (1991); (2) From Hawkins et al. (1991).

of the atomic interactions occurring at the electron and proton levels of the hydrogen and oxygen atoms.

CARBON ATOMS BUILD ICOSAHEDRAL BALLS

Carbon atoms associate in two forms: the hard and sparkling diamond and the soft and dull graphite (used in pencils). Both minerals are made of only carbon atoms. Different pressures and temperatures present at the time of their formation, led to opposing properties. In diamond, the carbon atoms build three-dimensional tetrahedral structures, while in graphite the atoms form sheets of contiguous hexagons (see Chapter 30, Figure 5). More recently, a third molecule of pure carbon has been discovered (Smalley, 1991). Using a laser machine, it was found that carbon atoms bind to one another in a closed spherical network building great balls called *fullerenes*. The molecule is very stable and of the type present in interstellar space, being considered to be most ancient, having participated in the first generation of stars. The balls have an icosahedral configuration (Figure 2). This property of the carbon atom helps to explain many spherical biological forms which hitherto could not be related to simple atom patterns.

SILICON ATOMS MAINTAIN RADIAL CONFIGURATIONS AS THEIR NUMBER INCREASES

Silicon does not occur free and because of its great affinity for oxygen it is found in nature chiefly as SiO_4. The biological significance of this atom group, with its tetrahedral form, lies in the fact that it can be linked into chains, rings and three-dimensional frameworks. When each SiO_4 shares two oxygens with contiguous tetrahedra, cyclic metasilicates are formed, with 3, 4, 6, and 8 linked tetrahedra (Greenwood and Earnshaw, 1989).

The remarkable feature is that the SiO_4 atom groups distribute themselves in the form of a ring as their number increases.

1. The tetrahedra are maintained united by their vertices.
2. By increasing their internal distance they form larger rings.
3. They do this without abandoning their ordered distribution.
4. The most common pattern which is found in asbestos is built by rings of six tetrahedra (Figure 3).

It is worth noting that these properties are an indication that simple atoms, such as silicon and oxygen, have the capacity to determine the circular and ordered arrangement of components that is also found in the distribution of petals in flowers and of arms in starfish.

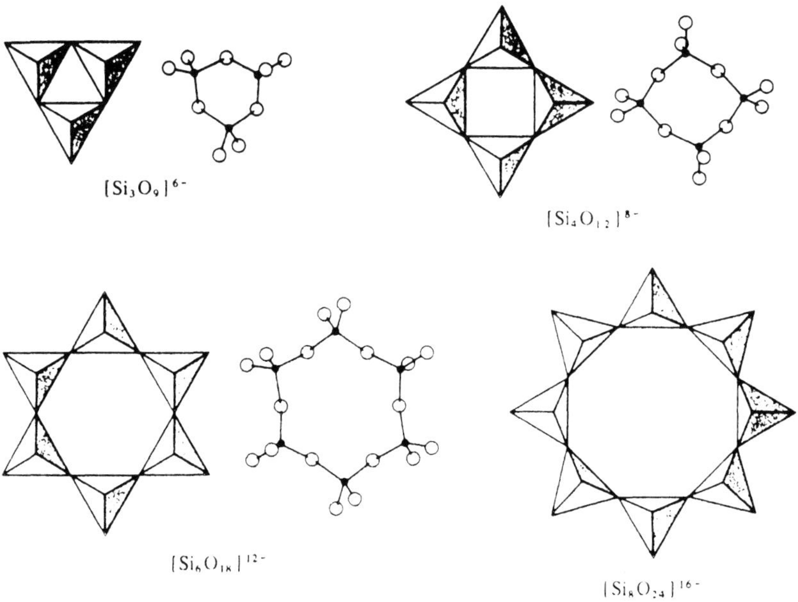

Figure 3. Combinations of the atoms of silicon and oxygen build specific regular patterns. The silicon atom (Si) combines usually with 4 oxygen atoms (O) building a tetrahedron (SiO_4). These can combine by corner sharing of O atoms into larger units which have a cyclic configuration. The structures of cyclic metasilicates built by 3, 4, 6, and 8 tetrahedrons are shown. These orient in a regular fashion, producing symmetric and radial structures.
Source: (From Greenwood and Earnshaw, 1989).

NUCLEIC ACIDS MAINTAIN ORDERED CONFIGURATIONS AS THEIR NUMBER OF COMPONENTS INCREASES

DNA crystallizes in distinct hexagonal structures that are seen in the electron microscope. Besides, it can occur in several forms called: A, B, C, D, E, and Z depending on the water content of the medium and the presence of metal ions. DNA-RNA hybrids and polynucleotides form, like DNA, helical structures.

When these are analyzed by X-ray diffraction the computer illustrations reveal the distribution of their components (Figure 4) (Saenger, 1988). The constituent atoms of these macromolecules, which are carbon, hydrogen, oxygen, phosphorous, and nitrogen, enable them to form ordered patterns. When viewed parallel to the helix axis, these structures show a distribution that maintains its order as the number of components increases from 6 to 22.

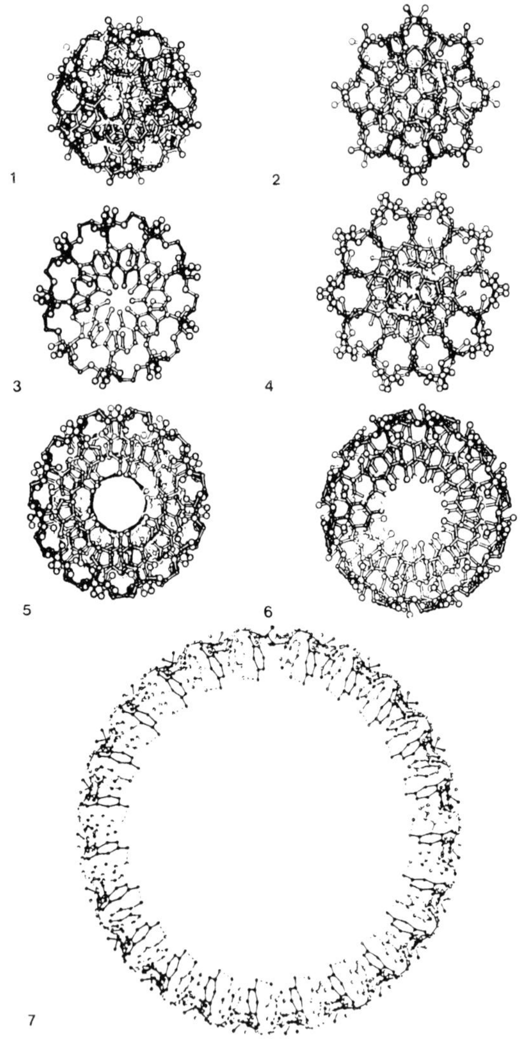

Figure 4. Structure of DNAs, DNA-RNA hybrids and polynucleotides, exhibiting radial configurations involving 6, 8, 9, 10, 11, 12, and 22 components. Computer illustrations of nucleic acid structures constructed on the basis of the coordinates obtained by means of X-ray diffraction analyses. The views of the structures are parallel to the helix axis. (1) Molecular structure of left-handed, double-helical poly (dG-dC)·poly (dG-dC). (2) Structure of D-DNA. (3) poly(A) single helix. (4) Structure of B-DNA. (5) Structure of A-DNA. (6) Structure of DNA-RNA hybrid. (7) Double helix constructed for the poly(A)·poly(U) duplex. The poly(U) chain is drawn heavier than the poly(A) chain.

Abbreviations: d = deoxyribonucleotide, G = guanine, C = cytidine, A = adenine, U = uracil.

Chapter 27

ORDERED TRANSFORMATION IN MINERALS

ASSOCIATION PROCESSES IN CRYSTALS

Most minerals form single crystals that belong to a system of symmetry. Under special conditions several crystals are formed together. This produces an arrangement that is called twinning, two cases being gypsum and calcite. These crystals have a shape and size that are not accidental. The twins are usually of: (1) the same form, (2) the same length, and (3) tend to be mirror images of one another. Their emergence, as left-handed and right-handed forms, that can combine into a single structure is especially evident in gypsum in which the twins have a pointed shape and form what mineralogists call a *swallow's tail.*

Other minerals produce three crystals whose arrangement also follows order. The three units are: (1) radially distributed, (2) located at equal distances from one another, (3) of the same shape, (4) of equal length. Moreover, the three crystals are of the same color, namely, metallic gray. Argirodite is an example of such a crystal combination (see Chapter 28, Figure 1).

When four and six crystals associate the same phenomenon takes place. They are: (1) radially distributed, (2) located at the same average distance from one another, (3) of the same shape, (4) of equal length. The six crystals that build the cyclic twins in chrysoberyl are usually green (Chapter 28, Figure 8). Another example of such crystal association is found in staurolite (Chapter 28, Figure 1).

When more than six crystals come together, they tend to build a spacial configuration in which they irradiate in many directions from a center. The result is a multiarmed star configuration. This is, for instance, the case in aragonite and pharmacolite.

When crystals of different chemical composition are superimposed, their location is also not accidental. They tend to occupy positions which are in accordance with one another. When the mineral staurolite forms crystals together with those of kyanite, the overgrowth is directed. Each mineral maintains its different structure and unit cell size, but the crystals build faces which are parallel to one another. The same occurs in the overgrowth of plagioclase on microline in which the crystals of the deposed mineral follow the orientation of the faces of the microline. Mineralogists term this phenomenon epitaxis which means order derived from overgrowth (Whitten and Brooks, 1988).

THE MECHANISM OF TWINNING IN MINERALS — ENERGY DETERMINES ORDER

At the atomic level, twinning arises when the regular arrangement of the atoms is interrupted on the outside of a growing crystal.

However, the altered arrangement of the atoms must be ordered, leading to the maintenance of a plane common to both crystals; otherwise, no twinning can be formed. The transition from a single crystal to twin formation is accompanied by an increase in internal atomic energy.

The overgrowth of several crystals into large groups does not result in an accidental structure. Rather, the crystals tend to aggregate with their axes and faces in a parallel way.

The distribution is regarded as being the result of a lower potential energy that the atoms use when an ordered pattern rather than an accidental one is formed.

THE PHYSICO-CHEMICAL PROPERTIES OF CRYSTALS OPEN THE WAY TO OUR UNDERSTANDING OF BIOLOGICAL TRANSFORMATION

The association of atoms and of crystals into ordered configurations allows us to better understand the regular association of components in biological structures. Moreover, crystals are known to have an internally guided pattern and a long range process that determines the emergence of their faces and shape (Klein and Hurlbut, 1985). These physico-chemical properties of atoms are critical to an understanding of biological transformation.

This leads to the analysis of structure at the next level of organization, namely, that of plants. Flowers are the plant organs that convey in the most direct way the types of symmetry and the organization of parts known from the minerals. However, all other plant organs ranging from the roots to the fruits, turn out to follow both externally and internally, the same rigid transformation process.

Chapter 28

ORDERED TRANSFORMATION IN PLANTS

MINERAL AND FLOWER TRANSFORMATION

Figures 1 through 3 show the assembly of minerals and flower parts with an increased number of components. The bilateral or twofold symmetry of the flower of the orchid *Orchis Morio* is compared with a crystal of the mineral argirodite. Not only do they have the same symmetry but the many faces that form the crystal have a distribution and organization that is partly repeated in the flower.

The three petals of *Alisma plantago-aquatica* irradiate from a central region where the six anthers (2 × 3) are located and from which three sepals also irradiate. Three crystals in the mineral argirodite irradiate from a central region which has also a threefold symmetry. The three crystals are of equal size just as the three petals are of the same length. The crystals terminate in a series of facets that give them a partially rounded shape; this terminal form is also evident in the petals of *Alisma*.

Staurolite is a complex mineral composed of iron, aluminum, silicon, oxygen, and hydrogen ($FeAl_4SiO_{10}(OH)_2$) which forms associations of crystals involving four components. Four crystals grow from a central point just as the four petals and the four stamens of *Ilex aquifolium* do. Moreover, in both cases they produce a regular cross and all components, mineral or biological, assume a specific shape and all parts stop growing at the same length (Figure 1).

The symmetry of five is not found among the crystallographic systems that prevail in minerals, but crystals can grow together, thereby forming a regular association of five units. This is significant because it shows that a symmetry forbidden at one level may become established when a more complex situation arises. The association of five crystals occurs in marcasite, a mineral composed of iron and sulphur atoms (FeS_2). The flower of *Cynanchum Vincetoxicum*, as well as

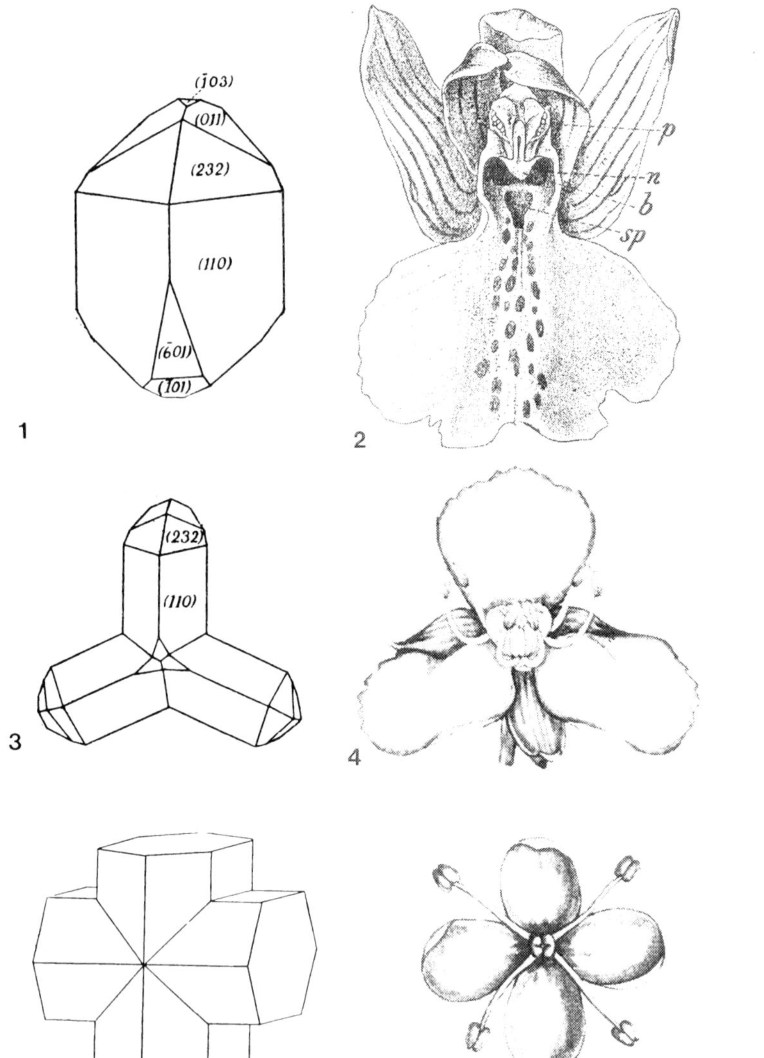

Figure 1. Incorporation of 2 to 4 components into minerals and flowers. (1) Single crystal of argirodite displaying bilateral symmetry (Ag_8GeS_6). (2) Flower of the orchid *Orchis Morio* whose symmetry is also bilateral. (3) Combination of three crystals around a center in argirodite. (4) Flower with three petals of *Alisma plantago aquatica.* (5) Staurolite ($FeAl_4Si_2O_{10}(OH)_2$) cruciform twin crystals. (6) Male flower of *Ilex aquifolium* with four petals and four stamens.

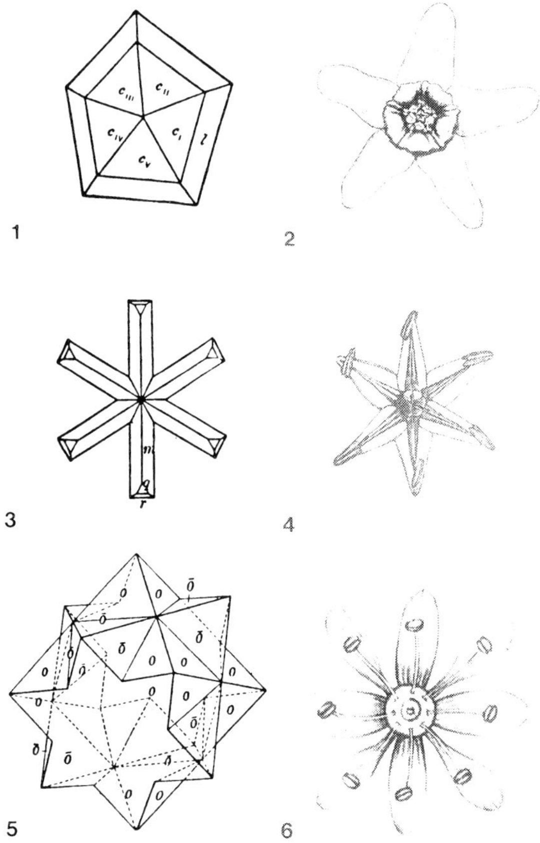

Figure 2. Incorporation of 5 to 8 components into minerals and flowers. (1) The association of five crystals in marcasite (FeS_2). (2) Flower with 5 parts of *Cynanchum Vincetoxicum* (Family Asclepiadaceae). (3) Six crystals formed around a center in arsenopyrite (FeAsS). (4) Flower of *Scilla autumnalis* with six petals. (5) Association of crystals in galena (PbS) consisting of two octahedrons building 16 vertices. (6) Male flower of *Acer platanoides* with 16 parts (8 petals and 8 stamens).

that of many other species, belonging to the same or other families, displays five petals. These irradiate from a central region that comprises five parts which have a shape and distribution similar to that of the crystals of marcasite. In both cases intermediate solutions between different symmetries seem to be excluded. Moreover, within each structure, be it mineral or plant, the parts have the same shape, length, and average distance from one another.

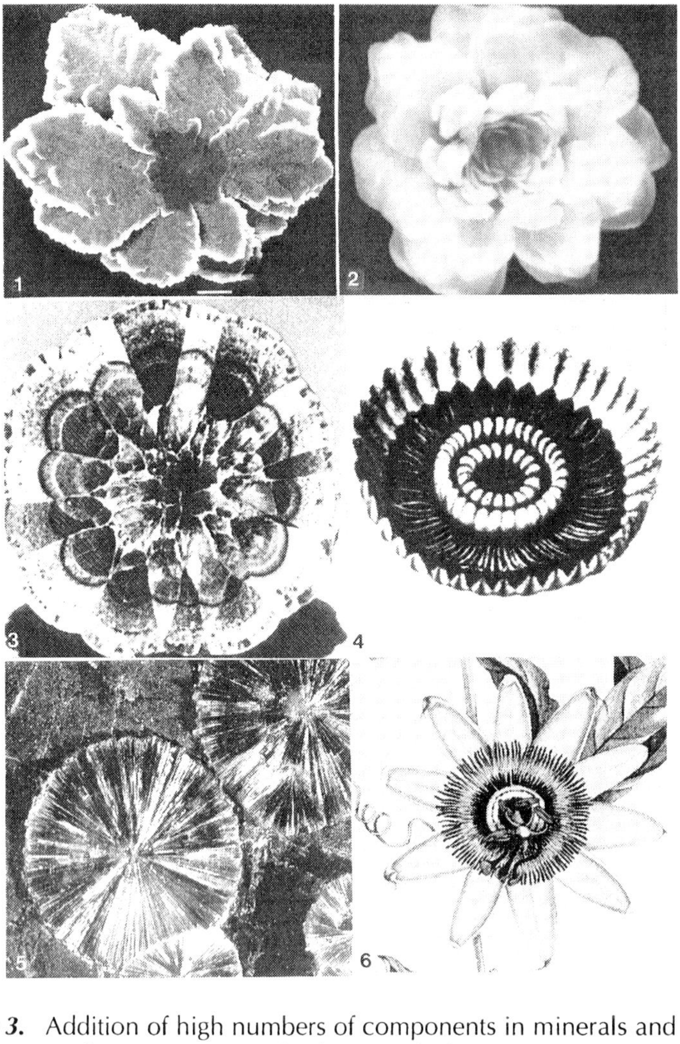

Figure 3. Addition of high numbers of components in minerals and plants. (1) Scanning electron micrograph of a crystal of vaterite ($CaCO_3$) with over 20 components arranged regularly, and in the shape of a flower (the bar is 10 microns). (2) Double mutant flower of *Arabidopsis thaliana.* (3) The copper carbonate malachite ($Cu_2(OH)_2CO_3$) showing radial distribution of 17 main sectors which increase in size towards the periphery. (4) Flower of *Napoleona imperialis* which is petal-less and formed by 38 modified stamens building a corolla-like structure. (5) The mineral wavellite ($Al_3(PO_4)_2(OH)_3{\cdot}5H_2O$) showing the regular arrangement of over 100 components. (6) Flower of *Passiflora caerulea* with corona consisting of over 100 filaments disposed regularly around its center and surrounded by 10 bracts.

As the number of components increases the same rules of organization apply, because the plant is a prisoner of the atomic organization of the mineral. The flower of *Scilla autumnalis* with its six petals and six stamens is a copy of the six crystals that have grown from a center of arsenopyrite (a compound of iron, arsenic, and sulphur, FeAsS). Again, the composing parts of each structure dispose themselves at equal distances from one another, have the same length and shape.

As a rule, crystallographic systems do not have symmetries higher than six, and more complex arrangements in minerals follow less rigid constructions. A prime example is galena, a widespread ore of lead (PbS) that builds associations of octahedral crystals. That is, the two combined crystals have each eight vertices. A similar solution is found in the male flower of *Acer platanoides* in which eight petals and eight stamens associate. This is taken as one piece of evidence that the higher orders of symmetry found in living organisms have their antecedents in mineral associations.

A recent photograph obtained with the scanning electron microscope reveals that a crystal of vaterite ($CaCO_3$) builds a pattern that cannot be easily distinguished from that of many flowers. (1) There is a central region that is darker and less differentiated than the rest, a feature characteristic of flower buds. (2) The peripheral components of the crystal have the shape that the corolla parts usually present. (3) They are also distributed in a circular fashion, as is the case in petals and sepals. (4) They are partly superimposed on one another as in several plant species. (5) The crystal parts situated below are bigger than those situated in the upper part near the dark central region, which is another feature well known from plant structure. For comparison the flower of a double mutant of *Arabidopsis thaliana* is shown (Figure 3).

Malachite is a copper carbonate, $Cu_2(OH)_2CO_3$. Seventeen components are distributed radially and form four concentric differentiations emerging from a central vortex. A similar radial distribution of parts is found in the petal-less flower of *Napoleona imperialis*, which shows 38 components (stamens and staminodes) forming a corolla-like structure that also consists of four concentric differentiations emerging from a central vortex.

Wavellite ($Al_3[PO_4]_2[OH]_3{\cdot}5H_2O$) is a mineral known for its radiating structures that consist of over 100 components. These are distributed with high regularity around a central region that has the form of a small circle. The flower of *Passiflora caerulea* has a corona consisting of over 100 filaments, irradiating with equal regularity from a central region that has also the configuration of a circle.

MINERAL AND ROOT ORGANIZATION

The configurations built by vascular tissues of young and old roots, can be seen in transverse sections. They often follow a combination of three plus three elements. This is reminiscent of water crystals, since they produce configurations in which

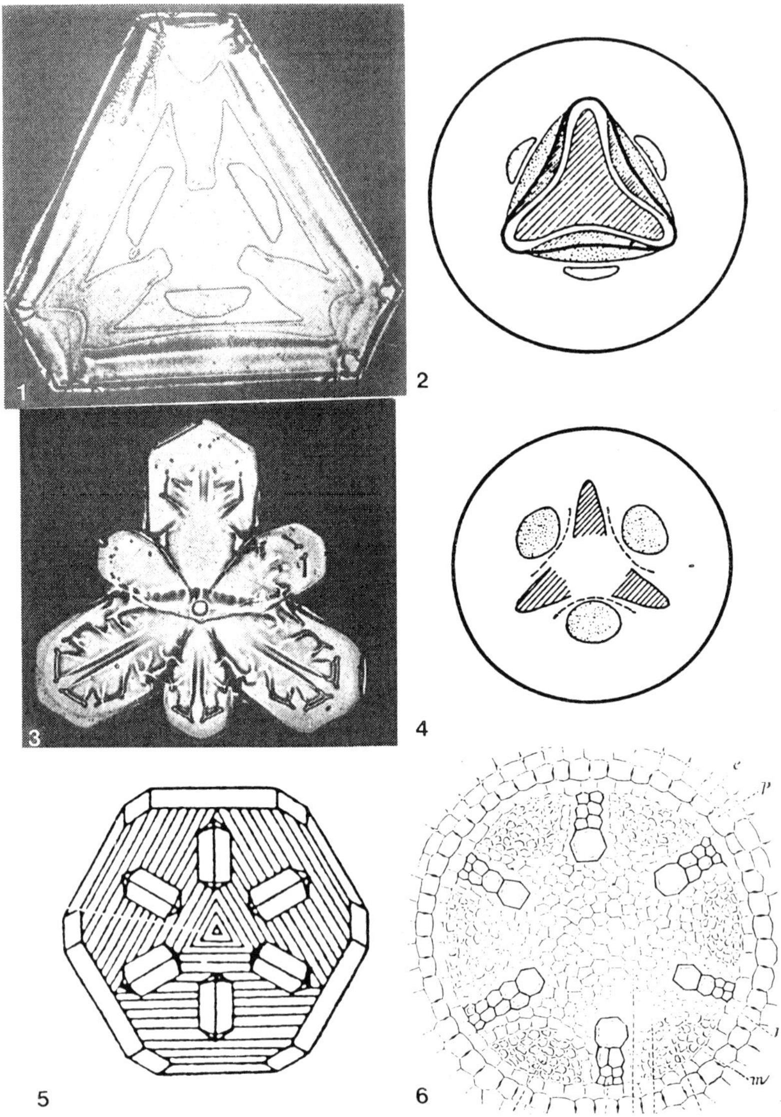

Figure 4. Incorporation of components into crystals and plant roots. (1) and (3) Snow crystals growing under different conditions of humidity and temperature showing variations of 6-ray symmetry. (5) Combination of the mineral rutile (TiO_2) with the mineral hematite (Fe_2O_3) building 6 components distributed radially. (2), (4) and (6) Transverse sections of young and older roots displaying variations of a 6-ray symmetry.

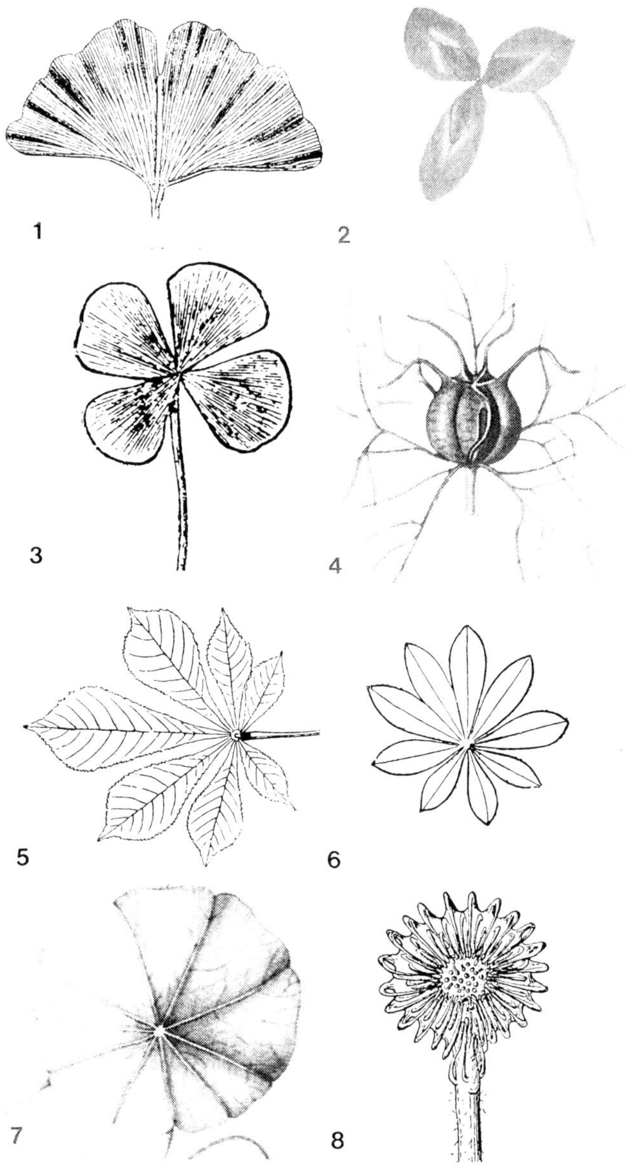

Figure 5. Leaves with 2 to 24 components. (1) A leaf from *Ginkgo biloba* (two). (2) Leaf of *Trifolium repens* (three). (3) Leaf of *Marsilea quadrifolia* (four). (4) Fruit of *Nigella damascena* surrounded by feathery leaves (five). (5) Leaf of *Aesculus hippocastanum* (seven); in this case, the radial symmetry tends to become bilateral, a feature common in leaves. (6) Leaf of *Lupinus* (nine). (7) Leaf of *Tropaeolum majus* (ten). (8) A leaf of the carnivorous plant *Drosera rotundifolia* (24).

three plus three components are distributed in a similar arrangement. When the titanium mineral rutile (TiO_2) combines with hematite, an iron mineral (Fe_2O_3), they do not mix, but the two different molecules grow into a pattern of six components distributed radially and at regular intervals. Two different tissues combine in the root following the same distribution in which six vascular bundles are located radially and at regular intervals (Figure 4).

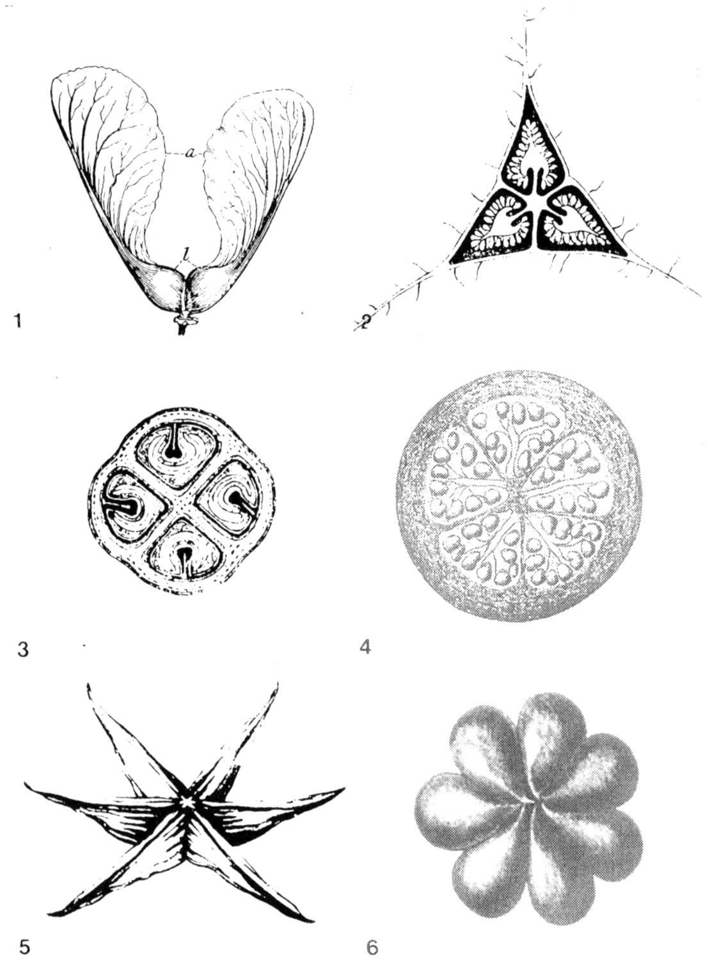

Figure 6. Fruits with 2 to 7 components. (1) Fruit of *Acer pseudoplatanus* (two). (2) Transverse section of ovaries of the plant Begonia (three). (3) Transverse section of fruit of *Rhamnus cathartica* (four). (4) Cross-section of the fruit of *Psidium Guajava* (five). (5) Group of united fruits in *Damasonium alisma* (six). (6) Fleshy fruit of *Phytolacca clavigera* (seven).

THE INCREASE IN THE NUMBER OF COMPONENTS IN LEAVES AND FRUITS FOLLOWS THE ATOMIC ORDER OF MINERALS

Leaves usually consist of several components that can be at least as many as 24. As the number of the components increases, order continues to be maintained, as occurs in minerals. The associated parts form bilateral or radial symmetries, or combinations of both, and are located at regular intervals. This is evident in leaves with few or many elements (Figure 5).

Fruits can have as many as 22 components, yet they show organization in accordance with the principles that minerals followed before the advent of plants.

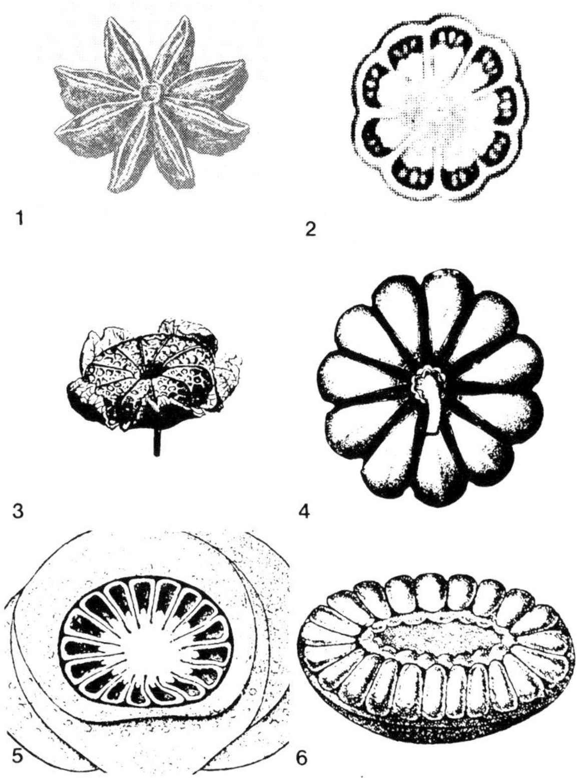

Figure 7. Fruits with 8 to 22 components. (1) Collective fruit of *Illicium verum* (eight). (2) Cross-section of ovary of *Marcgravia nepenthoides* (nine). (3) Fruit of *Malva silvestris* (ten). (4) View from below of the fruit of *Hura crepitans* (11). (5) Cross-section of the fruit of *Dillenia indica* (17). (6) Lower part of the fruit of *Ammobroma Sonorae* (22).

The components are found: (1) to be distributed radially, (2) to maintain the same average distance, (3) to have the same shape, and (4) to be of the same size (Figures 6 and 7).

THE COMPOSITAE ARE AN EXAMPLE OF THE RIGID ORDER THAT DETERMINES FLOWER ORGANIZATION

Many botanists have studied in detail the organization of the flowers of the Compositae (daisy family). This comprises: *Aster* (daisy), *Lactuca* (lettuce), *Taraxacum* (dandelion), *Helianthus* (sunflower) and many other species. Their flowers are not true flowers but large assemblies of single flowers, i.e., inflorescences.

What has puzzled biologists for generations is the fact that these inflorescences mimic single flowers both structurally and functionally: (1) The bracts may function as sepals. (2) The outer small flowers of the inflorescence imitate the outer petals of solitary flowers. This is due to the corolla of the outer flowers being extremely asymmetric. It is very large only on the outer side and brightly colored which makes it resemble a petal. (3) The small flowers of the central part of the inflorescence resemble the central part of solitary flowers. Their corollas are so short that they are hardly seen and instead the stamens and the styles protrude. This results in an overall picture in which one mainly sees stamens in the central part of the inflorescence, thereby resembling the central part of a single flower in which only stamens and styles are located.

What actually occurs is that these "false flowers imitate true flowers," as Leppik (1977) aptly remarks. The pseudoflowers of the Compositae simply follow the basic construction of the single and independent flowers of totally unrelated families, such as the Ranunculaceae (Figure 8).

PSEUDOFLOWERS HAVE ALSO THE PROPERTIES OF TRUE FLOWERS

There is not only a drastic change in the shape of the peripheral flowers but they also display the properties of solitary flowers in many other ways: (1) They are all of the same size, just as is the case with the petals of solitary flowers. (2) They have a shape that imitates that of petals. (3) Their colors are also bright and variegated like those of single flowers. (4) They are distributed from the center of the inflorescence in a regular radial symmetry. (5) They produce pleasant odors. In every respect, then, they have the characteristics of true flowers.

THE FLOWER ARRANGEMENT FOLLOWS SPECIAL NUMERIC SERIES AND LOGARITHMIC SPIRALS

It has been recognized for a long time that the flower distribution within the inflorescence of the Compositae follows a numeric order. The small flowers of the

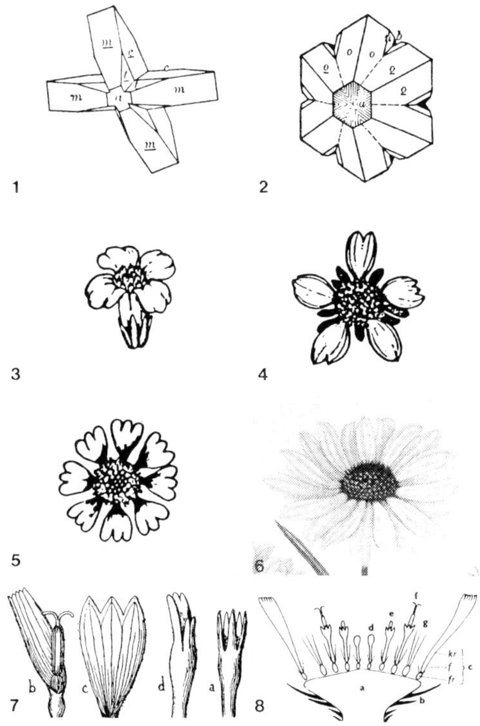

Figure 8. Differentiation into a central and a peripheral region in the association of minerals and the association of flowers. Single flowers of Compositae, when put together in inflorescences change their form. The whole inflorescence appears as a single flower. The inflorescences of the Compositae family consist of minute single flowers grouped into a central region, and of single flowers situated at the periphery having a large petal that protrudes conspicuously. (1) Association of four crystals in thenardite leads to differentiation into a center and protruding units. (2) Association of six crystals in chrysoberyl leads to differentiation into a central hexagonal region and six radiating units. (3) Inflorescence of *Bryomorphe zeyheri* with many centrally located minute flowers and with four peripheral flowers possessing protruding petals. (4) Inflorescence of *Bidens bipinnata* with many minute central flowers and five peripheral flowers with large petals. (5) Inflorescence of *Actinella odorata* with eight peripheral flowers possessing large protruding petals in addition to the central ones. (6) A typical inflorescence of the Compositae family, *Helianthus angustifolius,* with a large number of minute single flowers constituting a central dark region and many flowers situated at the periphery with large protruding petals. (7) The four types of flowers that normally occur in an inflorescence. (8) Diagrammatic section of an inflorescence to illustrate the distribution of the different types of single flowers.

central part of the inflorescence may be many but those that occupy the periphery, and resemble petals, are usually few. Their number tends to be: 8, 5, 3, and 2. Number 7 does not seem to occur and 6 and 1 are rare. Five turns out to be the most frequent number, as is the case with the number of petals in solitary flowers of other families such as the Rosaceae.

The sunflower (*Helianthus annuus*) has inflorescences consisting of over 100 central flowers. These are arranged in a radial symmetry and follow logarithmic spirals that have been the object of extensive mathematical studies. The most frequent patterns of their arrangement fall within a series of divergences: 2/5, 3/8, 5/13, 8/21 and 13/34. There are 21 long spirals intersecting 34 short ones. The peripheral flowers that look like petals also follow the geometrical construction of the central ones and occur only at the ends of these long logarithmic spirals.

THE SAME SOLUTION OCCURRED IN DIFFERENT SUBFAMILIES AND IN OTHER ORGANS

The inflorescence of the Compositae is the result of the aggregation of single solitary flowers that lost their individual shapes as they became part of a new organization. Different subfamilies of the Compositae (Tubuliflorae and Liguliflorae) have followed the same evolutionary trend in the construction of their inflorescences. This indicates how rigid evolutionary determination has been.

Other plant organs such as the stems also sought the same solution. Pseudoflowers started to occur at the end of single stems, while different stems associated later. However, instead of occupying accidental positions they formed ordered structures resembling the large inflorescences of other families such as the Umbelliferae (carrot family).

FEATURES OF FLOWER TRANSFORMATION IN COMPOSITAE

The following features are notable:

1. Flowers that build an inflorescence distribute themselves radially from a center.
2. The inflorescence as a whole has a polar orientation: an upper part with the flowers, and a lower part covered by bracts.
3. The flowers that build the central disk are disposed according to logarithmic spirals in several species.
4. The flowers situated at the periphery are located there in accordance with the position of the internal spirals.
5. The flowers lying in the periphery maintain approximately the same length. Also, instead of having different shapes and colors, they usually have the same shape and color.

6. The outer flowers have their large petals directed to the periphery.
7. When the number of flowers located in the periphery increases (from 2 to over 30), they do not aggregate into irregular groups, but usually maintain the same average distance. This results in a regular radial distribution.
8. Although the outer flowers can occur in different numbers some are optimal. Five is a common number.

Chapter 29

INTEGRATION OF COMPONENTS IN INVERTEBRATES

THE NUMBER OF ARMS IN LIVING AND FOSSIL STARFISH

Most living species of sea stars (echinoderms) have five arms. However, a minority have more than 40 arms, as in *Heliaster* (of the American coast). Within a single species the number of arms may vary from 7 to 14, as in the European sun star *Crossaster papposus*. The variation in number is quite extensive (Figures 1 and 2).

What is of particular interest is that in all the species, the adults have a symmetry which differs from that of the larvae. With arms on both sides of the slender body, the larvae have perfect two-fold symmetry. When metamorphosis begins, the anterior end of the larva degenerates, and the adult arms appear as new extensions of the body. The left and right side of the larva become transformed into the lower and upper part of the adult, which acquires polar and radial symmetry. The transformation could hardly be more drastic and more rapid.

The study of fossil starfish reveals that species with three, four, and five arms were abundant in earlier periods (Raff and Kaufman, 1983). This means that they had the ability to add components and at the same time maintain their body form. The number of arms in fossil and living species constitute a series, represented by the following numbers: 2, 3, 4, 5, 6, 7, 9, 10, 11, 13, 14, 19, 40, and over.

CHARACTERISTICS OF STRUCTURAL TRANSFORMATION WITHIN THE SAME ANIMAL GROUP

The body transformations that occur in starfish follow specific pathways.

1. Bilateral and radial symmetries can be produced by the same individual. This implies that no change in genetic constitution is necessary to produce totally different symmetry and body orientation.

2. The arms of an adult starfish radiate from a well-defined center.
3. The body has also a polarity. The mouth is located on the ventral side.
4. There is large variation in the number of units, contributing to body symmetry.
5. Whether few or many, the arms are all of the same length and shape (broad at the base and pointed at the end). An exception to this rule is *Luidia ciliaris* which has 5 large and 2 small arms. This shows that the change in length can occur but is generally precluded.
6. When the number of arms increases they do not aggregate into irregular groups (as they could do), but maintain the same average distance. The result is conservation of a regular radial distribution.
7. The most frequent number of arms among the 1,600 species so far described is five.
8. There is also a preference for color, which is usually yellow. Sea stars can be of several hues: red, blue, purple, and green, or combinations thereof, but the most common color is drab yellow.
9. The general radial symmetry of the adult can revert to the bilateral symmetry found in the larva. For example, the 5 large and the two small arms of *Luidia ciliaris* divide its body into a partly bilateral condition.
10. Intermediate solutions or extreme departures from this ground pattern have not been allowed (Figures 1 and 2).

Other groups of echinoderms display a similar transformation of their symmetries. Both brittle stars (Ophiuroidea) and the sea-lilies (Crinoidea) have radially located arms that are of equal length and shape, and distributed at the same average distance from each other. Moreover, if the number of arms increases from 5 to 10, as it does in these classes, they do not form accidental groups or assume different sizes and shapes, but rather they preserve the initial order.

DISTRIBUTION OF THE TENTACLES IN COELENTERATES

The Coelenterates include the hydras, jellyfish, sea anemones, corals, and comb jellies. Due to their striking radial symmetry they are also known as Radiata.

1. The body of these animals has polarity. The mouth occupies one of the ends of the body axis.
2. A circle of tentacles surrounds the mouth, thereby producing radial symmetry.
3. Most groups have a typical number of tentacles which is 6 in hydras, 8 in the larva of *Tubularia*, 10 in *Aglaophenia*, and 18 in *Obelia* (a hydroid). In the hydromedusae they are usually 4 (but there are species with one, two, or more than four tentacles). Octocorals are so called, because they always possess 8 tentacles. The comb jellies have the lowest number of tentacles:

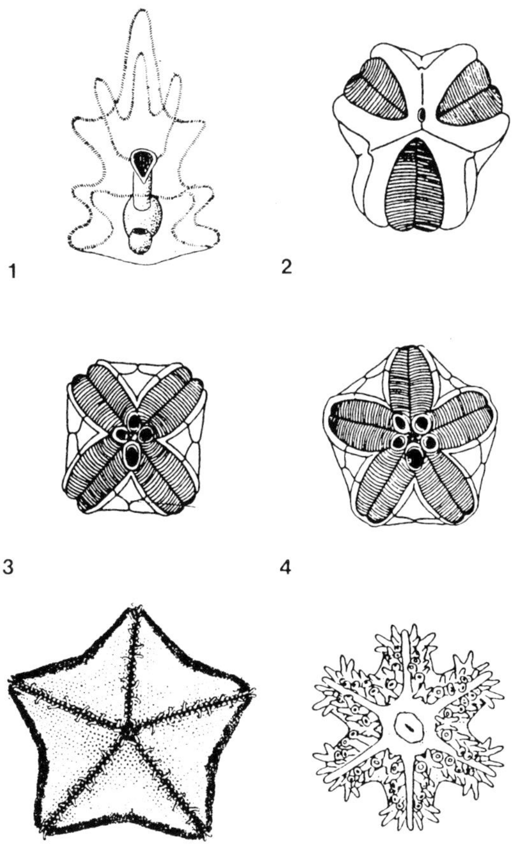

Figure 1. Starfish and related echinoderms with 2 to 6 components. (1) The bipinnaria type of larva characteristic of starfishes (class Asteroidea) with its bilateral symmetry. (2) Fossil echinoderm of the class Blastoidea (three arms). (3) Another fossil of the Blastoidea (four arms). (4) A third fossil of the same class (five arms). (5) Living starfish *Palmipes membranaceus* (five arms). (6) Living starfish *Leptasterias hexactis* (six arms).

only two. Two groups deviate from a constant number. The number of tentacles in jellyfish varies from 4 to over 400 (in *Aurelia* there are young forms with 8, 9, and 15 tentacles). Sea anemones have a tentacle number varying from 8 to several hundred.

4. In most species all the tentacles of an animal are of the same length, shape, and color.

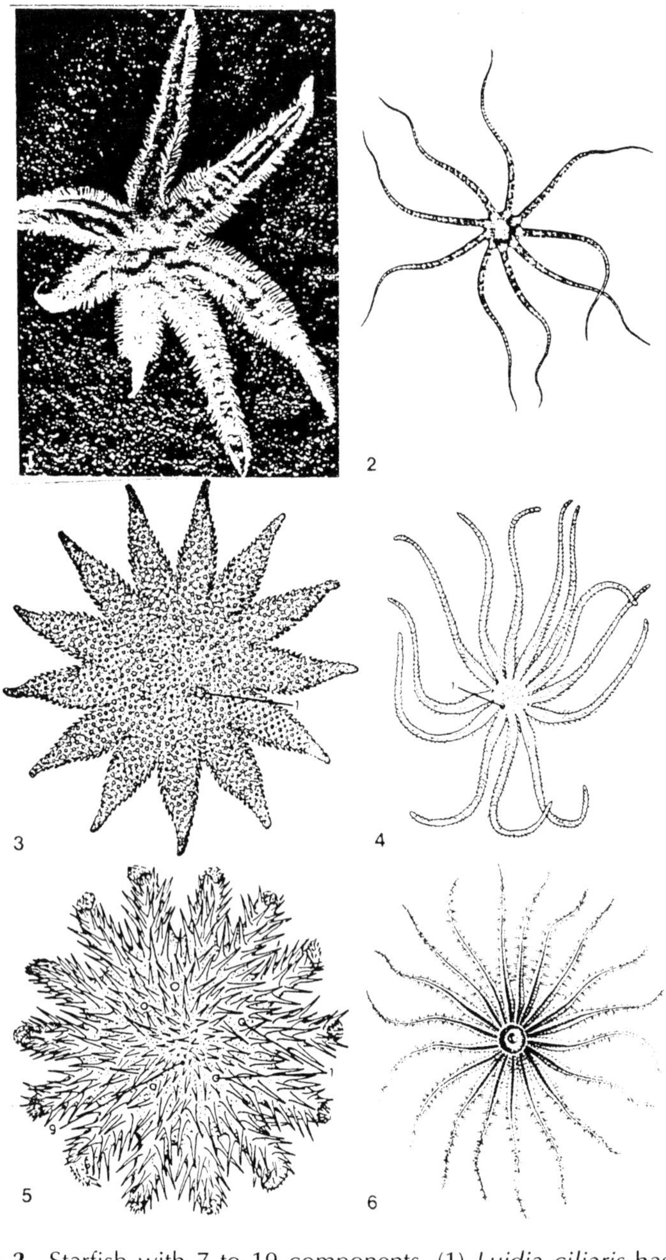

Figure 2. Starfish with 7 to 19 components. (1) *Luidia ciliaris* has partial bilateral symmetry (7 arms). (2) *Brisinga mediterranea* (9 arms). (3) *Crossaster papposus* (13 arms). (4) *Freyella sp.* with 13 slender arms. (5) *Acanthaster* (14 arms). (6) *Odinia elegans* (19 arms).

5. All tentacles are distributed at regular intervals. When their number increases the average distance between them decreases.
6. When the number of tentacles increases to several hundreds, these become much thinner and shorter, but they continue to have the same length, shape, and color (as is the case in the adult of the medusa *Aurelia*).

THE PARTITIONS OF THE BODY CAVITY FOLLOW TYPICAL NUMBERS

Not only the tentacles but also the partitions of the body cavity in sea anemones manifest a radial pattern.

1. The typical number is 12.
2. The increase in partitions occurs also by multiples of 12. In different species it is: 12, 24, 48, and so on.
3. The addition of a new cycle of 12 units occupies the space between the previous one with the result that as the number of units increases, the distance between them diminishes, and at the same time, the average distance between each unit is maintained.
4. There are variations of this numerical symmetry, but they represent exceptions.

The same phenomenon is found in other groups, such as the octocorals. There exists in them a correlation between the number of tentacles and body partitions. Both structures are always eight in number (Barnes, 1980).

THE LEGS OF SPIDERS AND INSECTS HAVE FIXED NUMBERS

The tendency to have a typical number of arms, as in starfish, coelenterates and octopods extends one step further, resulting in a number that is fixed for a whole class or order. This is the case with the legs of other invertebrates.

1. All spiders have 8 legs although there are not less than 32,000 species.
2. All insects have 6 legs. They are the most abundant group of living organisms, numbering over 1 million species.
3. In spite of an enormous species diversity, all Crustacea have two pairs of antennae. Within this class, the crabs have all four pairs of walking legs and comprise not less than 4,500 species.

CHAPTER 30

ADDITION OF COMPONENTS IN VERTEBRATES

IN VERTEBRATES THE NUMBER OF ADDED COMPONENTS MAY ALSO REACH 100

Over 100 flower parts are disposed regularly in the plant *Passiflora* (Chapter 28, Figure 3). The number of tentacles in invertebrates may be over 400. In vertebrates the number of added parts that follow a similar distribution reaches 100. This is the case in the fossil fish *Cyclobatis* in which the pectoral fins consist of about 100 spines (Figure 2). In other animals, the added components may be: 2, 3, 4, 5, 6, 8, 15, 16, 17, 29, 54, or some other value (Figures 1 and 2).

THE RADIAL DISTRIBUTION OF EYE PLATES IN FOSSIL FISHES AND REPTILES

There are indications that vertebrates followed the same rules of construction as those that directed the organization of invertebrates, plants and minerals.

Ichthyosaurs were the most specialized of the marine reptiles that lived mainly in the Triassic (230 million years ago). The earliest known species, *Utatsusaurus*, had eyes consisting of seven skeletal plates located radially around the center of the eye. Each plate was shorter near the center and got broader toward the periphery. All seven plates were of the same size and shape, forming a fan-like structure. Other species, such as *Mixosaurus*, had 14 eye plates. These were also radially distributed and of the same size and shape (Figure 3). *Plateosaurus* had 18 plates, all distributed radially, but two were larger and occupied opposite positions, thus giving the circle of plates a secondary bilateral symmetry (Figure 3). Among these species 14 plates were the most common, all of which were of the same size and shape (Boule and Piveteau, 1935; Carroll, 1987).

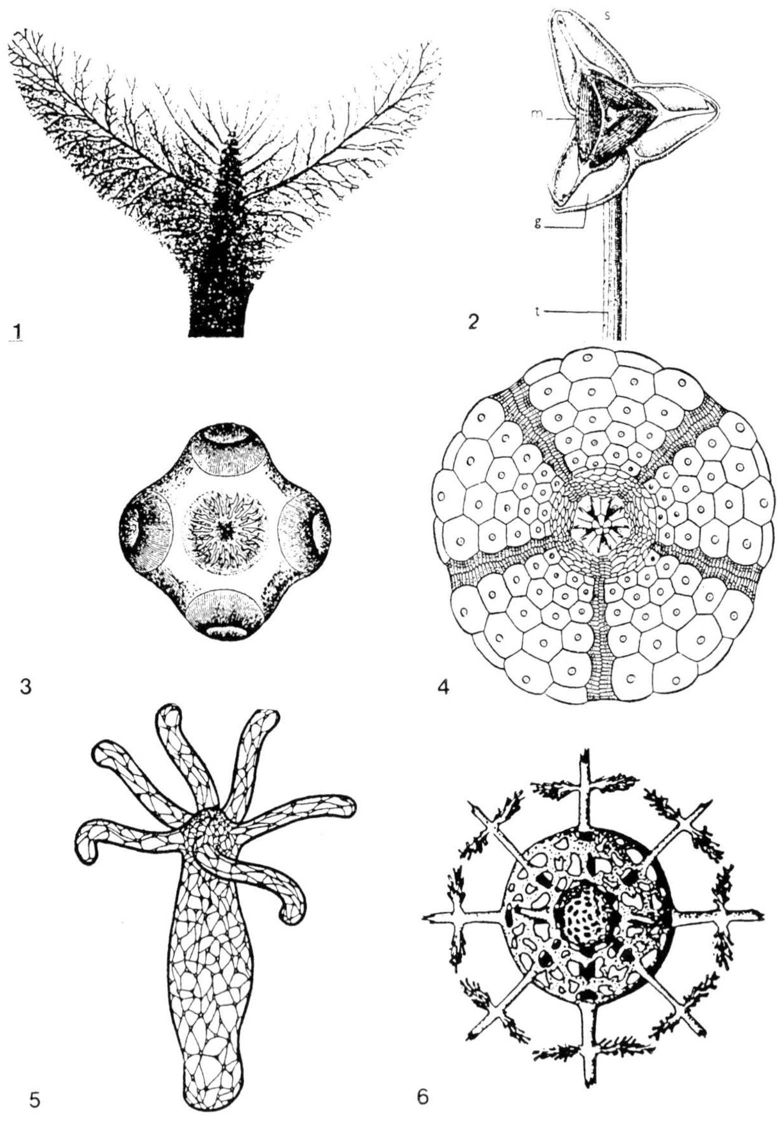

Figure 1. Animals or animal parts with 2 to 8 components. (1) X-ray picture of the arteries of the tail of the whale *Phocaena phocaena* (two). (2) Pincer-like structure of the echinoderm *Echinus esculentus* (three). (3) Head of the worm *Taenia saginata* (four). (4) The fossil sea urchin *Archaeocidaris Wortheni* (five). (5) The nerve net of a hydra showing the tentacles around the mouth (six). (6) The shell of the protozoan *Phractopelta tessarapsis* (eight).

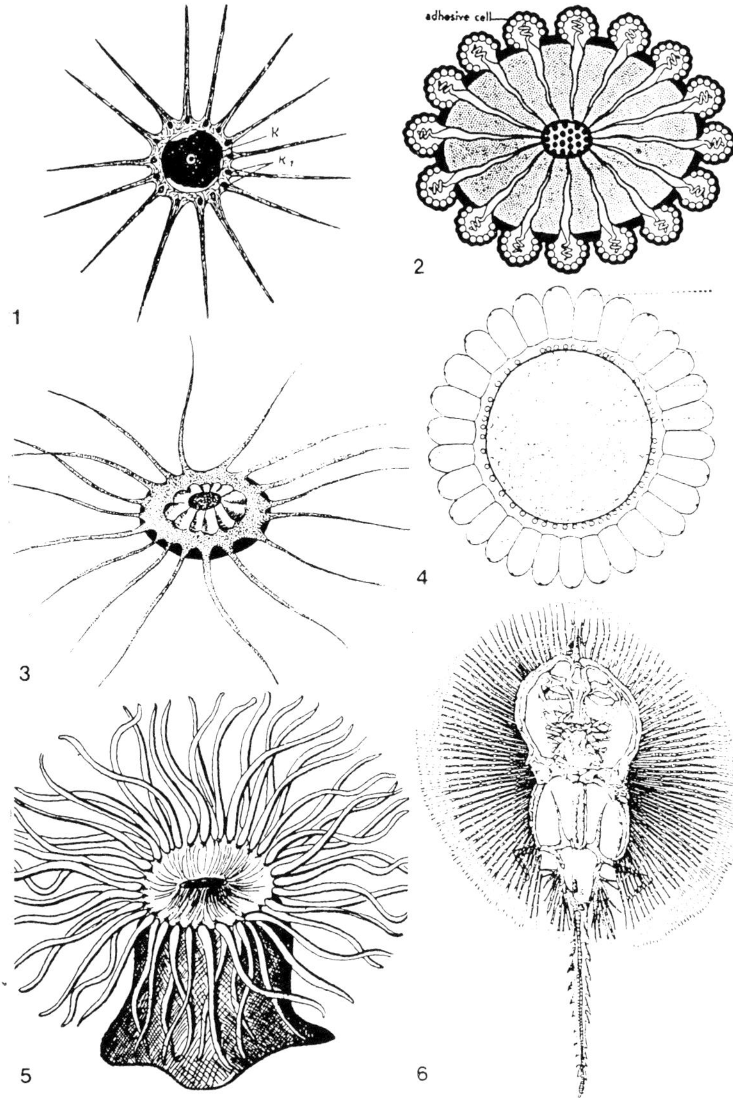

Figure 2. Addition of components in animal cells and animals numbering 15 to about 100. (1) Pigment cell of mollusc with stellar form (15). (2) Cross-section of a tentacle of a comb jelly showing adhesive cells at the periphery (16). (3) Spermatozoon of crustacean (17). (4) Egg, with its envelopes, of the tunicate *Ascidiella aspersa* (29). (5) The sea anemone *Actinia* with its tentacles (54). (6) The fossil fish *Cyclobatis* with highly expanded pectoral fins consisting of many spines (about 100).

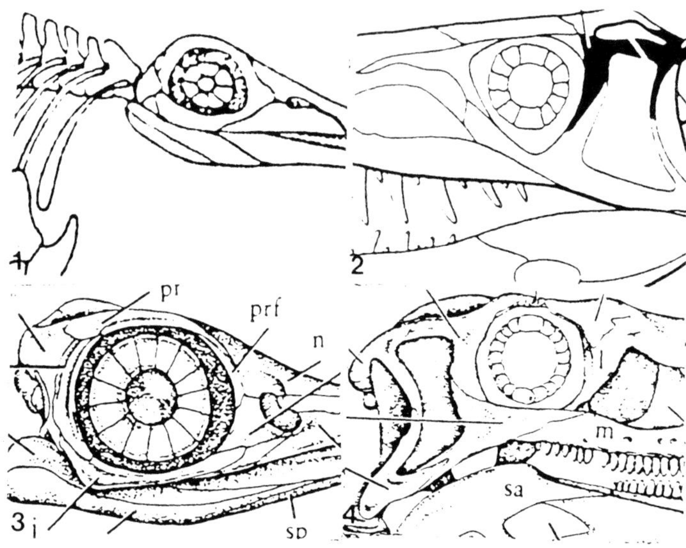

Figure 3. Addition of 7 to 18 plates in the eye of fossil reptiles. (1) The marine reptile *Utatsusaurus* (7). (2) Skull of the *Euparkeria* (12). (3) Skull of the marine species *Ophtalmosaurus* (14). (4) The herbivorous dinosaur *Plateosaurus* (18). In (2) and (4) there is a transition from radial to bilateral symmetry; two plates are slightly larger and located opposite each other.

A large eye of this type having ossified plates, existed in some of the earliest fish (which were the ancestors of the amphibians and the reptiles). They had 17 or 18 plates in their eyes, which were of the same size and shape, thus forming a regular radial structure (Babin, 1980). This pattern was maintained from the fishes to the reptiles. It was characterized by the preservation of several features. Chief among these were:

1. Radial symmetry.
2. Plates were of similar size and shape.
3. Though the number of components increased from 7 to 18, the structure was maintained.
4. The most common number was 14.

THE BILATERAL SYMMETRY OF A VERTEBRATE IS NOT DIRECTLY RELATED TO THE SYMMETRY OF ITS EGG

Embryologists have attempted to elucidate the mechanism underlying body symmetry. They performed a series of experiments to find out whether there was a relationship between the planes of egg cleavage and the symmetry of the future embryo. The use of vital dyes and other devices led them to the conclusion that there was no correspondence between the two phenomena. The median plane of the embryo did not always arise where the first cleavage plane of the egg was situated. They therefore concluded that cleavage cannot possibly be the specific factor through which bilateral symmetry of the embryo is laid down (Weiss, 1939). This gap in our knowledge needs to be bridged not only at the macromolecular level but also at the molecular and atomic levels.

THE HUMAN BODY IS BUILT ON THE ATOMIC PLAN OF A TWIN CRYSTAL

The body of a human, like that of any other mammal, is built according to a crystal plan. Bilateral symmetry in humans is indistinguishable from the twinning process in minerals. The two halves of the body are built just as contact crystal twins. These are characterized by the fusion of two structures that are identical. They have an axis in common, and one half has grown in a position, which corresponds to a rotation of 180°, thereby becoming a mirror image of the other.

Contact twins in minerals are not in intimate contact all over their structures. There is only a region along the common axis where they are united. This is also the case in humans. In the head and the trunk the two mirror structures are united; the common axis comprises the cranium and the vertebral column, but the arms and the legs are mirror structures in which contact is absent. A human, seen upside down with stretched legs, closely resembles the twin crystals of rutile that are only united at the base but are in the upper part separate as two independent units (Figure 4). In addition, the polarity of the human body is indistinguishable from that present in a crystal that has an apex and base.

Several other features are also derived from the mineral construction, one of which is the repetition of parts that is the basis of the body segmentation in invertebrates and the human column. Each vertebra is an added unit that has the same essential pattern, but vertebrae are not mirror images of one another. They are oriented in a single direction which is defined by body polarity. This is also the case in crystals; for example barite, in which there is a parallel growth of the crystals. The same type of unit is repeated many times, and oriented in the same direction.

Not only humans, but every invertebrate and vertebrate with bilateral symmetry is built like a twin crystal. The earliest fish have bilateral symmetry that dominates the whole body. They are built by the union of two mirror image structures that are kept together from the tip of the head to the end of the tail. The protruding fins also follow the general twin symmetry.

The same is true of plant organs. The leaves of flowering plants usually have strict bilateral symmetry. They are formed by two twin structures that are mirror images. The central nerve of the leaf is the common principal axis. In some species

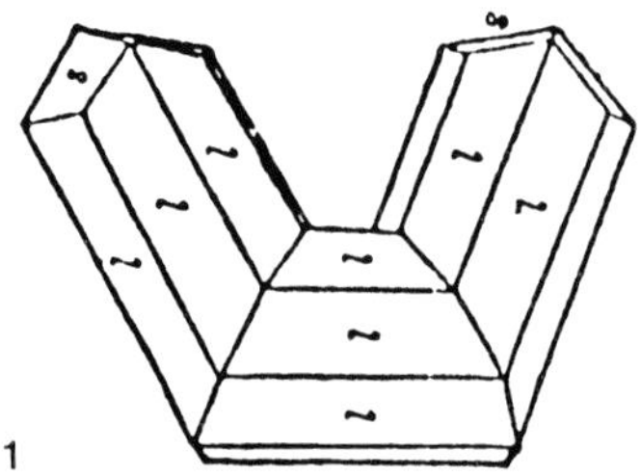

1

2

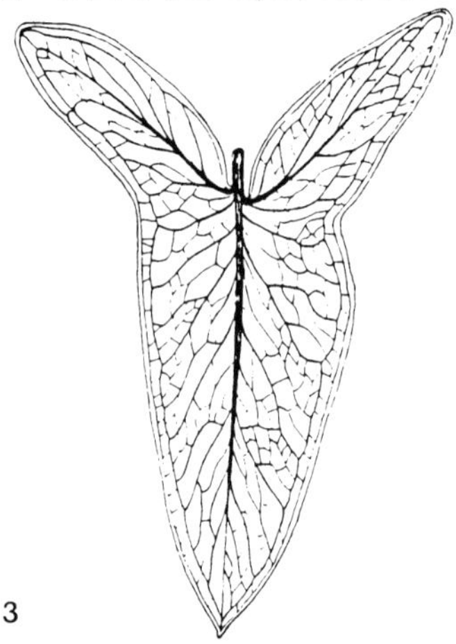

3

Figure 4. Twin crystal formation in minerals, and twin crystal-like organization in plants and vertebrates. (1) Two twin crystals of rutile, a titanium oxide (TiO_2). (2) Woman with open legs and arms doing gymnastics, seen upside down. (3) Leaf of *Arum hygrophilum*.

the twins are only in contact at the median region and tip of the leaf, and are separate in the posterior region. This occurs in the leaves of *Arum* and *Sagittaria* which are bifurcated at the base. In this respect, they are not only like the arms and the legs of the human body, but also like the contact twin crystals of gypsum that are only partially united.

THE PATTERN OF FLOWERS AND THAT OF HUMANS IS DETERMINED BY THE SAME TYPE OF GENES

Unexpectedly, recent genetic analysis of the embryonic development of *Drosophila*, has yielded information on the molecular processes leading to the formation of the pattern of flowers and that of the human body.

Homeotic genes, as they are called, specify structural differences between body segments along the head-to-rear axis of animals providing the different tissues with positional values. The genes are ordered along the chromosome and their expression on the body axis follows the chromosome order (Affolter et al., 1990; McGinnis and Kuziora, 1994).

Homologous homeotic genes have been found in worms, mollusks, mammals, humans and plants. The head-to-tail axis of humans turns out to be homologous to that of the mouse, the genes being expressed in order along two orthogonal axes: an anteroposterior and a proximodistal pattern.

Of relevance in connection with the similarity between the symmetry of humans and plants, is that the same type of homeotic genes specify the sequence of the parts of a flower. The precise order which leads to the emergence of sepals, petals, stamens, and carpels is determined by these genes in *Arabidopsis* (Jofuku et al., 1994; Meyerowitz, 1994). The genes encode regulatory proteins that control flower formation. These genes also occur in tomato, *Petunia*, and *Antirrhinum* (Coen et al., 1990; Sommer et al., 1990; Angenent et al., 1992).

Hence, the similarities between the symmetries found in plants and in humans can now be traced to the molecular level, since the same type of DNA sequences and proteins are involved in the formation of the body parts that establish these symmetries. These results together with those from molecular mimicry (in which different molecules result in the same pattern) are bringing us closer to an understanding of the pure atomic processes that appear to be responsible for the emergence of symmetries in living organisms. The gene is only the bearer, and the carrier, of the atomic order that already determined mineral symmetries (Lima-de-Faria, 1988).

THE ROLE OF GENES IN SYMMETRY CHANGES

What is actually the role of the structural gene in the formation and maintenance of the symmetry characteristic of a given species? The data coming from genetic experiments shed some light on this process. The best studied case is the mutation

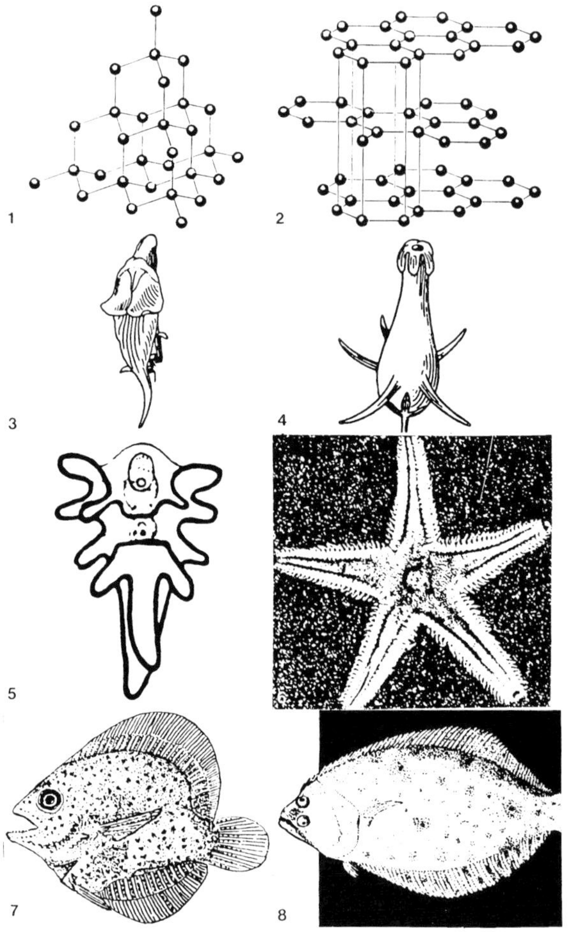

Figure 5. Changes in symmetry. Atomic structure of: (1) diamond, and (2) graphite. Both consist solely of carbon atoms that crystallize in the cubic system (diamond) or the hexagonal system (graphite). The change in the symmetric arrangement of the atoms occurs at a different temperature and pressure. (3) The flowers of the plant *Linaria vulgaris,* normal one-spurred flower with two-fold symmetry and (4) less common five-spurred form with radial symmetry. (5) The larva of a starfish with its two-fold symmetry and (6) an adult *Astropecten irregularis* with its radial symmetry. All starfish larvae have bilateral symmetry. (7) Young fish *Stromateoides* similar to a flat fish, but maintains bilateral symmetry with eyes on both sides of the body. (8) Adult of the flat fish *Paralichthys albiguttus* which when young has eyes on both sides but, as an adult, has both eyes on the upper side of the body. This change results in dorsiventral symmetry.

found in snapdragon (*Antirrhinum majus*) which transforms the normal bilateral flower directly into one with a radial five-fold symmetry. Though a single gene is involved, intermediate forms also arise (Baur, 1930; Stubbe, 1966). The same phenomenon was described long ago in a closely related species *Linaria vulgaris* (Figure 5).

De Vries, the founder of the mutation theory, described a mutant in *Chrysanthemum segetum* that accounts for the formation of the special arrangement of the flowers in the inflorescence of the Compositae. The normal plant of this *Chrysanthemum* species has an inflorescence consisting of a central disc in which the flowers have a rudimentary and symmetric corolla. The symmetry results from 5 small indentations at the periphery of the corolla that are distributed radially. These flowers are hardly visible because of their form and small size. Only the flowers at the periphery have one large petal that is formed on the outer side and sticks out quite distinctly. These last flowers are highly asymmetrical and vary normally in number from 8 to 21. De Vries found mutants in which all the flowers of the central disc had acquired large petals and were also oriented toward the outer side, thus mimicking the peripheral flowers. As many as over 200 flowers had changed and the character was transmitted to the progeny (Hertwig, 1929b). It is thus clear that a mutation transformed a rudimentary corolla with perfect radial symmetry into a highly asymmetrical structure formed by a single large petal oriented away from the center of the disc in all the flowers.

In the snapdragon the change in symmetry was from bilateral to radial, while in *Chrysanthemum* the modification was from radial to unilateral. This is practically the reverse process.

The larva of the sea urchin is found to change symmetry from a bilateral to a five-fold radial pattern in the adult. Genetic studies reveal mutants with a four-fold symmetry in the adults; these are fully functional. More than one gene is involved in the appearance of this character; crosses between these mutants yield two-, three-, four-, five- and six-sided progeny (Hinegardner, 1975).

The starfish larva possesses bilateral symmetry, which in the adult becomes transformed into a radial pattern. The meaning of this is clear: no change in genetic constitution is necessary to produce this radical transformation in body pattern and polarization. The simplest explanation of this phenomenon is that the DNA sequences change the expression of some of their genes during the initiation of the adult stage.

As one reaches the vertebrates similar changes are found to occur. The number of toes in the hindfoot of guinea pigs is normally three. Mutants with four toes were studied by Wright (1934a and b) who established that this trait is dependent on a set of alleles representing four genes. In this case a partly bilateral symmetry, consisting of the three toes oriented in a single direction, is changed into a four toe structure that is partly radial.

Mineral chemistry already shows that such transformations are the result of simple atomic processes. Graphite and diamond provide a prime example of two minerals that consist solely of carbon atoms but have totally different properties and symmetries (Figure 5). This line of thinking can be extended to living organisms where differences in symmetry are also expected to be due to different atomic configurations of the same or different protein molecules.

RULES OF TRANSFORMATION COMMON TO ATOMS, MINERALS, PLANTS AND ANIMALS

It is apparent from the comparisons that have been drawn of structural transformations that they follow the same path in living and non-living material. The rules applicable to the mineral, plant and animal kingdoms can be formulated as follows:

(1) When two units combine, they tend to be mirror images of one another, thereby forming bilateral symmetry. Examples are the twins of crystals, the symmetry of the bivalve molluscs, the leaves of plants, and the human body.

(2) When three or more units combine, they tend to emanate radially from a center and grow away from this center. This holds true when as many as 100 units combine in minerals, plants or animals.

(3) The units that combine radially tend to be of: (i) the same shape, (ii) the same length, and (iii) the same color. That the atomic number of a chemical element influences its color is a well-established fact.

(4) An increase in the number of units is accompanied by a change in the distance between each unit. The result is that the average distance between the units is maintained. As a consequence, the radial pattern is conserved.

(5) At every level of organization there is an optimal number of units that combine. This is: six in silicon atoms, six in water crystals, five in flower petals, five in the arms of starfish, six in insect legs, eight in spider legs, eight in the tentacles of octocorals, four in the tentacles of hydromedusae, and four in vertebrate limbs.

(6) Intermediate solutions are precluded or avoided. This is what one observes at the level of silicon atoms. At the other extreme, addition of units in the formation of the organism, is an all or none process. Limbs are either formed or they are not. Structures of other types or partial structures are blocked from emerging in minerals, plants and animals. For example, there are no starfishes with minute arms arising between the large ones. This is so in humans; no small arms appear between the large ones.

Hence, the order in the union of organs into large body structures, in plants and animals, does not derive from any accidental process, but has its origin in the atomic arrangements that were already at work at previous levels much earlier. When the atoms involved are different, the structural result can still be the same, because they are able to mimic the effect of others. The direct consequence is the emergence of structural periodicity at every level.

PART IX

MACROMOLECULAR AND ATOMIC MECHANISMS RESPONSIBLE FOR PEROIDICITY

Chapter 31

THE FORMATION OF NEW MOSAIC PROTEINS AND THE SUDDEN REAPPEARANCE OF OLD PROTEINS

PERIODICITY AND ITS MECHANISM

Periodicity as a phenomenon has been recognized only recently, and hence, information about its mechanism is still preliminary. Very likely a few decades will elapse before certain well-defined processes will be understood and these will, in turn, explain the sudden emergence of periodicity and its widespread occurrence. At the moment, the available information seems sufficient to provide guidelines in the planning and performance of new experiments.

PROTEINS HAVE AN ADDRESS IN THE CELL

Proteins are targeted to sites lying inside and outside the cell. This is made possible by targeting signals which are short segments of amino acids within each protein. Thus, proteins carry an address that is highly specific. Other proteins present in the cell membrane and other cellular structures have the ability to decode these signals. The outcome is a concerted interaction between proteins that gives the cell a coherent organization. There are also specific transport systems that allow the exit or entry of specific proteins. They carry signals that are recognized by receptors on cell surfaces and direct them to appropriate locations within the cell (Figure 1).

To reiterate, proteins synthesized in the endoplasmic reticulum, as well as mitochondria and chloroplasts, are provided with signals which tell them not only where they are to travel, but how they are to modify their structure and how they are to interact with other molecules once they arrive at their final station (Pugsley, 1989).

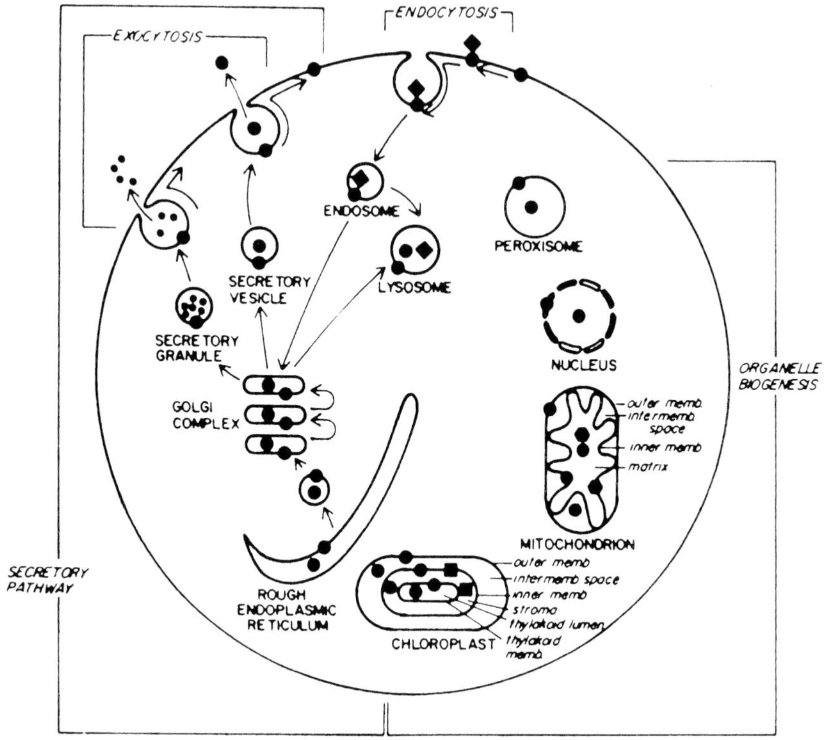

Figure 1. Diagram of eukaryotic cell showing the sites to which targeted proteins are directed. Nuclear encoded proteins are directed to a variety of cellular sites. They are sent into and through the endoplasmic reticulum to the Golgi complex where they are addressed to lysosomes and the plasma cell membrane (via secretory granules and vesicles). Chloroplasts and mitochondria code for a limited number of proteins. All other proteins are encoded by the nucleus and are imported from the cytoplasm.

Source: (From Pugsley, 1989).

Messenger RNAs have been found recently to also have specific addresses in the cell (Singer, 1993).

HUMAN PROTEINS CAN BE CLASSIFIED ACCORDING TO AGE

The human protein sequences can be divided into three classes, namely, (1) ancient, (2) middle-age, and (3) modern. Ancient proteins have not changed their function since prokaryotes and eukaryotes were separated from each other some 2 billion years ago. They are mainly enzymes involved in basic metabolic activities, such as triosephosphate isomerase. In regard to middle-age proteins, they are common in eukaryotes but have not been found in prokaryotes. Actin is an example of this group. And as for modern proteins, they are divided into three groups: (1) those present in animals or plants, but not in both, and absent from prokaryotes (example collagen), (2) those present only in vertebrates (e.g., plasma albumin), and (3) recent mosaic proteins, e.g., the low density lipoprotein (LDL)-receptor. This last class of mosaics is of particular interest relative to biological periodicity, since these proteins are viewed as being the result of intron-exon recombination (Doolittle et al., 1986).

THE NEW MOSAIC PROTEINS HAVE BEEN FORMED FROM OLD ONES BY INTRON-EXON RECOMBINATION

In the late 1970s it became evident that the genes of higher organisms are composed of two main types of DNA sequences: exons and introns. Although the whole DNA of the gene is transcribed into RNA, only a fraction of this DNA, contained in the exon sequences, is used for the formation of the messenger RNA, which is subsequently involved in the process of protein translation. The RNA of the intron regions, on the other hand, is cleaved and deleted. These findings prompted Gilbert (1978) and Blake (1978) to suggest that introns and exons could play an important part in evolution if they were able to change in number and position along the gene. They assumed that introns, due to their lengths of 50 to 50,000 base pairs, could participate in genetic recombination, and thereby, increase the rate at which exons can be recombined as independent structures. One way then, of creating new genes by the combination of exons into novel associations, would be by their duplication and migration along the gene during evolution. Much of the evidence coming from different laboratories supports this view (Figure 2).

Several examples of the recombination of exons in genes that are responsible for the evolution of plants, invertebrates and vertebrates can now be given. The first is the low-density lipoprotein (LDL)-receptor, which is a cholesterol transport protein. Since it performs multiple functions, each single function is carried out by a single protein domain. In order to establish the genetic basis of these domains and their relation to those of related proteins, Südhof et al. (1985 a, b) examined in detail the exon organization of this protein. Its gene was found to contain 18 exons and to be 45,000 bases long. The exons correspond to the functional domains, with thirteen of them encoding protein sequences that are present in other proteins. Five of these exons are similar to one present in human plasma (C9 complement component). Three exons encode a sequence similar to the one found in the epidermal growth factor (EGF) and in three other proteins of the blood clotting system (factor IX, factor X, and protein C). Such results provided evidence in support of the conclusion that there is a correlation between functional domains in the LDL-receptor protein, and the exon-intron organization of its gene, and that the

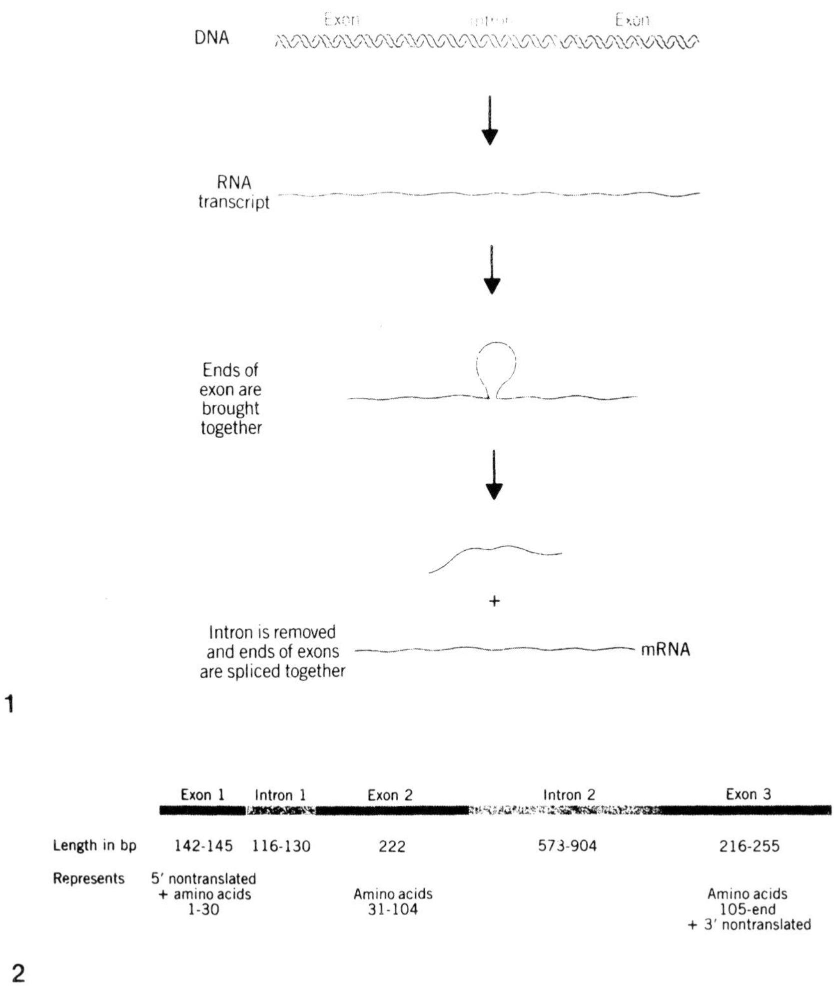

Figure 2. Many structural genes are interrupted structures consisting of intervening sequences (introns) and of sequences coding for proteins (exons). (1) Both are transcribed into RNA but only the exons become part of the messenger RNA that is translated into the different amino acid sequences that form the protein. (2) The mammalian beta globin gene with its three exons and the amino acids coded by each of them.

Abbreviations: (mRNA = messenger RNA; ; bp = base pairs in DNA).

Source: (From Lewin, 1983).

protein is the result of the association of exons shared with five different proteins. These exons are considered to have come from other genes.

Another example is the structural gene that is responsible for the formation of the human tissue-type plasminogen activator. This protein is one of the proteases involved in blood clots. The studies of Ny et al. (1984) showed that exons represent building blocks coding for discrete structural or functional protein domains, since the DNA sequence analysis of this gene revealed that the structural domains of this protein correlated well with the exon-intron pattern of the gene. There are, at least, 14 exons and 13 introns in this gene.

OTHER MECHANISMS RESPONSIBLE FOR PROTEIN DIVERSITY

Three other mechanisms need to be considered which have led to protein evolution; these are: DNA duplication, DNA rearrangement, and alternative RNA splicing (Smith et al., 1989; Benne and Van der Spek, 1992).

1. Gene duplication is the result of the addition of the complete sequence of a given gene, or of chromosomes that contain this gene, such as in polyploidy. Lactalbumin which occurs in milk is derived by this process from lysozyme (Doolittle et al., 1986).
2. DNA rearrangement in single genes leads to protein diversity. This is best exemplified by the immunoglobulin and T-cell receptor genes, which are a part of the immune system. Two segments of the gene in germ cells are separated by an intron. They become united in the somatic cells producing a protein (Hozumi and Tonegawa, 1976; Rabbitts, 1978).
3. Alternative RNA splicing does not alter the DNA content of the cell. Rather, it involves only modifications of the messenger RNA. Three examples may be cited. The first is the myosin heavy chain gene of *Drosophila*, which has 20 or more exons that can be alternatively spliced. This could give rise to a very large number of isomorphous proteins. The second is the GTP-binding protein Gs, the alpha component of which can generate a high diversity of splicing (Smith et al., 1989). And the third case are the more than 30 different exons or pairs of exons known to be alternatively spliced in sarcomeric contractile protein genes (Breitbart et al., 1987).

DIFFERENT PROTEINS EVOLVE AT DIFFERENT RATES ON ACCOUNT OF THEIR ATOMIC CONSTRUCTION

Of the more than 4000 amino acid sequences known, more than 200 represent human proteins. This information has made it possible to draw certain conclusions concerning the evolution of proteins. Chief among these are:

1. Various proteins change at different rates that are characteristic of each protein, the governing factor being amino acid replacement.
2. Various parts of a protein change at different rates, the exterior regions being modified more rapidly.
3. Binding sites and catalytic assemblages tend to be conserved.
4. The cysteine residues are the slowest to change when disulfide bonds are present in the protein.
5. The ancient proteins seem to be changing very slowly. Metabolic enzymes from humans and bacteria are about 40–50% identical.

The first coded proteins are supposed to have appeared on Earth some 3.5 billion years ago. Assuming a constant rate of change, then the most recent proteins are those produced by pseudogenes and fibrinopeptides. Lactalbumins, lysozymes and ribonucleases are considered intermediate in this time frame. The old ones are: hemoglobins, acid proteases, triosephosphate isomerase and others (Doolittle et al., 1986).

These rates of evolution depend on the structure of the constituent amino acids, and especially the nature of the atoms, such as sulfur that determines the behavior of the disulfide bonds, which also means that these atoms play a key role in determining the molecular clock of evolution. What is perhaps important to realize is that each protein goes its own evolutionary way (dependent on its atomic structure) regardless of whether the organism per se likes it or not.

PROTEIN DIVERSITY IN DEVELOPMENT AND EVOLUTION IS THE RESULT OF THE SAME MOLECULAR MECHANISM

Studies carried out on *Drosophila* (Boggs et al., 1987; Bell et al., 1988) have shown that alternative RNA splicing can also generate protein diversity during early development, as well as in the later stages of cell differentiation. Thus, alternative splicing participates both in organism development and evolution. This line of evidence reinforces the view that the two processes of development and evolution cannot be separated at the molecular level and that the variations they produce are similar in type. Alternative splicing provides proteins that, by and large, share extensive regions of identity; only a few of the domains differ.

THE NUMBER OF INTRONS INCREASED WITH EVOLUTION

Eubacteria are known to have no introns but there are some introns in both ribosomal RNA and transfer RNA genes in Archaebacteria (Pace et al., 1986). Fungi, such as *Aspergillus*, which are primitive eukaryotes, have few introns in their genes. In yeast 5 to 10% of the structural genes have introns. Human and mouse cytochrome c genes have introns but they are lacking in yeast. The alcohol dehydrogenase gene present in *Drosophila* has introns but not in yeast. Most lower eukaryotes have genes without introns.

As one reaches the invertebrates few introns are found and those that occur tend to be short. In insects, the number and type of introns are variable, as is the case in *Drosophila* ribosomal DNA (Engberg, 1985).

With the emergence of the vertebrates, and especially mammals, one notices the formation of genes with a vast number of introns. The ovalbumin gene in the chicken consists of 7 introns (Chambon, 1981). The human tissue-type plasminogen activator gene has 13 introns (Ny et al., 1984). The low density lipoprotein receptor gene contains 17 introns (Südhof et al., 1985a). The human gene that encodes the blood-clotting factor VIII consists of 25 introns. This gene spans 186,000 base pairs, while the coding sequence for its protein is only 9,000 base pairs (Gitschier et al., 1984). Such huge genes and the extreme number of introns they have are unknown in prokaryotes and invertebrates.

THE NUMBER OF INTRONS ALSO DECREASED IN SOME GROUPS

In general, introns are not easily lost but there is evidence that its number in some groups of organisms has decreased. In fact, some protein coding genes do not have any introns. Genes encoding histones — the main constituents of nucleosomes — only rarely have introns. This is also the case with interferons, proteins produced by vertebrates in response to viral infections.

Furthermore, in eukaryotes, several cases of intron deletion have been described (Antoine and Niessing, 1984). Such results lend credence to the view that intron deletion is a viable evolutionary mechanism (Green, 1986).

PRODUCTION OF NEW PROTEINS BY THE FORMATION OF COMPOUND MESSENGER RNAs

The African trypanosomes, such as *Trypanosoma brucei* and related species, have been extensively studied at the molecular level. They are unicellular parasites that cause sleeping sickness by proliferating in the host's bloodstream. What has made them the focus of much attention is the periodic change of their proteins. These proteins build the coats that cover their cell surfaces. By continuously producing new proteins, these protozoa escape immune attack. The genes that code for these variant cell surface glycoproteins are called VSG genes. Hitherto several novel mechanisms of gene action have been elucidated, one of which is the trans-splicing of RNA (Figure 3).

Every messenger RNA of a VSG gene consists of information from two DNA components. Two exons are derived from separate genes located in different chromosomes. The first component is a mini-exon (that is non-coding) coupled to a donor-intron. This produces a mini-exon donor RNA (med RNA). The second

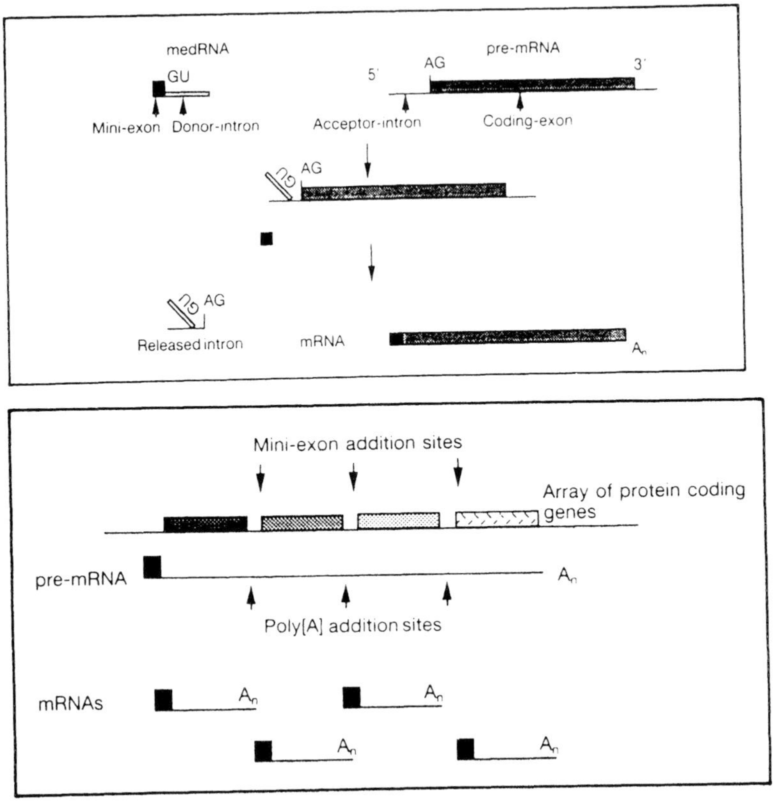

1

2

Figure 3. Production of compound messenger RNAs in eukaryotic cells. (1) Representation of the process of trans-splicing in the genome of the trypanosome *Trypanosoma brucei* (and related species) occurring in the genes that code for the variant cell surface glycoproteins (VSG genes). The trans-splicing leads to the joining together of different RNAs into a single messenger RNA. Two separately transcribed RNA precursors unite to form the final messenger (mRNA). The first precursor is transcribed from a mini-exon which is coupled to an intron that functions as a donor (med RNA). The second precursor derives from a coding-exon coupled to an acceptor-intron and is a pre-messenger RNA (pre-mRNA). The splicing which takes place at the terminal nucleotides (GU and AG), is accompanied by the release of the introns. The final product is a compound messenger RNA (mRNA) that is used in the translation of a new protein. (2) Trans-splicing occurring with the help of a donor-intron which adds mini-exon sites to an array of protein coding genes. This generates multiple mature messenger RNAs (mRNAs) from a single pre-messenger RNA (pre-mRNA). Each final messenger is a compound structure which contains, in addition to one of the segments of the pre-messenger RNA, the sequence from the mini-exon.

Source: (From Van der Ploeg, 1990).

component is an acceptor-intron followed by a coding-exon, which gives rise to a pre-messenger RNA (pre-mRNA). The joining of the mini-exon with the coding exon occurs through the trans-splicing of the independently transcribed precursor RNAs. The result is that precursor RNAs, from different chromosomes, are joined to form compound messenger RNAs that become translated into new proteins.

Another mechanism involving trans-splicing allows the generation of multiple mature messenger RNAs from a single precursor RNA (Figure 3).

A number of points which are important to an understanding of biological periodicity are the following. (1) By means of several molecular mechanisms the trypanosome is able to produce entirely different proteins without changing its genetic constitution. (2) The production of the new proteins is a periodic process. (3) This phenomenon is determined internally, independently of the environment to which the protozoan is exposed, the rate of transformation in *T. brucei* being about 10^{-6} to 10^{-7} per cell division (Van der Ploeg, 1990).

THE REVERSIBILITY OF ALTERNATIVE SPLICING AND OF EXON-INTRON RECOMBINATION ARE KEY PHENOMENA TO AN UNDERSTANDING OF THE EMERGENCE OF PERIODICITY

One of the most interesting properties of the alternative splicing of RNA is the reversibility of this process (Smith et al., 1989). The genetic constitution does not change and the splicing pathways need not be discarded in order to use different ones. This is due to the fact that the phenomenon does not occur at the level of DNA, but rather at the level of the transcribed RNA. It is the type of exon segments that are incorporated into the final messenger that determines the structure of the protein.

The order of the RNA segments can be easily reversed, thus leading to the production of a protein which had not been produced for some time. The decision to produce an alternative but different protein is inherent in the RNA sequences. As such, the protein emerges without being necessarily correlated with the phylogenetic position of the organism. Two organisms may look very different, and as a consequence may belong to quite different classes or phyla, yet if they happen to have a few genes in common, the old protein may suddenly reappear.

The same holds true of exon-intron recombinations at the DNA level. As mentioned above, new genes can be created by exon migration and recombination. This process is also reversible since exons can be included from other genes. This mechanism would allow the same type of protein to reappear suddenly.

It is noteworthy that each exon situated along the structural gene is responsible for the production of a separate domain that is only a part of the whole protein. For example, each of the five exons situated along the gene in fragment 1 of bovine prothrombin corresponds to a well-defined domain within the protein. This facili-

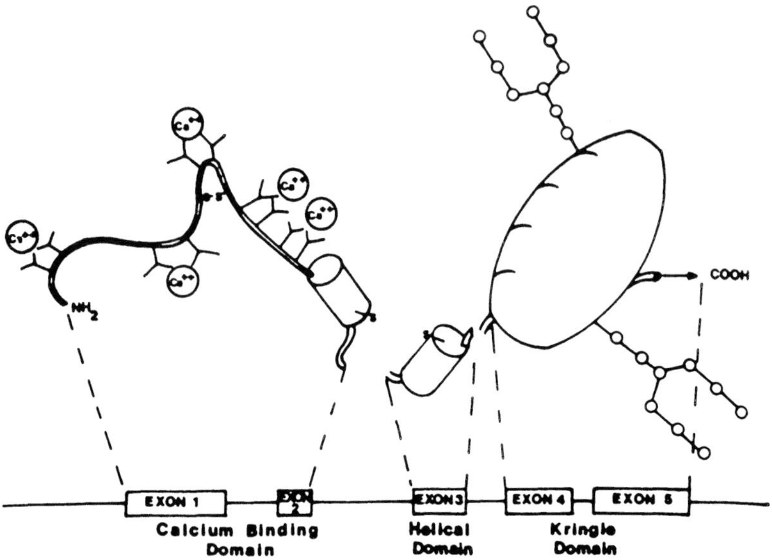

Figure 4. Diagram of the relationship between the three-dimensional structure of fragment 1 of bovine prothrombin and the fine structure of its gene. Each exon sequence of the DNA corresponds to a different structural domain of the protein.

Source: (From Blake et al., 1987).

tates the recombination of exons since individualized domains can be reassembled in an orderly manner (Figure 4) (Blake et al., 1987).

Recently, genes with over 60 introns, such as the DNA sequence for human Duchenne muscular dystrophy and with 117 introns, as is the case with the human type VII collagen gene, have been described (Koenig et al., 1987; Christiano et al., 1994). This association of an impressive number of small DNA sequences into functional genes, implies that the recombination of exons and introns may have been much more extensive during the evolution of higher organisms than it was anticipated.

Chapter 32

CASCADES OF MOLECULES AND OF GENE ACTIVATIONS ARE IRREVOCABLE EVENTS THAT PRODUCE INTEGRATED PACKAGES OF STRUCTURES AND FUNCTIONS

THE MOLECULAR CASCADE—A HORMONE ELICITS SIX DIFFERENT MOLECULES

The cell is not a "bag of enzymes," as was previously thought, but rather turns out to be multi-compartmented in which molecules participate in well defined sequences of chemical reactions. These sequences are so well determined within microdomains that the production of a molecule results in what is called a cascade.

A liver cell is stimulated by the hormones epinephrine and glucagon. Either is able to activate the production of glucose, but this chemical is the final product of six intermediate reactions that follow each other in a sequence that cannot be altered. Epinephrine and glucagon interact with specific receptors located at the outer side of the cell membrane. This triggers the formation of cAMP (cyclic adenosine 3′:5′ monophosphate) as the result of the catalytic breakdown of ATP (adenosine 5′-triphosphate) by activated adenylate cyclase. Cyclic AMP, in turn, activates cAMP-protein kinase which leads to a number of protein phosphorylation reactions including glycogen phosphorylase. The next step is the break-down of glycogen to glucose-1-phosphate which is transformed into glucose (Figure 1).

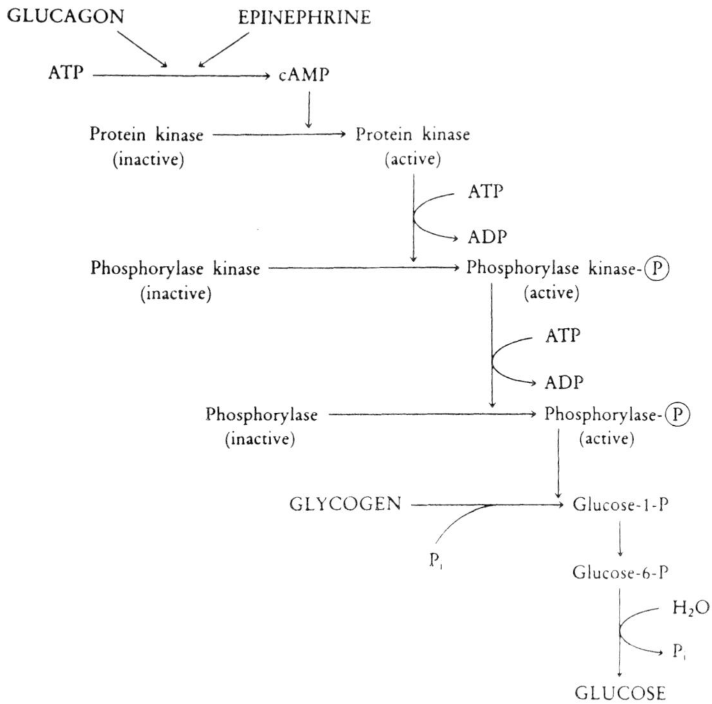

Figure 1. The stimulation of a liver cell by the hormones glucagon and epinephrine results in a cascade of molecules. ATP (adenosine 5′-triphosphate) is transformed into cyclic AMP (adenosine 5′-monophosphate) by membrane adenylate cyclase. Cyclic AMP, in turn, initiates a number of phosphorylation reactions that finally lead to the conversion of glycogen into glucose.
Source: (From de Duve, 1984b).

The above cascade has the following features:

1. Virtually all the steps involve enzyme catalysis.
2. Every macromolecule has its exact place in the sequence which cannot be altered.
3. Once one of the specific hormones acts, six different molecules are produced as seen in Figure 1. This occurs in a way that the organism, as a whole, cannot stop or modify.
4. Moreover, the cascade involves amplification of the initial molecular signal: for every hormone bound to its receptor one expects 100 million molecules of glucose to be produced. It is thus clear that minor alterations in hormone concentration can have far-reaching developmental consequences (de Duve, 1984b).

BLOOD CLOTTING OCCURS BY A CASCADE INVOLVING MORE THAN 10 PROTEINS

After inflicting a wound in any part of the body, a clot is formed that stops bleeding. More than 10 different proteins are involved in this process and they follow each other in a cascade. The activated form of one protein catalyses the activation of the next protein. The principal cascade results from three minor ones that comprise an intrinsic and an extrinsic pathway in which proactivator and activator proteins are involved (Figure 2). Both the proenzymes participating in blood coagulation and the fibrinolytic steps have polypeptide units which are common to many of them. The inference thus drawn is that the cascade is determined by intrinsic molecular events that are inherent in the evolution of the proteins themselves (Figure 2). The result is a package of proteins that are functionally closely connected and inseparable. This produces a canalized sequence which precludes deviation (Blake et al., 1987; Stryer, 1988). What also characterizes the cascade is the requirement for very

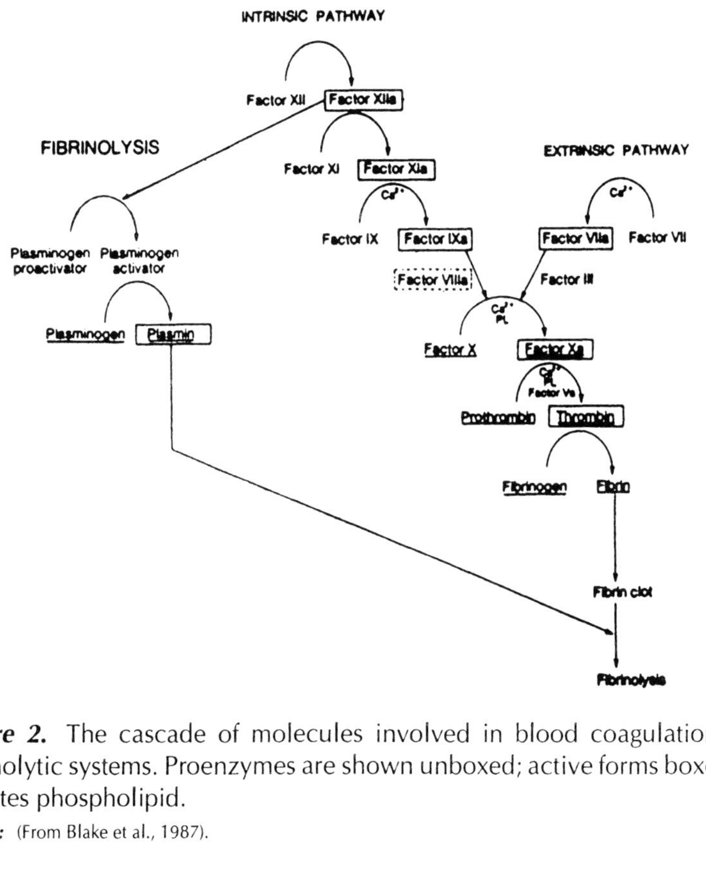

Figure 2. The cascade of molecules involved in blood coagulation and fibrinolytic systems. Proenzymes are shown unboxed; active forms boxed; *PL* denotes phospholipid.

Source: (From Blake et al., 1987).

small amounts of the initial enzymes because of the catalytic nature of the activation process. As a consequence, the response to trauma is very rapid. Hyper-canalization and speed are the two main distinguishing features (Ryden and Hunt, 1993).

OTHER TYPES OF CASCADE — SIX HORMONES FORMED FROM A SINGLE POLYPEPTIDE

More molecular cascades directing cell activities are being discovered as more information becomes available. The anterior pituitary gland, situated in the middle of the brain, produces a signal-directed secretory polypeptide called melanocorticotropin. In a first step this polypeptide is cleaved into two segments that are the hormones ACTH (adrenocorticotropin) and beta-LPH (lipotropin). In a second step these peptides are carved once more producing alpha-MSH, beta-MSH (melanocyte stimulating hormones) and beta-endorphin. In a final cleavage, enkephalin is released from endorphin (Figure 3).

1. The six hormones are all amino acid sequences of various lengths. The smallest of all consists only of five peptides (enkephalin).
2. The cleavage is a cascade process that occurs at specific sites along the macromolecules and irreversibly leads to the formation of these hormones.
3. The hormones act on quite different organs: (i) They stimulate the adrenal cortex adjacent to the kidneys. (ii) Act on lipid metabolism. (iii) Affect the pigmented cells of the skin. (iv) Influence brain function in a way similar to morphine.

Taken together, this means that an integrated package of structures and functions is produced as a result of the initial formation of a single polypeptide (de Duve, 1984b).

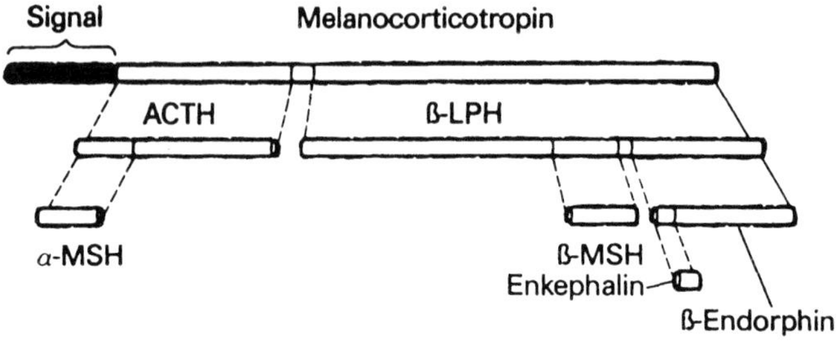

Figure 3. One polypeptide gives rise to six hormones that activate different organs. The melanocorticotropic hormone is carved into adrenocorticotropin (ACTH) and beta-LPH (lipotropin). A second cleavage leads to the release of alpha- and beta-MSH (melanocyte stimulating hormones) and beta-endorphin. The third cleavage produces enkephalin.

Source: (From de Duve, 1984b).

THE CELL HAS ITS OWN SAFEGUARDS

A second source of order derives from a series of safeguards. In the event that a particular cascade and its amplification mechanism were to get out of control, then the molecular apparatus of the cell would respond as follows: (1) destroy the hormone, (2) isolate the hormone in a compartment, (3) introduce antagonistic hormones, (4) inactivate the receptor, and (5) hydrolyse cyclic AMP through phosphodiesterase.

In the next section a third source of order, which resides in sequential activation, as well as the repression of genes along the chromosomes, and within the genome as a whole, will be touched upon.

CASCADES OF GENE PRODUCTS ARE A MECHANISM RESPONSIBLE FOR ORDERED DEVELOPMENT

Genes do not function independently of each other. Detailed molecular analysis carried out during the last few years reveals that certain sets of genes code for proteins that are necessary for the transcription of the next class of genes. This interdependence creates what has also been called a cascade, since groups of genes are turned on (or turned off) in a well defined sequence and at specific times during development. The gene cascades are found at several levels of organization: these include viruses, animal embryos and endocrine control in higher organisms.

CASCADE OF GENE ACTION DURING PHAGE INFECTION

A phage is a virus that infects bacteria. On infection it subverts bacterial functions, thus directing them to the production of its progeny.

The production of virus progeny in T4, lambda and other phages, follows a three phase cascade: (1) A first group of genes is transcribed. (2) One or several of these genes is a regulator that controls the expression of the second phase. (3) One or more of these last genes is also a regulator necessary for the expression of the genes of the third phase.

The regulation of transcription is carried out by proteins that bind specifically to regions of the virus DNA and which involve interactions between amino acids and bases (Campbell, 1971; Ptashne, 1967 a and b, 1978, 1986; Friedman and Gottesman, 1983; Wharton et al., 1984; Lewin, 1990).

CASCADES IN EMBRYO DEVELOPMENT

The gene cascades that occur, during the development of the embryo of *Drosophila* are not unrelated to those of viruses. In both cases DNA-binding proteins are employed as regulatory factors.

Genes involved in the regulation of development are identified by mutations that are lethal, or cause disturbances in differentiation. These mutations fall into three groups: (1) Maternal genes which are expressed (before embryo formation) in the body of the mother during oocyte formation. They affect polarity and large regions of the embryo. (2) Segmentation genes are expressed after fertilization and determine the number of body segments. And (3) homeotic genes control the function of a given segment.

The genes of these three groups act in a cascade that restricts successively the properties of the different parts of the embryo. For example, the fate of the cells in the anterior part of the embryo is determined by the *bicoid* protein which turns on the transcription of another gene *hunchback*. Other parts of the cascade are manifested by *gap* genes which are necessary for the expression of the *pair-rule* genes. The expression of genes involved in segmentation and cell differentiation is regulated in a hierarchical and sequential way (Scott and Caroll, 1987; Ingham, 1988; Graham et al., 1989; Lewin, 1990).

NEW GENES ARE PRODUCED DURING DEVELOPMENT IN HUMANS

Immunoglobulin genes in mammals, including humans, are assembled from several DNA sequences in blood cells (lymphocytes). Each immunoglobulin is an antibody consisting of a variable region and a constant region. The genes responsible for these regions of the protein are located apart at considerable distance in the germ line of humans, and are not expressed. It is only when the genes are closer to each other, as in lymphocytes (of the B type), that the antibody is formed. But at birth the organism does not possess the functional genes for producing a particular antibody. It is only later during somatic differentiation that the rearrangement of DNA sequences results in new gene formation and expression (Tonegawa, 1983; Flavell et al., 1986; Alt et al., 1987).

The rearrangements have two characteristics: One is that they are an orderly process involving well-defined DNA sequences. The other is that they change the expression of specific DNA regions. Thus, they represent a clear-cut example of the production of new functions by simple modification of the internal chromosome sequence.

CASCADES OF REGULATORY EVENTS IN PLANTS

Plants, e.g., *Arabidopsis*, are able to detect environmental changes, including rain and wind, and to respond to them by altering structural features, such as body elongation and stem thickness. Briefly, some 10 to 30 minutes after stimulation, messenger RNA levels increase by as much as 100-fold and four genes are expressed. Calmodulin and related proteins, as well as calcium ions, are involved. The regulatory events represent a cascade in which calcium ions bind to calmodulin and this in turn acts on a series of enzymes which function as key regulators of a variety of cellular processes (Braam and Davis, 1990).

Chapter 33

THE RELATIONSHIP BETWEEN CHEMICAL AND BIOLOGICAL PERIODICITY

THE PRINCIPLES FOLLOWED BY FORM AND FUNCTION ARE THE SAME AT EVERY LEVEL OF ORGANIZATION

(1) The initial determinant is internal. Every component is built from within.

Energy and matter have concretized and continue to concretize into elementary particles. Atoms emerge from nucleons that trap electrons, which become associated in peripheral orbits. Crystals grow from nuclei of atoms that function as seeding centers. DNA replicates when an initial primer molecule is available. Multicellular organisms develop from a single initial fertilized egg that functions as the center from which the large mass of the organism's cells are derived.

(2) Self-assembly is the mechanism involved in the production of more complex components.

Elementary particles have the ability to organize themselves spontaneously. The assembly of atoms and macromolecules is also intrinsic and spontaneous. At the cell level the self-assembly of ribosomes which has been studied in great detail arises automatically from the isolated proteins and nucleic acids (Nomura, 1977; Röhl and Nierhaus, 1982; Spirin, 1986). Viruses self-assemble from their separated macromolecular components, thereby producing infectious particles (Fraenkel-Conrat, 1962; Mindich et al., 1982). Cells have also the capacity to self-assemble into tissues and organs, because of chemical receptors and messengers. Whole organisms such as hydras can self-assemble from their dispersed cells (Bonner, 1952; Gierer, 1974). Human skin has been produced by allowing separated cells to self-assemble (Dubertret et al., 1987) (Table 1).

(3) Particle and atom congruence determine the channels prevailing in self-assembly.

When elementary particles self-assemble, their symmetries and energies govern what types are formed, as is the case when quarks and antiquarks build mesons.

When the 53 proteins that constitute a ribosome are isolated and allowed to self-assemble, they follow a specific sequence. Moreover, the addition of one protein enhances the incorporation of the next (Nomura, 1977). This is also the case in the self-assembly of cells in the slime mould *Dictyostelium*. The separated cells secrete a messenger substance, called acrasin, that is recognized by the surface protein receptors of other cells. The order of assembly determines the function that the different cells perform in the newly built organism (Sussman, 1964; Bonner, 1983).

(4) Self-disassembly is the counteracting process which leads to decay and destruction of self-organized structures and functions.

Elementary particles decay into other particles. The same phenomenon is present at the atomic level, and has the well known name of radioactivity, in which atoms disintegrate with the emission of particles. Crystal growth in a solution is a process that involves simultaneously self-assembly and self-disassembly (Pauling, 1987). DNA is a molecule that self-assembles but also disassembles when it is heated: the two strands of the double helix separate from each other. This is a reversible process (Watson et al., 1987). Proteins are broken down into minor polypeptide chains with specific functions, as exemplified by pro-insulin (de Duve, 1984b). Cells self-assemble into tissues but can also disassemble under the influence of various chemicals. This is an event that is regularly encountered in embryonal development (Moscona and Hausman, 1977). An obligatory event in living organisms is death whose immediate outcome is cell disassembly.

(5) The stabilization and perpetuation of a given combination is mainly governed by energy levels.

Among the elementary particles only 20 are long-lived, four of which are stable (photon, electron, neutrino and proton). The neutron is only stable when bound to an atom nucleus. The energy state is the factor governing stabilization (Pitt, 1988). In minerals, crystal stability is directed by the chemical bonding mediated by the electrons of the constituent atoms (Jaffe, 1988). Stability in higher organisms, such as a human, is totally dependent on the amount of energy made available through food. Elimination of this energy source leads to immediate decay and destruction. In mammalian cell cultures the cells are made to self-assemble or self-disassemble by means of chemical substances that change the energy state of their protein receptors (Alberts et al., 1983).

The forms and functions that are most frequent in organism evolution, and those that predominate in development, ought to be the most energetically stable. Potential energy thresholds limit the number and type of conformations that can occur

Table 1. Self-Assembly Extends from Elementary Particles to Living Organisms

Level	*Example*	*Reference*
Elementary particles	Mesons result from union of pairs of quarks and antiquarks.	Pagels, 1982
Atoms	Protons, neutrons, and electrons self-assemble into atoms.	Pitt, 1988
Crystals	Atoms self-assemble into crystals by ionic, covalent, metallic and molecular bonding.	Jaffe, 1988
Proteins	Aspartate transcarbamoylase; the 12 units, which have been dissociated, are able to reconstitute the enzyme.	Bothwell and Schachman, 1974
Membranes	Nuclear membranes self-assemble from cell-free extracts of eggs.	Dabauvalle et al., 1991
Ribosomes	The 3 RNAs and 53 proteins of ribosomes have been isolated and reassemble into active particles. The information for correct self-assembly is present in the structure of the molecules.	Spirin, 1986
Nucleosomes	Purified DNA and histones self-assemble into the essential component of chromosomes, called nucleosome.	Oudet et al., 1975
Viruses	The isolated RNA and proteins of tobacco mosaic virus self-assemble, thus producing an infectious virus.	Mindich et al., 1982
Organisms	Self-assembly of the dispersed amoebae of *Dictyostelium discoideum* results in a complete slime mold.	Bonner, 1983
Organisms	Hydras can self-assemble from their dispersed cells.	Gierer, 1974
Organisms	After dissociation, marine sponges produce a whole organism from their cells.	Müller et al., 1976
Organs of Mammals	Dispersed liver cells, kept in tissue culture, self-assemble. The resulting liver reacquires its function.	Moscona, 1959
Organs of Mammals	Dissociated cells of rat gonads reorganize into testes and ovaries.	Ohno et al., 1978
Human tissues	Human skin has been produced by allowing dispersed cells to self-assemble.	Dubertret et al., 1987
Human Organs	Human capillary endothelial cells from tissue cultures are able to self-assemble into blood capillaries.	Folkman and Haudenschild, 1980

in nucleic acids. They limit the rotation of atoms, and as such, the type of changes in molecular configuration that can arise (Saenger, 1988).

(6) Internal processes determine which particles and atoms are incorporated from the external environment.

Elementary particles are incorporated from the outer environment according to the internal energy state and symmetry of the molecular cell components. Consider, for example, in the ampler sense, the unique properties of water. These properties are primary determinants of cell construction. The simple fact is that it is mainly the molecules that have been, and can be accepted by water, which are allowed to participate in the evolution of the cell. One feature is that water concentration in the cell determines the configuration of DNA (Rich et al., 1984). The uptake of elementary particles, such as photons, is determined by the cell's internal construction. Chloroplasts within a plant cell align themselves along the outer membrane when few photons reach a leaf, but when light intensity increases they move to the side walls of the cell. By displacing themselves chloroplasts are able to adjust the amount of photons that they absorb (Denffer et al., 1971; Björkman, 1992). Also one cannot overlook the fact that mammalian cells, whether in culture, or *in vivo*, when subjected to noxious chemical agents, react by increasing the number of specific genes (amplification), thereby blocking or minimizing the entry of such toxic agents into the cell (Stark et al., 1990).

BIOLOGICAL PERIODICITY HAS FOLLOWED THE FEATURES THAT CHARACTERIZE PERIODICITY AT THE CHEMICAL LEVEL

What led to the concept of periodicity in the chemical elements was the finding that the same properties appeared suddenly in atoms that, from the beginning, seemed to have no structural relationship and that these properties emerged at regular intervals. When the elements were distributed according to their atomic weights there was a periodic recurrence of characteristic properties as shown, for instance, by the elements with atomic numbers, 2, 10, 18, 36, 54 and 86 which are inert gases. The Periodic Table contains, among other sequences: two short periods of eight elements each, two long periods of 18 elements, followed by a very long period of 32 elements (Pauling, 1949; Greenwood and Earnshaw, 1989).

The chemical periodicity has several features among which are:

1. The same property is repeated.
2. The property emerges at relatively regular intervals.
3. The periodicity was initially recognized in the absence of any structural relationship to a predecessor with the same property. At that time no mechanism was known which explained the recurrence of similar features.
4. Partly related elements do not have the property.
5. Not just one, but a whole set of similar properties characterizes a given group of atoms.
6. Large groups of elements constitute separate series that do not fit well into the Periodic Table.
7. The regularity of the periodicity is not total, that is, there are exceptional situations.
8. Both simple and complex properties show periodic trends.

9. The different atom components follow a hierarchy in the determination of the periodic properties.
10. The simplest, as well as the most complex atoms, display periodicity.

These features turn out to be essentially similar at the biological level.

1) The same property is repeated, as attested by the emergence, in quite different groups, of similar structural and functional sets of properties.

2) There is an emergence at relatively regular intervals. This is, for instance, the case in flight. The length of the interval is not yet easy to establish. This is mainly due to the present ignorance of the main mechanisms directing gene regulation and gene recombination. It is to be expected that different structures, as well as different functions, may have different intervals.

3) Initially, there was no apparent structural relationship between the atoms that exhibited the same properties. This was only revealed when their electronic orbitals were determined. This is also the case, at present, at the biological level. It remains to be disclosed which type of electronic orbitals, in key atoms, are responsible for the mimicry of the different molecules (proteins or others) that produce the same combinations of structures and functions.

4) Chemical elements located nearby in the Periodic Table have different properties. This is, for instance, the case of the elements of Group I which are at the side of Group II. The situation is comparable to that of vision within the molluscs, in which closely related species, placed side by side in the same class, have among the simplest and the most complex eyes known to occur in the Animal Kingdom e.g. *Nautilus* and *Octopus*.

5) There is a whole set of properties that accompanies the groups of the Periodic Table of elements. The alkali metals (Group I) are: (i) very reactive, (ii) soft metals, and (iii) have low melting points. The alkaline-earth metals (Group II) are adjacent to Group I but have a different package of properties. They are: (i) less reactive, (ii) harder, and (iii) have higher melting points than the alkali metals (Group I). The same is true of the biological properties. The wing by itself would never allow a bat or an insect to fly; it is only when correct articulations, muscles, blood flow, energy source and brain coordination are present that the periodicity of flight becomes a reality.

6) The lanthanide elements or "rare earth metals" (Nos. 58 to 71) have similar properties constituting the lanthanum series. The actinide elements or "uranium metals" (Nos. 90 to 103) also have similar properties constituting the actinium series. The lanthanides are silvery in appearance and are based on typically metallic, close-packed arrangements. Their chemistry is dominated by the + 3 oxidation state, and they are very reactive metals. The actinides are also silvery in appearance, and are based on metallic, close-packed arrangements. Their chemistry is similarly dominated by the + 3 oxidation state. They also react with most non-metals. Both

series tarnish in air (Chapter 1, Fig. 1). These two series are prominent in the Periodic Table and their elements are placed side by side because they have properties in common. For instance, cerium (Ce) has properties similar to those of thorium (Th). These analogies extend to 14 elements. Consider now the placental and the marsupial groups: they show relationships that are comparable to those of these two series of chemical elements. These mammals form two parallel series of animal species. They are a case apart in biological evolution. The striking fact is that they show equivalence of the same characters, the result being that practically similar species occur in the two series. Both the placental and marsupial groups have mammalian characters. Yet each group has its own typical feature: all placentals have properties in common (e.g. the placenta) and most marsupials have properties in common (e.g. the pouch). However, as in the case of the two atom groups, they also comprise animal species having a perfect correspondence in both series, which includes sabretooth cats, lions, mice, moles and others (Table 2).

7) The regularity of biological periodicity seems at present to be only partial. This is not unexpected. When the chemical elements with similar properties were aligned, this was only possible by leaving empty spaces in the Periodic Table. Several atoms were considered to be missing. This was one of the initial merits of the chart. The recognition of these exceptions led to the prediction of the missing elements which were later discovered and had the properties predicted. As more fossil species are bound to be discovered, and as new organisms are created with the help of gene manipulation, biological periodicity is bound to become better understood, and a clearer pattern will gradually emerge. However, as in the case of the atoms, one should not expect biological periodicity to be totally regular.

8) The most characteristic property of an element is its valence. This is a simple property. Other properties such as the bond type are complex. Both show periodic trends. In living organisms, periodicity is evident not only in simple properties such as luminescence (which already appears in minerals), but also in flight that depends on brain activity which is considered as being the most complex body function.

9) Not all electrons which form a cloud around the atom nucleus, are equal in importance in determining the properties of the elements. It is the electrons of the outer shell that are the main determinants of the atom's chemical properties. The same is true of living organisms. Not all molecules in the cell are of equal importance. Its major functions are determined by nucleic acids and proteins. Moreover, it is the bases within the nucleic acids that are the significant components. The hierarchy found in the cell is already present within the atom.

10) Periodicity is not directly dependent on atom or living organism complexity. Atoms may have only one proton in their nucleus, as is the case with hydrogen, or may possess as many as 103 protons as in lawrencium. Consider helium: it has 2 protons and essentially the same properties as xenon which has 54 protons. The same is true of animals. Their genomes may consist of little DNA, as in invertebrates, or contain large amounts of DNA as in most vertebrates. Thus, there is no

Table 2. Chart of Functional Periodicity

	Functions						
	Bioluminescence	*Advanced vision in invertebrates*	*Placenta*	*Penis*	*Return to aquatic life*	*Flight*	*High mental ability*
	Photobacteria	Cubomedusae (Cnidaria)	Onychophorans	Platyhelminthes (Flatworms)	Fresh water Gastropods	Insects	Insects
	Dinoflagellates (Protozoa)	Prosobranchia (Mollusca)	Sharks [3] (Fish)	Gastrotrichs	Fresh water (Insects)	Pterosaurs (Reptiles)	Cephalopoda (Mollusca)
	Cephalopoda (Mollusca)	Pecten bivalvia (Mollusca)	Amphibians	Acanthocephala	Stereospondyls (Aquatic amphibians)	Teleosts (Fish)	Mammals
	Fireflies Beetles (Insects)	Cephalopoda (Mollusca)	Reptiles	Gnathostomulida	Ichthyosaurs (Aquatic Reptiles)	Birds	
General Increase in Organism complexity	Asteroidea (Echinodermata)	Alciopidae (Annelida)	Teleosts (Fish)	Gastropoda (Mollusca)	Plesiosaurs (Aquatic reptiles)	Bats	
	Fish[1]	Decapoda[2] (Crustacea)	Flowering[5] plants	Cirripedia (Crustacea)	Sea Cows (Aquatic mammals)		
		Amphipoda[2] (Crustacea)	Placental Mammals	Odonata (Insects)	Whales (Aquatic mammals)		
		Neuroptera (Insects[2])		Marsupials	Seals (Mainly Aquatic Mammals)		
				Placentals	Hippopotamuses (Mainly Aquatic Mammals)		

Table 2. (Continued)

Placental Series[4]													
Sabre-tooth	Lion	Hyena	Otter	Wolf	Cat	Mouse	Gliding Squirrel	Gliding colugo	Mole	Shrew	Anteater	Rhino-ceros	Wild boar
Smil-odon	*Pant-hera*	*Cro-cuta*	*Lutra*	*Canis*	*Felis*	*Mus*	*Anoma-lurus*	*Cynoce-phalus*	*Talpa*	*Sorex*	*Taman-dua*	*Dice-ros*	*Sus*
Marsupial Series													
Sabre-tooth	Lion	Hyena	Water opposum	Wolf	Cat	Mouse	Greater Glider	Sugar Glider	Mole	Phasco-gale	Anteater	Rhino-ceros	Pig-Footed Bandi-coot
Thyla-cosmy-lus	*Thyla-coleo*	*Borhy-aena*	*Chiro-nectes*	*Thyla-cinus*	*Sarco-philus*	*Darcy-cercus*	*Petau-roides*	*Petau-rus*	*Notor-yctes*	*Phasco-gale*	*Myrme-cobius*	*Dipro-todon*	*Chaer-opus*

Notes: To avoid overloading of this table the following restrictions were introduced:

1 The distribution of bioluminescence is found in Chapter 6, Table 1. Only included here are the groups in which it is most prominent.

2 Only some Decapoda and Amphipoda have advanced eyes. Neuroptera are only mentioned as an example; other insect orders have advanced eyes.

3 Some functions, such as the placenta, are only present in some species of a group. Detailed information is found in the text.

4 The equivalence between the placentals and the marsupials is displayed separately as two parallel series, because their parallelism is comparable to that of the chemical elements that represent the Lanthanum and Actinium series (Chapter 1, Figure 1). Both at the chemical and the biological levels, the components within each series have common characteristics, yet they show properties common to both series that have an exact correspondence.

5 The flowering plants are less complex than the fish, amphibians and reptiles, yet they were located in the chart after them, because they arose in evolution after these groups had established themselves.

6 The examples given of high mental ability are only a few. Others could have been added to the Table.

obvious correlation between the recurrent appearance of the properties studied and DNA content.

The fact that a series of main features are common to atomic and biological periodicities implies that the phenomenon has been maintained, despite the existence of an enormous gap between the two levels. Such evidence lends credence to the concept of autoevolution, according to which biological evolution has been conditioned by the initial transformation processes occurring at the atom level (Lima-de-Faria, 1983, 1988).

When periodicity was recognized in the properties of atoms, it seemed elusive at first, and the exceptions were too numerous for the phenomenon to be taken seriously. J. Newlands's paper on the "law of octaves" was sent for publication in 1883 but was rejected and ridiculed. Few chemists at that time could conceive such a high degree of order among the chemical elements (Pauling, 1949). However, as knowledge increased, periodicity became established as a central phenomenon that explains most chemical processes (Sanderson, 1967). To date, there are still exceptions to the Periodic Table of chemical elements. It is therefore reasonable to expect that the task of understanding periodicity in living organisms is going to be a difficult one and that less well defined groups and more exceptions will be found.

THE MOLECULAR MIMICRY THAT IS RESPONSIBLE FOR BIOLOGICAL PERIODICITY MAY BE DECIDED BY KEY ATOMS WITH SIMILAR ELECTRONIC CONFIGURATIONS

Molecular mimicry is found in crystals that are of different chemical composition, yet display the same form and belong to the same crystallographic system. But the phenomenon of chemical mimicry is present at an even earlier level, i.e. that of single atoms. Though the chemical elements have been classified into a periodic system on the basis of the similarity of their properties, they consist of different numbers of protons, neutrons and electrons. The significance of this is that it is the number of electrons in the outer shell that is mainly responsible for the similarity of the properties within a given group (Jaffe, 1988). This means that the mimicry is the result of similar electronic properties. At the more complex level of crystal formation, it turns out that single atom groups are responsible for this phenomenon. The carbonates of magnesium and iron have the same crystal form; but it is the carbonate part of the molecule that is responsible for the maintenance of the pattern. In other words, it is a given part of the molecule that is mainly responsible for mimicry. Remarkably, the same situation prevails in living organisms. Very different macromolecules may lead to the same pattern or function.

The snow crystal with six radiating rays, the six crystals formed around a center in arsenopyrite, the corolla of a lily flower with its six petals, the starfish with six arms, and the reptile with twelve (2×6) eye plates radially distributed, are all expected to be the expression of a similar type of atomic process that is located in different molecular configurations. Such a particular atom, or group of key atoms,

Table 3. Chart of Structural Periodicity

	Number of Components Associated Orderly								
	2	*3*	*4*	*5*	*6*	*7*	*8*	*9*	*10*
Chemicals		Cyclic metasilicate (3 SiO_4 molecules)	Cyclic meta-silicate (4 SiO_4 molecules)		Cyclic meta-silicate (6 SiO_4 molecules)		Cyclic meta-silicate (8 SiO_4 molecules)		
Minerals	Argirodite (Single crystal with bilateral symmetry)	Argirodite (3 crystals, twinning)	Staurolite (Cruciform twins)	Marcasite (5 crystals, twinning)	Arsenopyrite (6 crystals, twinning)				
Nucleotides					Poly (dG-dC) Poly (dG-dC) (6 components)		D-DNA (8 components)	Poly (A) (9 components)	B-DNA (10 components)
Flowers	*Orchis Morio* (Dorsal petal with bilateral symmetry)	*Alisma plantago* (3 petals)	*Ilex aquifolium* (4 petals)	*Rosa pendulina* (5 petals)	*Scilla autumnalis* (6 petals)	*Paeonia tenuifolia* (7 petals)	*Acer platanoides* (8 petals)	*Drimys winteri* (9 petals)	*Passiflora caerulea* (10 bracts)
Fruits	*Acer pseudo-platanus* (2 units)	*Begonia* (3 units)	*Rhamnus cathartica* (4 units)	*Psidium guajava* (5 units)	*Damasonium alisma* (6 units)	*Phytolacca clavigera* (7 units)	*Illicium verum* (8 parts)	*Marcgravia nepenthoides* (9 units)	*Malva silvestris* (10 units)
Leaves	*Ginkgo biloba* (bilateral symmetry)	*Trifolium repens* (3 parts)	*Marsilea quadrifolia* (4 parts)	*Nigella damascena* (5 parts)		*Aesculus hippocastanum* (7 parts)		*Lupinus* (9 parts)	*Tropaeolum majus* (10 parts)

Table 3. (Continued)

Fossil and living starfish	Bipinnaria larva of starfish (bilateral symmetry)	Fossil Echinoderm Blastoidea (3 arms)	Fossil Echinoderm Blastoidea (4 arms)	Starfish *Palmipes membranaceus* (5 arms)	Starfish *Leptasterias hexactis* (6 arms)	Starfish *Luidia ciliaris* (7 arms)		Starfish *Brisinga mediterranea* (9 arms)	Starfish *Solaster papposus* (10 arms)
Other invertebrates	Body of many invertebrates (bilateral symmetry)	Echinoderm *Echinus* Pincerlike structure (3 parts)	Worm head *Taenia saginata* (4 parts)	Echinoderm Ophiuroides *Ophionereis* (5 arms)	Coelenterate *Hydra* (6 tentacles)		Protozoa *Phractopelta tessarapsis* (8 parts)	Schyphozoa Jellyfish *Aurelia* larva (9 tentacles)	Coelenterate *Aglaophenia* (10 tentacles)
Vertebrates	Body of most vertebrates (bilateral symmetry)		Body of most vertebrates (4 limbs)	Foot of *Gecko techadactylus* (5 toes, radial to bilateral symmetry)		Fossil Reptile *Utatsusaurus* (7 eye plates)		Flagellar Axoneme (9 outer microtubule doublets)	

Note: The empty boxes do not need to imply that the missing numbers of associated components do not occur in nature. Some may actually not be formed but others may be only less frequent, and as such, difficult to detect.

that are responsible for producing the recurrent pattern, constitute what may be termed, the "periodicity center" of the molecule (Table 3). This might be better understood by considering the following example. The six branched crystal structure of water is produced by a totally different molecule, viz. iodoform. This carries a far-reaching implication for our understanding of biological periodicity. Clearly, one does not need to produce exactly the same chemical to have a similar form or function. Iodoform consists of hydrogen, carbon and iodine. The first two atoms are found in every protein. It is thus not surprising that these macromolecules may be able to produce in plants and animals the same type of properties present in simpler atomic combinations. Since mimicry among atoms is mainly determined by the number of electrons present in their outer shell, it follows that molecules that mimic one another ought to have key atoms in their "periodicity centers" that possess the same numbers of electrons in their outer shells.

The gene by producing a specific type of molecule that mimics an earlier atomic configuration, makes it necessary for the same structural pattern to reappear. Both exon-intron recombination and RNA-splicing can lead to the emergence of both new and old proteins. This represents one of the primary mechanisms by which periodicity may originate, irrespective of the location of the organism in the evolutionary scale.

The fact that bilateral symmetry has remained unaltered during the evolution of the body of the vertebrates, that is, from fishes to humans, is evidence that agrees with the concept of the determination of biological evolution by atomic order. The permanence of bilateral symmetry could hardly have occurred outside the rules that guide atomic behavior.

As pointed out in Chapter 12 animals or plants may even possess a series of different DNA sequences, as found in unrelated organisms; yet such different genes can still produce the same structure or function if the proteins they give rise to (or other molecules), have comparable effects. An example is found among the four proteins: staphylococcal nuclease, yeast asp-tRNA synthetase, cholera-like cytotoxin and Shiga-like cytotoxin. The amino acid sequences show no significant similarity. However, all four proteins perform the same function, e.g. they bind oligonucleotides or oligosaccharides (Murzin, 1993).

Another example is provided by 15 proteins, showing completely different amino acid sequences that all form the eight-stranded alpha-beta-barrel structure, a most regular protein feature. This is an outstanding case of structural mimicry (Branden and Tooze, 1991).

In essence, then, if the molecules contain pertinent atoms that have similar effects, one can expect "carbon copies" to emerge. As pointed out by Schwabe and Travers (1993) in proteins, very similar structures can he formed from widely differing amino acid sequences, and the same biological function has more than one structural solution.

BIOLOGICAL PERIODICITY FOLLOWS A GENERAL RULE

The information assembled provides an overall picture of the phenomenon of biological periodicity. No matter the property studied, periodicity follows the same pattern. This allows the formulation of a general rule of biological periodicity which can be stated as follows: An integrated package of forms and functions arises in an organism independently of the main properties of its immediate predecessors, recurs at relatively regular intervals irrespective of the organism's phylogenetic position, and is not necessarily directly related to the major factors prevailing in the environment.

This rule demands qualification.

Qualification 1: Both structures and functions emerge as an integrated package. This indicates that they are the result of a biochemical cascade, especially since they appear at the same time and are obligatory components of a functionally interrelated unit.

Qualification 2: The properties of the immediate predecessors show no direct correlation with the emergence of periodicity. Nonetheless, the properties must be such that they facilitate its formation. Whilst certain cellular and genetic constitutions tend to favor its emergence, others do not.

Qualification 3: The regularity of the recurrence of most of the properties studied thus far is impressive. But since the phenomenon has come to light only recently, the length of the interval cannot be determined with exactitude at present. This is not surprising since the length of the interval already varies in chemical elements.

Qualification 4: The phylogenetic position is not a limiting factor in the emergence of a given form or function. However, limitations on the degree of complexity of the emerging structure are imposed by the organism. A protozoan is not in a position to produce a multicellular organ, but once one reaches the invertebrate stage, the limit to complexity vanishes, at least in many cases.

Qualification 5: The internal and external environments are not necessarily decisive factors in the emergence of periodicity, but they may serve as contributory factors e.g. they may trigger it or canalize it in a specific direction.

The periodic law of the chemical elements was originally enunciated in the following way: "The properties of the elements are a periodic function of their atomic weights" (Greenwood and Earnshaw, 1989). Later the weights were related to the number of protons, neutrons and electrons that form the atom. Thus, the biological periodic rule can be reformulated in a simplified way. The properties of living organisms are a periodic function of the atomic constellations of their major molecules. These are in the last analysis expected to depend on the number of electrons in the outer shell of their key atoms. Thus, although periodicity results from the configuration of all cell molecules, it is expected to be largely derived from that of nucleic acids and proteins.

Chapter 34

A CONSEQUENCE OF PERIODICITY IS THE PREDICTION OF NOVEL BIOLOGICAL TRANSFORMATIONS

SCIENTIFIC PREDICTIONS ARE PROVISIONAL

Predictions in science do not represent visions through a crystal ball. Rather, they are the outcome of the assembly of data that have been obtained following a certain scheme. What then emerges is a coherent picture.

Nothing is predictable in the true sense of the term. One simply uses pieces of information that are already available and only combines them in a new way. The views thus put forward reflect the evidence and the technology available at a given time, and as such can only be considered as provisional guidelines in future research.

At first, it was difficult to establish the periodicity of the chemical elements because there were missing elements. It was this recognition of "black holes" in the system that led to the prediction of the existence of 10 unknown elements. These were discovered afterwards. However, other elements were found that had not been foreseen. It was not possible to predict the existence of the noble gases or the number of lanthanide elements which were recognized subsequently (Greenwood and Earnshaw, 1989). Thus, even in the case of atoms, predictions were limited. There is no question that periodicity at the biological level allows fewer predictions to be made and that there will be more surprises in biological research than can now be anticipated.

THE PRESSING DEMAND FOR GENE MANIPULATION IN MEDICINE, ANIMAL HUSBANDRY AND AGRICULTURE LEADS TO BIOLOGICAL TRANSFORMATIONS

There is a general public demand for more effective treatments of cancer and other diseases, including those that involve serious viral infections. Research grants in pure and applied research are being actively channeled into these areas in an effort to obtain rapid results (Varmus, 1988). A prime example is AIDS research. In addition, the destruction of natural environments is directing the research of multinational companies into animal husbandry and agriculture (Lichtenstein and Draper, 1986; Hiatt et al., 1989).

Gene manipulation and gene therapy have become standard procedures at molecular biology laboratories. Segments of DNA that harbor the desired genes can be targeted to specific tissues or organs and incorporated into the cell's chromosomes, thus leading to a change in the genetic constitution in a predictable way (Palmiter and Brinster, 1986). Biotechnology is being used as a main weapon in the search for the cure or amelioration of many human diseases. It is also being directed to the production of animal and plant species with new combinations of valuable properties (Van der Putten et al., 1985).

By pressing for an urgent solution of these problems, we are at the same time opening the way to a biological revolution. This will lead to the creation of a multibiological society in which the limits between humans, animals, and plants will become difficult to define, simply as a result of the transfer of genes between these groups. Unexpected and serious ethical problems are bound to arise. Although it may seem surprising, we are already in the middle of this revolution (Lima-de-Faria, 1991c; Lee, 1991).

As an example, one may cite the transfer of a specific gene from an invertebrate to a flowering plant. As described in Chapter 6, bioluminescence is displayed in organisms ranging from bacteria to the vertebrates. However, the emission of light is absent from whole classes such as the mammals. No flowering plant is known to exhibit luminescence. However, some insect species are strongly luminous, such as the firefly. The luciferase gene from the insect species, *Photinus pyralis*, was introduced into plants of *Nicotiana tabacum*. Briefly, a DNA clone of the gene was connected with a promoter and a plasmid, thereby enabling it to be carried into tobacco seedlings. The transgenic plants became luminescent and the light emission was organ specific, being localized mainly to the cotyledons (Ow et al., 1986; Millar et al., 1992).

SEALS AND WHALES MAY BE RECREATED BY GENE TECHNOLOGY

Both seals and whales are placental mammals that have a streamlined body and are partly or wholly aquatic. They are not derived from ancestors that lived in water; rather, they originated from animals that were fully terrestrial. Whales are regarded as having originated from the ancestors of the deer (ungulates), while seals from those of the otter and the dog (carnivores). This view is based on evidence coming from various studies, most notably DNA–DNA hybridization (Janvier, 1984; Macdonald, 1984; Arnason and Widegren, 1986).

The return of these species to water was a modification that did not demand many novelties. Mammals at birth go through a period of partial asphyxia. Studies carried out in humans and seals show that the changes in blood flow and the heart rate, resulting from a decrease in the available air, involve the same hormone in both species (Lagerkrantz and Slotkin, 1986). This mechanism allows the human infant to survive a prolonged birth and the seal to remain submerged for periods exceeding 50 minutes. The hydrodynamic form of the seal had also been produced earlier, not only in the fishes, but also in the aquatic reptiles. Hence, the transformation of the terrestrial placentals into aquatic forms may well have been a relatively rapid process, since some of the main structural and functional solutions were already available or had been produced at an early stage of evolution. The first fossil whales are fairly similar to today's living species which indicates that the change was rather abrupt. Thus, one may be in a position to change a deer into a whale, or a dog into a seal, by means of few and relatively rapid gene manipulations (Lima-de-Faria, 1988; 1991b).

MAKING THE MARSUPIALS AQUATIC

The parallel evolution of the placentals and the marsupials is an impressive example of periodicity, yet the equivalence is not quite total. No seals or whales, or comparable animals, exist among the marsupials. This represents a real "black box" in the periodicity chart. One reason could be the timing of birth in these mammals. Marsupials are born at an early embryonic stage and complete their development in the pouch of the mother. They may lack, for this reason, the ability to tolerate a long period of asphyxia, which is a pre-requisite for an aquatic life — a feature uncommon among marsupials.

However, the kangaroos swim well. Moreover, the water opossum (a marsupial) is partly aquatic and has the body features of the placental otter. The opossum closes its pouch under water while the otter closes its ears and nostrils when it dives (Macdonald, 1984). This raises the possibility that a marsupial seal or a marsupial whale might be created by using appropriate gene technology.

MARSUPIALS GLIDE BUT THERE ARE NO MARSUPIAL BATS

There are gliding marsupials like the greater glider *Petauroides volans* and the sugar glider *Petaurus brevicepts*. Both have membranes that extend between the forearms and the hindfeet. They glide as efficiently as placental species, such as the scaly-tailed squirrel or the Philippine colugo. Yet there are no flying species among the marsupials that are comparable to the placental bats. This is another "black box" in the equivalence between the two groups.

The gap between gliding and flying seems at first sight not to be very large. However, flight represents the combination of a series of new structures and functions. It demands the intervention of different bones, new articulations, new muscles, brain coordination and large amounts of energy that must be delivered rapidly. Yet, it is not excluded that more advanced knowledge of gene splicing and exon-intron recombination may provide the means of producing a marsupial bat.

FILLING THE EMPTY BOXES IN THE CHART OF BIOLOGICAL PERIODICITY

Humans and bats have in common DNA sequences, as is known from DNA–DNA hybridization between different mammalian species (McCarthy, 1969; King and Wilson, 1975; Davidson and Britten, 1979; Nagl, 1986; Lima-de-Faria et al., 1986; Loomis, 1988). The wings of bats were not just added to the body of a terrestrial species, for they represent two well defined events. These are: (1) an emergence that was rapid and direct. Intermediate fossils between bats and other mammals are not available, and (2) flight is not regarded as an isolated biological event but rather the outcome of a series of interrelated processes. These include a whole set of structures and functions that are interdependent.

The decision by the National Academy of Sciences of the USA to sequence the DNA of human chromosomes and other mammalian species (Roberts, 1988a and b) has led to an unprecedented international effort to map and sequence genomes. The complete DNA sequence of yeast chromosome III has already been established (Oliver et al., 1992) and success in the sequencing of the human genome is being obtained much earlier than expected. The entire human chromosome 21q has been sequenced (Chumakov et al., 1992). Thus, there are reasons for thinking that when the DNA of bats will be sequenced, it will then be possible to find out which structural genes and regulatory DNA sequences are involved in the production of the wing package.

Artificial yeast chromosomes have been produced in the laboratory, and yeast chromosomes that carry human genes have also been obtained (Murray and Szostak, 1987; Burke et al., 1987). As for the organization of human chromosomes, it has become much better understood in recent years and the production of human artificial chromosomes is the main goal of several laboratories (Lima-de-Faria, 1991a). It may not be an insuperable task to create a human with wings by employing gene manipulation based on sufficient knowledge of the organization of chromosomes in bats and humans. The relationships among bats and primates, including humans, are being studied actively at the molecular level (Sarich, 1993).

Paleontologists have pointed out, time and again, that there are in the fossil record many unsettled questions arising from lack of fossils of organisms that represent transition stages. Another prominent aspect of the fossil record is that all the animal phyla that have hard parts appeared some 600 million years ago in the Cambrian in a sudden way. Few major body plans seem to have emerged since that time. Subsequent evolution has consisted mainly of variations of those basic themes (Levinton, 1992).

As a conclusion, one could state that biotechnology, guided by information relating to periodicity, will help to clarify the problem of the emergence of new groups, as well as of organisms that participated in their transition stages. Such a state of affairs will fill some of the empty boxes that exist in the periodic chart of both fossil and living organisms.

Chapter 35

SUMMARY AND CONCLUSIONS

When chemical elements were aligned after their increasing atomic weight, they turned out to exhibit properties that recurred at relatively regular intervals. It was this unexpected periodicity that led to the establishment of the periodic chart of the elements. This phenomenon took over 100 years to establish. The first indication appeared in 1772 when de Morveau made a table of chemically simple substances. This was followed by a series of descriptions of the recurrence of similar properties among elements that had quite different atomic weights. One had to await until 1871 for Mendeleev to enunciate the periodic law. He then predicted the existence of 6 elements that were missing in his improved atomic tables. These elements were soon discovered and as a result, the ordered periodicity among chemical elements became established.

However, a mechanism that explained the phenomenon was not available. Again, one had to wait until 1913 and 1922 when Bohr described the groups of atoms in the periodic chart on the basis of his theory of atomic structure. It is the number of electrons present in the outer shell of the atom that is mainly responsible for the similarity of the properties. As more information became available the atomic weight was substituted by the atomic number. The periodicity is especially evident in: helium (He), neon (Ne), argon (Ar), krypton (Kr), xenon (Xe), and radon (Rn) that have the atomic numbers: 2, 10, 18, 36, 54, and 86 respectively, yet they all have the same properties, i.e., they are inert gases. This is so in spite of the fact that helium is one of the simplest atoms and radon one of the most complex. The central finding is that irrespective of the degree of complexity of the atoms, the same properties reappear. The length of the period that marks the repetition of a certain property may vary from 2 to 32 chemical elements. Chemical periodicity has turned out to be a most fruitful finding that allows us to explain many of the phenomena encountered today in chemistry. At present, exceptions to the Periodic Table still exist that cannot be easily explained. The most regular phenomena contain deviating components; otherwise, no evolution would have occurred.

The periodicity of the elements has determined the periodicity of properties in minerals. The electrons governed not only the behavior of atoms but also how they were packed into minerals. This led to the transfer of the periodicity to the next level of organization. The structural periodicity is evidenced by the crystallization in the same system of minerals with unrelated chemical composition, i.e., the same pattern re-emerges in different mineral classes. Moreover, the same mineral proto-function, such as luminescence, is also present in minerals with different chemical compositions. Thus, a primary form of structural and functional periodicity is already evident at this intermediate level of organization.

Since the cell was formed by the same atoms that are present in the minerals, and continues to receive many of them from this source, it is not surprising that periodicity is present at the biological level. The cell has no genetic code for iron, magnesium, zinc, or cobalt. Yet these elements are an integral part of many macromolecules active in the cell: iron in hemoglobin, magnesium in chlorophyll, zinc in gene regulatory proteins, and cobalt in vitamin B_{12}. It is not the long chains of amino acids, or of other molecules that are involved; rather it is these single atoms, originating from the minerals, that are responsible for the main functions of these molecules. It is also noteworthy that the individual atoms present in the cell retain the properties they had both in the minerals and as primeval elements.

Biological periodicity is characterized by an equally unexpected recurrence of groups of properties that emerge at relatively regular intervals, as the biological complexity increases throughout the evolutionary scale. An example is flight. This function originates in the insects, does not show up in any other invertebrate group, reappears in the fossil reptiles, later becomes well established in birds; it occurs in certain fishes, and finally reappears suddenly in the bats but not in any other mammalian order. Periodicity elucidates the sudden emergence of functions that had been absent from many species for millions of years of evolution. Organisms belonging to remote groups of invertebrates happen to have the same function found in the most advanced mammals, such as humans, yet in between, there are many species that do not possess the corresponding properties.

In living organisms, as in the case of chemical elements, it is not just one property but a whole group of several integrated properties that are repeated. Flight is not the sole product of the formation of a wing but emerges as the result of a series of functions and structures that are highly coordinated. This is put in evidence by comparing the flight of insects with that of birds. They are strikingly similar although they involve different tissues and organs. Some of the common features are: (1) Flying animals have a streamlined body. (2) Feathers in birds and bristles in insects smoothen the exterior of the body. (3) Wings are located above the center of gravity in both cases. (4) Bones in bird wings are air filled and the skeletal parts in wings of insects contain air tubes. (5) Particular muscles are involved in the movement of both types of wings. (6) Articulations are also present in the two groups; otherwise, the wings could not fold. (7) High amounts of energy derived from glycogen are used by both insects and birds. (8) Air sacs exist in birds and insects, thereby allowing continuous air flow. (9) The speed of flight is detected through nostrils in birds and in insects by hairs in the head and by the antennae. (10) Birds and insects use highly complex types of flight. (11) The nervous system plays a major role in the flight process through: vision, balance and muscle movements. This list could be extended. In the flying reptiles, flying fishes and bats similar solutions were produced which have led to the reemergence of this function.

Several other functions show the same periodic occurrence and also involve packages of integrated tissues and organs. These are: (1) The formation of complex organs of vision in invertebrates, which are found in organisms as simple as the cubomedusae and get their highest expression in the cephalopods. (2) The formation of the placenta, which already appears among invertebrates, such as the onychophorans, and which reappears in fish, amphibians, and reptiles to finally establish itself in the placental mammals. (3) The existence of bioluminescence from bacteria to high eukaryotes, which is present in many groups but totally absent in others. (4) The function of the penis that is as highly developed in early invertebrates and other groups as it is in humans. (5) The features associated with the return to aquatic life in plants, invertebrates and mammals. (6) The occurrence among placental mammals of a series of structural and physiological "carbon copies" of the marsupial species. And (7) The emergence of high mental ability in distant groups.

The reemergence of these functions throughout the evolutionary process has several features in common that led to the establishment of biological periodicity. These are: (1) The function is not necessarily exhibited by the immediate predecessors or by closely related species, as is the case in the molluscs in which there are species with very simple and highly complex eyes. (2) The emergence occurs with few intermediate forms or suddenly (e.g., the emergence of flight in bats). (3) The group of properties displayed is essentially the same even in the most distantly related groups (e.g., return to water in invertebrates and mammals). (4) It tends to appear in a well defined and restricted group (e.g., bioluminescence in fishes but not in other vertebrates). (5) It occurs at relatively regular intervals, but the length of the interval varies as it already does in the chemical elements. (6) The degree of complexity that a given function exhibits is not necessarily related to the degree of complexity of the organism (e.g., insects have two and four wings, whereas birds have only two). And (7) There is no obligatory relationship with the general environment; in some cases, the environment may trigger the recurrence of a function (as is the case of an aquatic medium) but in others, it represents an adverse situation (e.g., fish that fly in air). (8) The same organ may be formed from quite different tissues (e.g., the placenta has its origin in unrelated cell types). (9) In some of these functions, it has been established that the emergence is decided by a cascade of hormones. (10) Some of these functions not only go back to the simplest living organisms, such as the bacteria, but are already present in minerals (e.g., luminescence). (11) No change in genetic constitution is necessary to produce a novel structure or function; this occurs, for instance, when different parts of the same plant show hydrodynamic shapes and functional modifications; and when bilateral symmetry is transformed into radial symmetry within the same individual sea urchin. (12) The reemergence of a given function occurred within the same group not only once, but twice and even three times (e.g., gliding marsupials). (13) Both simple and complex biological functions display periodicity: luminescence is one of the simplest while mental ability is one of the most complex.

The wing of an insect and that of a bird have been considered by zoologists to be non-homologous. The reason being that the two structures were so different that they seemed to have no relationship. The genetic analysis of the embryonic development in *Drosophila* and the chicken, has shown that the same type of genes determine the wing structure of the fly and of the bird. Moreover, the molecular analysis of these genes discloses that they have maintained the same position along the chromosome in invertebrates and vertebrates. Hence, the analogy became a homology since it is the product of the ordered expression of similar DNA sequences.

The same turned out to be the case in what concerns the eyes of invertebrates and those of humans. No relationship was believed to exist, at the genetic level, between the compound eye of an insect, with its thousands of facets, and the single spherical structure that builds the human eye. Yet the recent genetic analysis in *Drosophila*, worms, cephalopods, mice, and humans, reveals that also in this case the same genes are responsible for the formation of the eye in all metazoa. The DNA sequences of these genes, as well as their proteins, have been analyzed, establishing the basic similarity of eye formation from worms to humans.

Function cannot be separated from structure, but neither can structure be separated from function. Like matter and energy they are dual aspects of the same phenomenon. Yet, for the sake of simplicity, one treats them separately. Periodicity is still more evident in the patterns found in plants and animals that turn out to be an extension of those present earlier in crystals. The different types of symmetries, and the arrangement of components, were transferred intact to the biological level. Until recently, these striking similarities between the two levels of organization were considered to be accidental and without evolutionary significance. Although the mineral and biological patterns were described long ago, as following related mathematical transformations, these results were disposed of as irrelevant. There was no material basis that could be pinpointed as being responsible for the mechanism behind the similarity. However, at present we face a different situation. The reemergence of the patterns is now intelligible in atomic terms as a result of the discovery of molecular mimicry in minerals and in the cell's most important macromolecules. At both levels, molecules with quite different atomic compositions result in the same pattern or even the same function, i.e., they mimic one another. One example of this are the 15 proteins with completely different amino acid sequences. All form the eight-stranded alpha-beta-barrel structure, a most regular protein feature. Another example is furnished by the proteins of the staphylococcal nuclease group that were found to have different amino acid sequences, yet perform the same function. The molecular mimicry turns out to be the result of key atoms or of specific groups of atoms, both in the case of small and large molecules, that compel the same structure or function to reemerge. This means that organisms with different genetic constitutions can suddenly display the same solutions, and that very different molecules such as carbonates, cellulose, chitin and proteins can result in the emergence of similar patterns. This is especially evident in the symmetries that were transferred from the minerals to the living organisms, as shown by the following examples.

Bilateral symmetry occurs in: (1) the mineral argiorite, (2) the flower of *Orchis Morio*, (3) the leaf of *Ginkgo biloba*, (4) the fruit of *Acer pseudoplatanus*, (5) the larva of starfishes (class Asteroidea), and (6) the body of vertebrates including humans. The three-fold symmetry is present in: (1) the associations of rutile crystals, (2) the flower of *Alisma plantago-aquatica*, (3) the leaf of *Trifolium repens*, (4) the fruit of *Begonia*, and (5) the pincer-like structure of *Echinus esculentus* (echinoderms). Four-fold symmetry is found in: (1) the mineral staurolite, (2) the flower of *Ilex aquifolium*, (3) the leaf of *Marsilea quadrifolia*, (4) the fruit of *Rhamnus cathartica*, (5) the head of the worm *Taenia saginata*. Five-fold symmetry appears in: (1) the association of five crystals in marcasite, (2) the flower of *Cynanchum Vincetoxicum*, (3) the feathery leaves of the fruit of *Nigella damascena*, (4) the fruit of *Psidium Guajava*, and (5) the sea urchin *Archaeocidaris wortheni*. The six-fold symmetry may be observed in: (1) the crystals of arsenopyrite, (2) the flower of *Scilla autumnalis*, (3) the fruits of *Damasonium alisma*, (4) the tentacles of *Hydra*, (5) the arms of *Leptasterias hexactis* (starfish).

The biological and mineral structures not only have the same type of symmetry but also consist of components that are associated following similar rules: (1) In most cases the different units irradiate from a central region. (2) They are usually of equal length. (3) Tend to have the same shape. (4) Tend to have the same color. (5) The increase in the number of units leads to a rearrangement that maintains the same average distance between all units. (6) What is significant is that intermediate solutions between the different symmetries seem to be excluded, which is a further indication of strong mineral canalization. (7) At every level of organization there is an optimal number of units that combine: six in silicon atoms, six in water crystals, five in flower petals, five in the arms of sea stars, six in insect legs, eight in spider legs, eight in the tentacles of octocorals, and four in vertebrate limbs. It may be noted that there are more insect species (over one million) than all the other living organisms combined, and that insects have had at their disposal 425 million years to evolve, yet they did not depart from the six leg pattern, just like water crystals are locked in a six-ray configuration. These data reinforce the notion that structural periodicity is the result of the transfer of the same type of atomic order to successive levels of organization.

Recently the body pattern of humans and that of flowers has been found to be determined by the same type of genes. These are the homeotic genes which have been analyzed in detail at the molecular level and which have turned out to be conserved from invertebrates to humans as well as in the flowering plants. These studies are bringing us closer to an understanding of the pure atomic processes that appear to be responsible for the emergence of symmetries in living organisms. The gene is only the bearer, and the carrier, of the atomic order that already determined mineral symmetries. Moreover, due to the occurrence of molecular mimicry an organism may not even need to have the same genes to produce the same structural pattern.

Another aspect to be considered is the role of embryonic development in periodicity. This is revealed by the fact that no sharp demarcation can be established between development and evolution in many biological situations. An example of this is *Trichoplax*, an organism that was considered to be a larva of a Coelenterate until it was discovered that it actually represented an adult of a separate phylum. The onset of reproduction is not the sole component disclosing the artificial character of this demarcation. More important is that there is no distinction at the molecular level between the two processes. The same molecular machinery of replication, transcription, translation and molecular recognition is used in development and evolution. The cell could have created separate compartments, and separate molecules since it has a plethora of RNAs and an array of specialized organelles separated by membranes. Yet all of them participate in both processes. What is called metamorphosis is evolution occurring within the individual. The same type of larva, such as the trochophore, can actually give rise to four different phyla: Annelida, Echiura, Sipuncula, and Mollusca. Development may contribute to the establishment of periodicity by the introduction of reproduction at comparable stages of morphogenesis.

Not only molecular mimicry and morphogenesis are mechanisms to be taken into account, but other processes at the chromosome level must also be considered, if the mechanisms involved in periodicity are to be elucidated. Molecular cytogenetics and molecular biology have advanced so rapidly in the last decade that genetic phenomena that were difficult to conceive were suddenly disclosed.

Human proteins can be divided into three classes, viz. ancient, middle-age and modern. Among the last ones there is a class of mosaic proteins that is of special interest. These proteins are viewed as the result of intron-exon recombination, i.e. they are formed by the rearrangement of the non-coding and coding DNA sequences that can change in number and position in split genes. An example is the low-density lipoprotein (LDL)-receptor. Detailed analysis of its 18 exons has shown that it results from the association of exons that have come from other split genes. This means that different types of proteins can be recreated by repeating the initial intron-exon combination.

Another mechanism participating in protein evolution is alternative RNA splicing. This type of rearrangement does not alter the DNA content of the cell. Rather, it involves only modifications of the messenger RNA. One such case is the myosin heavy chain gene; another is the sarcomeric contractile protein gene in which more than 30 different exons, or pairs of exons, are known to be alternatively spliced. Hence, organisms do not need to change their genetic constitution to produce different proteins.

In African trypanosomes, which are unicellular parasites, the messenger RNA of a cell surface glycoprotein gene, consists of information from two DNA components located in different chromosomes. Through intron donors and acceptors, and exon-intron coupling, new compound messenger RNAs are formed that become translated into proteins. This process is determined internally, independently of the environment to which the protozoan is exposed, the rate of transformation in *Trypanosoma brucei* being about 10^{-6} to 10^{-7} per cell division.

Another mechanism that needs to be clarified is the one involving the reemergence of integrated packages of structures and functions found in several organs. The cascades of molecules and of gene activations provide now a preliminary explanation of such biological processes. Hormones e.g. epinephrine and glucagon activate the liver cell to produce glucose. But this chemical is the final product of six intermediate reactions that follow each other in a sequence that cannot be altered. In this cascade every molecule has its exact place and most steps involve enzyme catalysis. The cascade also results in amplification of the initial molecular signal. One hormone molecule can lead to 100 molecules of glucose. In blood clotting 10 proteins are produced in a strictly canalized sequence. But the best example comes from melanocorticotropin, the polypeptide produced by the anterior pituitary gland situated in the brain. The polypeptide is cleaved into six hormones that are amino acid sequences, the smallest consisting only of five peptides. The cleavages occur at specific sites along the molecules. What is significant for the understanding of periodicity is that the hormones act on very different organs: (1) They activate the adrenal cortex adjacent to the kidneys. (2) Act on lipid metabolism. (3) Affect the cells of the skin. And (4) Influence brain function in a way similar to that of morphine. It is thus not surprising that the over-all process is integrated and coherent, since it has its origin in a single molecule that by means of a rigid hormone cascade affects the function of four different organs. Cascades of gene action have also been studied extensively. The expression of genes participating in body segmentation and cell differentiation is regulated in a hierarchical and sequential way. This adds extra order to the distribution in the body of structures and functions. Taken together these results make it easier to understand the periodic reemergence of phenomena, such as flight, which results from the coordinated interaction of wing frameworks, blood vessels, muscles, glycogen derived energy and brain control.

Every component of the cell is built from within as a result of self-assembly. Elementary particles have the ability to organize themselves spontaneously. The assembly of atoms and macromolecules is also intrinsic and spontaneous. Viruses self-assemble from the combination of their nucleic acids and proteins, producing infectious particles. Most cell organelles self-assemble from their molecular components e.g. ribosomes, nucleosomes and membranes. Whole organisms self-assemble from their dispersed cells. Examples are hydras, sponges and slime molds. Human skin has been produced by allowing separate cells to self-assemble. Moreover, there is increasing evidence that indicates that the evolution of proteins and of nucleic acids is dictated by their atomic configurations.

What at first sight seems surprising is that biological periodicity has followed certain features that characterize periodicity at the chemical level. The enormous gap between the two levels has not erased the initial phenomenon. The biological and chemical periodicity have not less than 10 features in common. The reason for this is actually simple: one is dealing at the cellular level, with the same atoms that were present at the dawn of the formation of matter.

A significant finding is that nucleic acids and proteins consist of atoms which are not located at random in the Periodic Table. DNA and RNA are built only by: hydrogen, oxygen, carbon, nitrogen and phosphorus (H, O, C, N, and P). The polypeptides of proteins consist only of H, O, C, N, and S (sulfur). These atoms have the following characteristics: (1) They are few. (2) Four of them are the same in nucleic acids and proteins. (3) The six atoms are not spread over the Periodic Table, but are located only on its right side. (4) They are all non-metals. (5) They actually occupy a "niche" in the periodic system. These properties reveal how the periodicity of the atomic level has canalized the key molecules that direct the cell level. Thus, the biological periodicity is not an accidental phenomenon, but is anchored to the order which already determined the atomic periodicity.

The length of the period at the biological level is difficult to determine at present. This is due to the vast ignorance of the molecular mechanisms responsible for many biological phenomena, such as embryonic development and chromosome organization. Not even the chromosome number is known in whole groups of animals. A series of factors may be involved in the determination of the period length. These may be: atomic, genetic and environmental. The length of the period already varies at the simple atomic level. As a higher complexity is reached in living organisms, more variation is expected to occur, but not disorder. This is because the same atoms participate in the building of both levels, and as a consequence, they do not allow any other basic solutions than those resulting from their electronic configurations.

The information presented above enables the formulation of a general rule of biological periodicity: An integrated package of forms and functions arises in an organism independently of the main properties of its immediate predecessors, recurs at relatively regular intervals, irrespective of the organism's phylogenetic position, and is not necessarily directly related to the major factors prevailing in the environment.

Because of the occurrence of molecular mimicry, organisms with different genetic constitutions are able to suddenly display the same solutions. The emergence of biological periodicity is not necessarily dependent on the organisms having the same genes but rests mainly on their ability to produce molecules with the same type of key atoms.

The establishment of the concept of periodicity leads to the prediction of novel biological transformations, and to the planning of experiments that can be based on the existence of molecular order in evolution.

REFERENCES

Abercrombie, M., Hickman, C.J., & Johnson, M.L. (1951). A Dictionary of Biology. Penguin Books, Harmondsworth, Middlesex.

Affolter, M., Schier, A., & Gehring, W.J. (1990). Homeodomain proteins and the regulation of gene expression. Curr. Opin. Cell Biol. 2, 485–495.

Alberts, B., Barry, J., Bittner, M., Davies, M., & Hama-Inaba, H., Liu, C.C., Mace, D., Moran, L., Morris, C.F., Piperno, J., & Sinha, N.K. (1977). *In vitro* DNA replication catalyzed by six purified T4 bacteriophage proteins. In: Nucleic Acid-Protein Recognition (Vogel, H.J., Ed.), pp. 31–63. Academic Press, New York, San Francisco, London.

Alberts, B., Bray, D., Lewis, J., Raff, M., Roberts, K., & Watson, J.D. (1983). Molecular Biology of the Cell. Garland Publishing, Inc., New York.

Alberts, B., Bray, D., Lewis, J., Raff, M., Roberts, K., & Watson, J.D. (1989). Molecular Biology of The Cell, 2nd ed. Garland Publishing, Inc., New York & London.

Allen, O.E. (1984). Planeten Jorden—Atmosfären. Bokorama.

Alt, F.W., Blackwell, T.K., & Yancopoulos, G.D. (1987). Development of the primary antibody repertoire. Science 238, 1079–1087.

Amaldi, G. (1966). The Nature of Matter. The University of Chicago Press, Chicago, London.

Anderson, K.V., Bokla, L., & Nüsslein-Volhard, C. (1985). Establishment of dorsal-ventral polarity in the *Drosophila* embryo: the induction of polarity by the *Toll* gene product. Cell 42, 791–798.

Andrèsen, N. (1942). Cytoplasmic components in the amoeba *Chaos chaos* Linné. Compt. Rend. Trav. Lab. Carlsberg, Sér. Chim. 24, 138–184.

Angenent, G.C., Busscher, M., Franken, J., Mol, J.N.M., & van Tunen, A.J. (1992). Differential expression of two MADS box genes in wild-type and mutant Petunia flowers. The Plant Cell 4, 983–993.

Antoine, M. & Niessing, J. (1984). Intronless globin genes in the insect *Chironomus thummi thummi*. Nature 310, 795–798.

Arnason, U. (1985). Valen—mera hjort än lejon? Forskning och Framsteg 1985(6), 16–21.

Arnason, U. & Widegren, B. (1986). Pinniped phylogeny enlightened by molecular hybridizations using highly repetitive DNA. Mol. Biol. Evol. 3(4), 356–365.

Arnott, S., Chandrasekaran, R., Day, A.W., Puigjaner, L.C., & Watts, L. (1981). Double helical structures for polyxanthylic acid. J. Mol. Biol. 149, 489–505.

Arnott, S., Chandrasekaran, R., Hukins, D.W.L., Smith, P.J.C., & Watts, L. (1974). Structural details of a double-helix observed for DNA's containing alternating purine-pyrimidine sequences. J. Mol. Biol. 88, 523–533.

Arnott, S. & Hukins, D.W.L. (1972). Optimised parameters for A-DNA and B-DNA. Biochem. Biophys. Res. Commun. 47, 1504–1510.

Arnott, S. & Hukins, D.W.L. (1973). Refinement of the structure of B-DNA and implications for the analysis of X-ray diffraction data from fibers of biopolymers. J. Mol. Biol. 81, 93–105.

Aron, M. & Grassé, P. (1939). Précis de Biologie Animale. Masson et Cie, Editeurs, Paris.

Asimov, I. (1992). Atom: Journey Across the Subatomic Cosmos. Truman Talley Books, Plume, New York.

Atkins, P.W. (1987). Molecules. Scientific American Library, New York.

Attardi, B. & Ohno, S. (1978). Physical properties of androgen receptors in brain cytosol from normal and testicular feminized (Tfm/y♂) mice. Endocrinology 103, 760–770.

Auber, J. (1969). La myofibrillogenèse du muscle strié. I. Insects. J. de Microsc. 8(2), 197–232.

Azorin, F. & Rich, A. (1985). Isolation of Z-DNA binding proteins from SV40 minichromosomes: Evidence for binding to the viral control region. Cell 41, 365–374.

Babin, C. (1980). Elements of Palaeontology. John Wiley & Sons, Chichester, New York, Brisbane, Toronto.

Ballhausen, C.J. (1962). Introduction to Ligand Field Theory. McGraw-Hill, New York.

Barnes, R.D. (1980). Invertebrate Zoology, 4th ed. Saunders College, Philadelphia.

Bateson, W. (1894). Materials for the Study of Variation. University Press, Cambridge.

Baur, E. (1930). Einführung in die Vererbungslehre. Verlag von Gebrüder Borntraeger, Berlin.

Beatty, R.A. (1967). Parthenogenesis in vertebrates. In: Fertilization, Vol. I (Metz, C.B. & Monroy, A., Eds.), pp. 413–440. Academic Press, New York.

Beaver, H.H. (1967). Morphology. In: Treatise on Invertebrate Paleontology. Part S, Echinodermata I, Vol. II (Moore, R.C., Ed.). Courtesy of the Geological Society of America and University of Kansas.

Beazley, M. (Ed.) (1974). The World Atlas of Birds. Mitchell Beazley Publishers Limited, London.

Beazley, M. (1980). The Atlas of World Wildlife. Rand McNally & Company, The Netherlands.

Beçak, W., Beçak, M.L., & Ruiz, I.R.G. (1978). Gene regulation in polyploid amphibians. In: XIV International Congress of Genetics, Contributed Paper Sessions, Abstracts, Part II, Moscow 1978, p. 147. Nauka, Moscow.

Becker, R.D. (1974). The significance of bioelectric potentials. Bioelectrochemistry and Bioenergetics 1, 187–199.

Bell, L.R., Maine, E.M., Schedl, P., & Cline, T.W. (1988). Sex-lethal, a drosophila sex determination switch gene, inhibits sex-specific RNA splicing and sequence similarity to RNA binding proteins. Cell 55, 1037–1046.

Bell, P.R. & Woodcock, C.L.F. (1971). The Diversity of Green Plants, 2nd ed. Edward Arnold, London.

Bender, W., Akam, M., Karch, F., Beachy, P.A., Peifer, M., Spierer, P., Lewis, E.B., & Hogness, D.S. (1983). Molecular genetics of the bithorax complex in *Drosophila melanogaster*. Science 221, 23–29.

Benfy, T. (1964). Spiral periodic chart. In: Chemistry (Benfy, T., Ed.), p. 14. Amer. Chem. Soc., Washington, D.C., 37(6), 14.

Benne, R. & Van der Spek, H. (1992). L'Editing des messages génétiques. La Recherche—245 Juillet-Août 23, 846–854.

Bentley, W.A. & Humphreys, W.J. (1962). Snow Crystals. Dover Publications, Inc., New York.

Benzer, S. (1955). Fine structure of a genetic region in bacteriophage. Proc. Natl. Acad. Sci. USA 41, 344–354.

Benzer, S. & Freese, E. (1958). Induction of specific mutations with 5-bromouracil. Proc. Natl. Acad. Sci. USA 44, 112–119.

Berg, J.M. (1988). Proposed structure for the zinc-binding domains from transcription factor IIIA and related proteins. Proc. Natl. Acad. Sci. USA 85, 99–102.

Bergeron, R.J. (1986). Iron: a controlling nutrient in proliferating processes. Trends Biochem. Sci. 11, 133–136.

Berleth, T., Burri, M., Thoma, G., Bopp, D., Richstein, S., Frigerio, G., Noll, M., & Nüsslein-Volhard, C. (1988). The role of localization of *bicoid* RNA in organizing the anterior pattern of the *Drosophila* embryo. EMBO J. 7, 1749–1756.

Bernal, J.D. (1965). Molecular matrices for living systems. In: The Origins of Prebiological Systems (Fox, S.W., Ed.), pp. 65–88. Academic Press, New York and London.

Bernardi, G. & Bernardi, G. (1986). Compositional constraints and genome evolution. J. Mol. Evol. 24, 1–11.

Berrill, N.J. & Karp, G. (1976). Development. McGraw-Hill, New York.

Bienfait, H.F. (1985). Regulated redox processes at the plasmalemma of plant root cells and their function in iron uptake. J. Bioenerg. Biomembr. 17, 73–83.

Björkman, O. (1992). Protective chloroplast movement in high light. In: Carnegie Institution of Washington. Year Book 91, The President's Report, July 1991–June 1992. p. 119, Baltimore.

Blake, C.C.F. (1978). Do genes-in-pieces imply proteins-in-pieces? Nature 273, 267.

Blake, C.C.F., Harlos, K., & Holland, S.K. (1987). Exon and domain evolution in the proenzymes of blood coagulation and fibrinolysis. Cold Spring Harbor Symposia on Quantitative Biology, Vol. LII. Evolution of Catalytic Function, pp. 925–931. Cold Spring Harbor Laboratory.

Blaringhem, L. (1923). Pasteur et le Transformisme. Masson et Cie, Editeurs, Libraires de l'Académie de Médecine, Paris.

Bloss, F.D. (1971). Crystallography and Crystal Chemistry. An Introduction. Holt, Rinehart and Winston, Inc., New York.

Blumenthal, A.B., Kriegstein, H.J., & Hogness, D.S. (1973). The units of DNA replication in *Drosophila melanogaster* chromosomes. Cold Spring Harbor Symp. Quant. Biol. 38, 205–223.

Boggs, R.T., Gregor, P., Idriss, S., Belote, J.M., & McKeown, M. (1987). Regulation of sexual differentiation in *D. melanogaster* via alternative splicing of RNA from the transformer gene. Cell 50, 739–747.

Böhmer, V. (1974). Skön Motion. Lunds Stadsbibliotek, Wahlström & Widstrand.

Bohn, G. (1934a). Leçons de Zoologie et Biologie Générale. III, Les Invertébrés (Coelentérés et Vers). Actualités Scientifiques et Industrielles 133. Hermann et Cie, Editeurs, Paris.

Bohn, G. (1934b). Leçons de Zoologie et Biologie Générale. VI. Vertébrés Inférieurs (Poissons, Batraciens, Reptiles). Actualités Scientifiques et Industrielles 183. Hermann et Cie, Editeurs, Paris.

Bohn, G. (1935). Leçons de Zoologie et Biologie Générale. IV. Les Invertébrés. Arthropodes, Mollusques et Echinodermes. Actualités Scientifiques et Industrielles 242. Hermann et Cie, Paris.

Bohr, V.A. & Wasserman, K. (1988). DNA repair at the level of the gene. Trends Biochem. Sci. 13, 429–432.

Bolk, L. (1926). On the problem of anthropogenesis. Proc. Section Sci. Kon. Akad. Wetens. Amsterdam 29, 465–475.

Bonnardel, R., Bull, L., Combes, R., Fessard, A., Laugier, H., Magne, H., Piéron, H., Plantefol, L., Rabaud, E., & Mlle Veil, C. (1938). La Biologie. Palais de la Découverte, Masson et Cie, Paris.

Bonner, J.T. (1952). Morphogenesis. An Essay on Development. Princeton University Press, Princeton, NJ.

Bonner, J.T. (1983). Chemical signals of social amoebae. Sci. Am. 248(4), 106–112.

Borror, D.J., DeLong, D.M., & Triplehorn, C.A. (1976). An Introduction to the Study of Insects. 4th Ed. Holt, Rinehart and Winston, New York.

Borst, P. (1986). Discontinuous transcription and antigenic variation in trypanosomes. Ann. Rev. Biochem. 55, 701–732.

Bostock, C.J. & Sumner, A.T. (1978). The Eukaryotic Chromosome. North-Holland Publishing Company, Amsterdam.

Bothwell, M. & Schachman, H.K. (1974). Pathways of assembly of aspartate transcarbamoylase from catalytic and regulatory subunits. Proc. Natl. Acad. Sci. USA 71, 3221–3225.

Boule, M. & Piveteau, J. (1935). Les Fossiles. Eléments de Paléontologie. Masson & Cie, Editeurs, Paris.
Boulenger, E.G. (1937). World Natural History. B.T. Batsford Ltd. London.
Boutaric, A. (1938). Précis de Physique d'après les Théories Modernes. G. Doin & Cie, Paris.
Bowman, J.L., Sakai, H., Jack, T., Weigel, D., Mayer, U., & Meyerowitz, E.M. (1992). SUPERMAN, a regulator of floral homeotic genes in *Arabidopsis*. Development 114, 599–615.
Braam, J. & Davis, R.W. (1990). Rain-, wind-, and touch-induced expression of calmodulin and calmodulin-related genes in Arabidopsis. Cell 60, 357–364.
Brachet, J. (1957). Biochemical Cytology. Academic Press Inc., New York.
Brack, C., Hirama, M., Lenhard-Schuller, R., & Tonegawa, S. (1978). A complete immunoglobulin gene is created by somatic recombination. Cell 15, 1–14.
Bramwell, M. (Ed.) (1974). The World Atlas of Birds. Mitchell Beazley Publishers Limited, London.
Brand, A.H., Micklem, G., & Nasmyth, K. (1987). A yeast silencer contains sequences that can promote autonomous plasmid replication and transcriptional activation. Cell 51, 709.
Branden, C. & Tooze, J. (1991). Introduction to Protein Structure. Garland Publishing, Inc., New York and London.
Brash, D.E. & Haseltine, W.A. (1982). UV-induced mutation hotspots occur at DNA damage hotspots. Nature 298, 189–192.
Breitbart, R.E., Andreadis, A., & Nadal-Ginard, B. (1987). Alternative splicing: A ubiquitous mechanism for the generation of multiple protein isoforms from single genes. Annu. Rev. Biochem. 56, 467–495.
Britten, R.J. & Kohne, D.E. (1969). Repetition of nucleotide sequences in chromosomal DNA. In: Handbook of Molecular Cytology (Lima-de-Faria, A., Ed.), pp. 21–36. North Holland Publishing Company, Amsterdam, London.
Brønstedt, H.V. (1970). Planarian Regeneration. Pergamon Press, Elmsford, NY.
Brumpt, E. (1936). Précis de Parasitologie. II. Collection de Précis Médicaux, Masson & Cie.
Buchsbaum, R. (1951). Animals without Backbones. Vol. I. Penguin Books, Harmondsworth, Middlesex.
Bullock, T.H. & Horridge, G.A. (1965). Structure and Function in the Nervous Systems of Invertebrates, Vols. I and II. Freeman and Company, San Francisco.
Burke, D.T., Carle, G.F., & Olson, M.V. (1987). Cloning of large segments of exogenous DNA into yeast by means of artificial chromosome vectors. Science 236, 806–812.
Burnie, D. (1988). Träd. Fakta i Närbild. Bonniers Junior Förlag AB, Stockholm.
Burns, R.G. (1970). Mineralogical Applications of Crystal Field Theory. Cambridge University Press, London.
Burton, R. (1987). Egg. Nature's Miracle of Packaging. William Collins, London.
Bynum, W.F., Browne, E.J., & Porter, R. (1981). Dictionary of the History of Science. (Bynum, W.F., Browne, E.J., & Porter, R., Eds.), Macmillan, London.
Cabrera, A. (1935). Historia Natural. Vol. I. Zoología (Vertebrados). Instituto Gallach, Barcelona.
Cabrera, A. (1937). Historia Natural, Geologia, Vol. 4. Instituto Gallach, Barcelona.
Cairns, J., Overbaugh, J., & Miller, S. (1988). The origin of mutants. Nature 335, 142–145.
Calvin, M. (1983). The path of carbon: From stratosphere to cell. In: Advances in Gene Technology: Molecular Genetics of Plants and Animals (Downey, K., Voellmy, R.W., Ahmad, F., & Schultz, J., Eds.), pp. 1–35. Miami Winter Symposia, Vol. 20. Academic Press, New York.
Campbell, A. (1971). Genetic Structure. In: The Bacteriophage Lambda. (Hershey, A.D., Ed.), pp. 13–44. Cold Spring Harbor Laboratory, New York.
Campbell, A.K. (1988). Chemiluminescence. Principles andApplications in Biology and Medicine. Ellis Horwood, VCH.
Cantor, C.R. (1981). DNA choreography. Cell 25, 293–295.
Cantor, C.R. & Schimmel, P.R. (1980). Biophysical Chemistry. Part III: The Behavior of Biological Macromolecules. Freeman, San Francisco.

Carpenter, F.M. (1973). Geological history and evolution of insects. In: Syllabus. Introductory Entomology (Tipton, V.J., Ed.), pp. 78–88. Brigham Young University Press, Provo.

Carroll, R.L. (1987). Vertebrate Paleontology and Evolution. W.H. Freeman and Company, New York.

Catterall, W.A., Seagar, M.J., Takahashi, M., & Nunoki, K. (1989). Molecular properties of dihydropyridine-sensitive calcium channels. In: Calcium Channels: Structure and Function. Annals of the New York Academy of Sciences, Vol. 560 (Wray, D.W., Norman, R.I., & Hess, P., Eds.), pp. 1–14. The New York Academy of Sciences, New York.

Challis, J.R.G. (1979). Prostaglandins. In: Mechanisms of Hormone Action, Reproduction in Mammals, Book 7 (Austin, C.R. & Short, R.V., Eds.), pp. 81–116. Cambridge University Press, Cambridge.

Chambon, P. (1981). Split genes. Sci. Am. 244(5), 48–59.

Champoux, J.J. & Dulbecco, R. (1972). An activity from mammalian cells that untwists superhelical DNA—a possible swivel for DNA replication. Proc. Natl. Acad. Sci. USA 69, 139–146.

Changeux, J.-P. (1985). Neuronal Man. The Biology of Mind. Oxford University Press, New York, Oxford.

Chistyakov, I.G. & Pikin, S.A. (1988). Liquid crystals. In: Modern Crystallography IV. Physical Properties of Crystals (Shuvalov, L.A., Ed.), pp. 513–578. Springer-Verlag, Berlin, Heidelberg, New York.

Christiano, A.M., Hoffman, G.G., Chung-Honet, L.C., Lee, S., Cheng, W., Uitto, J., & Greenspan, D.S. (1994). Structural organization of the human type VII collagen gene (COL7A1), composed of more exons than any previously characterized gene. Genomics 21, 169–179.

Chumakov, I., Rigault, P., Guillou, S., et al. (1992). Continuum of overlapping clones spanning the entire human chromosome 21q. Nature 359, 380–387.

Cluzet, J. & Ponthus, P. (1939). Précis de Physique Médicale. G. Doin & Cie, Paris.

Coen, E.S., Romero, J.M., Doyle, S., Elliott, R., Murphy, G., & Carpenter, R. (1990). *Floricaula*: A homeotic gene required for flower development in *Antirrhinum majus*. Cell 63, 1311–1322.

Colbert, E.H. (1980). Evolution of the Vertebrates. A History of the Backboned Animals through Time. 3rd ed. John Wiley & Sons, New York.

Combes, R. (1938). La biologie végétale. In: La Biologie (Palais de la Découverte, Mason & Cie, M., Eds.), pp. 9–22. Paris.

Cormier, M.J. (1974). Flashing sea pansies and glowing midshipmen. Natural History, Vol. LXXXIII, No. 3.

Crouse, H.V. (1961). X-ray effects on sex of progeny in Sciara coprophila. Biol. Bull. Woods Hole 120, 8–10.

Crouse, H.V. (1965). Experimental alterations in the chromosome constitution of Sciara. Chromosoma 16, 391–410.

Crouse, H.V. (1979). X heterochromatin subdivision and cytogenetic analysis in Sciara coprophila (Diptera, Sciaridae). II. The controlling element. Chromosoma 74, 219–239.

Crow, J.F. (1979). Genes that violate Mendel's rules. Sci. Am. 240(2), 104–113.

Crozier, R.H. (1979). Genetics of sociality. In: Social Insects, Vol. I (Hermann, H.R., Ed.), pp. 223–286. Academic Press, New York.

Cruickshank, D.W. (1961). The role of 3d-orbitals in π-bonds between (a) silicon, phosphorus, sulfur, or chlorine and (b) oxygen or nitrogen. J. Chem. Soc. 1961, 5486–5504.

Cuénot, L. (1932). La Genèse des Espèces Animales. Librairie Félix Alcan, Paris.

Cullis, C.A. (1977). Molecular aspects of the environmental induction of heritable changes in flax. Heredity 38(2), 129–154.

Curry-Lindahl, K. & Tinggaard, K.A. (1965). Djuren i Färg. Däggdjur, Kräldjur, Groddjur. Almqvist & Wiksell, Stockholm.

Dabauvalle, M.-C., Loos, K., Merkert, H., & Scheer, U. (1991). Spontaneous assembly of pore complex-containing membranes ("annulate lamellae") in *Xenopus* egg extract in the absence of chromatin. J. Cell Bio. 112(6), 1073–1082.

Dahlberg, J.E. (1977). RNA primers for the reverse transcriptases of RNA tumor viruses. In: Nucleic Acid-Protein Recognition (Vogel, H.J., Ed.), pp. 345–358. Academic Press, New York, San Francisco, London.

Dale, J.E. (1982). The Growth of Leaves. The Institute of Biology's Studies in Biology no. 137. Edward Arnold, London.

Dana, E.S. (1955). A Textbook of Mineralogy with an extended treatise on Crystallography and Physical Mineralogy. 4th ed. John Wiley & Sons, Inc., New York, Chapman & Hall, Limited, London.

Darlington, C.D. (1937). Recent Advances in Cytology. 2nd Ed. J. & A. Churchill Ltd, London.

Darlington, C.D. (1953). The Facts of Life. George Allen & Unwin Ltd., London.

Darlington, C.D. & Ammal, E.K.J. (1945). Chromosome Atlas of Cultivated Plants. George Allen & Unwin Ltd., London.

Davidson, E.H. & Britten, R.J. (1979). Regulation of gene expression: possible role of repetitive sequences. Science 204, 1052–1059.

Dawydoff, C. (1928). Traité d'Embryologie Comparée des Invertébrés. Masson et Cie, Paris.

De Beer, G. (1958). Embryos and Ancestors. Oxford at the Clarendon Press.

De Duve, C. (1984a). A Guided Tour of the Living Cell, vol. I. Scientific American Library, Scientific American Books, Inc., New York.

De Duve, C. (1984b). A Guided Tour of the Living Cell, vol. II. Scientific American Library, Scientific American Books, Inc., New York.

Degens, E.T., Deuser, W.G., & Haedrich, R.L. (1969). Molecular structure and composition of fish otoliths. Marine Biol. 2, 105–113.

De Jussieu, M.A. (1873). Cours Elémentaire Histoire Naturelle. Botanique. Organes et Fonctions de la Végétation. Langlois et Leclercq, Fortin, Masson, Paris.

Denffer, D. von, Schumacher, W., Mägdefrau, K., & Ehrendorfer, F. (1971). Strasburger's Textbook of Botany, New English Edition. Longman, London and New York.

De Robertis, E.M., Oliver, G., & Wright, C.V.E. (1990). Homeobox genes and the vertebrate body plan. Sci. Am. 263(1), 26–32.

Desautels, P.E. (1968). The Mineral Kingdom. Ridge Press Book/Madison Square Press, Grosset and Dunlap, New York.

Dickerson, R.E. (1980). Cytochrome c and the evolution of energy metabolism. Sci. Am. 242(3), 99–110.

Dickerson, R.E. & Geis, I. (1979). Chemistry, Matter, and the Universe. An Integrated Approach to General Chemistry. Benjamin/Cummings, Menlo Park.

Dilworth, M.J. & Coventry, D.R. (1977). Stability of leghaemoglobin in yellow lupin nodules. In: Recent Developments in Nitrogen Fixation (Newton, W., Postgate, J.R., & Rodriguez-Barrueco, C., Eds.), pp. 431–442. Academic Press, New York.

Donelson, J.E. & Turner, M.J. (1985). How the trypanosome changes its coat. Sci. Am. 252(2), 32–39.

Doolittle, R.F., Feng, D.F., Johnson, M.S., & McClure, M.A. (1986). Relationships of human protein sequences to those of other organisms. Cold Spring Harbor Symposia on Quantitative Biology, vol. LI. Molecular Biology of *Homo sapiens*, pp. 447–455. Cold Spring Harbor Laboratory.

Doolittle, W.F. (1987). What introns have to tell us: Hierarchy in genome evolution. Cold Spring Harbor Symposia on Quantitative Biology, vol. LII. Evolution of Catalytic Function, pp. 907–913. Cold Spring Harbor Laboratory.

Dorrington, J. (1979). Pituitary and placental hormones. In: Mechanisms of Hormone Action, Reproduction in Mammals, Book 7 (Austin, C.R. & Short, R.V., Eds.), pp. 53–80. Cambridge University Press, Cambridge.

Dorst, J. & Dandelot, P. (1988). A Field Guide to the Larger Mammals of Africa. Collins, London.

Drake, J.W. (1969). Comparative rates of spontaneous mutation. Nature 221, 1132.

Dreller, C. & Kirchnert, W.H. (1993). How honeybees perceive the information of the dance language. Naturwissenschaften 80, 319–321.

Driever, W. & Nüsslein-Volhard, C. (1988a). A gradient of *bicoid* protein in Drosophila embryos. Cell 54, 83–93.

Driever, W. & Nüsslein-Volhard, C. (1988b). The *bicoid* protein determines position in the *Drosophila* embryo in a concentration-dependent manner. Cell 54, 95–104.

Drost-Hansen, W. & Singleton, J.L. (1989). Liquid asset. How the exotic properties of cell water enhance life. The Sciences, September/October 1989, 38–42.

Dubertret, L., Coulomb, B., Saiag, PH., & Touraine, R. (1987). Les peaux artificielles vivantes. La Recherche 18(185), 149–155.

DuBois, A.M. (1933). Chromosome behavior during cleavage in the eggs of Sciara coprophila (Diptera) in relation to the problem of sex determination. Z. Zellforsch. 19, 595–614.

Duclaux, J. (1936). Actualités Scientifiques et Industrielles 350. Leçons de Chimie Physique. Appliquée a la Biologie, VII. Diffusion dans les Gels et les Solides. Hermann & Cie, Paris.

Dujon, B. (1989). Group I introns as mobile genetic elements: facts and mechanistic speculations—A review. Gene 82, 91–114.

Duncan, I. (1987). The bithorax complex. Ann. Rev. Genet. 21, 285–319.

Dyer, A.F. (1979). Investigating Chromosomes. Edward Arnold, London.

Dyson, F.J. (1953). Field theory. In: Particles and Fields, Readings from Scientific American, 1980, pp. 17–21. W.H. Freeman and Company, San Francisco.

Eames, A.J. & MacDaniels, L.H. (1925). An Introduction to Plant Anatomy. McGraw-Hill Book Company, Inc., New York.

Eckert, R. & Randall, D. (1978). Animal Physiology. W.H. Freeman, San Francisco.

Edelstein, S.J. (1973). Introductory Biochemistry. Holden-Day, San Francisco.

Edenberg, H.J. & Huberman, J.A. (1975). Eukaryotic chromosome replication. Annu. Rev. Genet. 9, 245–284.

Eibl-Eibesfeldt, I. (1970). Amor e Odio, Historia Natural dos Padroes Elementares do Comportamento. Livraria Bertrand, Lisboa.

Eisenbeis, G. & Wichard, W. (1987). Atlas on the Biology of Soil Arthropods. Springer-Verlag, Berlin.

Eisenberg, J.M. (1981). A Collector's Guide to Sea Shells of the World. McGraw-Hill Book Company, New York, London.

Ellfolk, N. (1972). Leghaemoglobin, a plant hemoglobin. Endeavour 31, 139–142.

Elvers, I. (1965). Vår Flora i Färg. Almqvist & Wiksell, Stockholm.

Engberg, J. (1985). The ribosomal RNA genes of Tetrahymena: structure and function. Review Article. Eur. J. Cell Bio. 36, 133–151.

Engel, W., Schmidtke, J., & Wolf, U. (1975). Diploid-tetraploid relationships in teleostean fishes. In: Isozymes Vol. IV Genetics and Evolution (Markert, C.L., Ed.), pp. 449–462. Academic Press, New York.

Englund, P.T., Kelly, R.B., & Kornberg, A. (1969). Enzymatic synthesis of deoxyribonucleic acid. XXXI. Binding of deoxyribonucleic acid to deoxyribonucleic acid polymerase. J. Biol. Chem. 244, 3045–3052.

Engström, A. & Finean, J.B. (1958). Biological Ultrastructure. Academic Press Inc., New York.

Feder, J.L., Chilcote, C.A., & Bush, G.L. (1988). Genetic differentiation between sympatric host races of the apple maggot fly *Rhagoletis pomonella*. Nature 336, 61–64.

Fehilly, C.B., Willadsen, S.M., & Tucker, E.M. (1984). Interspecific chimaerism between sheep and goat. Nature 307, 634–636.

Fersht, A. (1977). Enzyme Structure and Mechanism. W.H. Freeman and Company, Reading, San Francisco.

Fieldes, M.A. (1994). Heritable effects of 5-azacytidine treatments on the growth and development of flax (*Linum usitatissimum*) genotrophs and genotypes. Genome 37, 1–11.

Fiorito, G. & Scotto, P. (1992). Observational learning in *Octopus vulgaris*. Science 256, 545–547.

Flavell, R.A., Allen, H., Burkly, L.C., Sherman, D.H., Waneck, G.L., & Widera, G. (1986). Molecular Biology of the H-2 histo-compatibility complex. Science 233, 437–443.

Flemming, W. (1881). Beiträge zur Kenntniss der Zelle und ihre Lebenserscheinungen I, II, III, Archiv für Mikroskopische Anatomie (Bonn) XVI, XIX, XX (1879, 1880, 1881).
Florkin, M. & Scheer, B.T. (1972). Chemical Zoology. Vol. 7. Academic Press, New York.
Folkman, J. & Haudenschild, C. (1980). Angiogenesis *in vitro*. Nature 288, 551–556.
Fox, H. (1981). Cytological and morphological changes during amphibian metamorphosis. In: Metamorphosis: A Problem in Developmental Biology (Gilbert, L.I. & Frieden, E., Eds.), pp. 327–362. Plenum Press, New York.
Fraenkel-Conrat, H. (1962). Design and Function at the Threshold of Life: The Viruses. Academic Press, New York.
Francé, R.H. (1943). La Maravillosa Vida de los Animales. Una Zoologia Para Todos. Editorial Labor, S.A., Barcelona.
Frankel, R.B., Blakemore, R.P., & Wolfe, R.S. (1979). Magnetite in freshwater magnetotactic bacteria. Science 203, 1355–1356.
Freeman, R. (1972). Classification of the Animal Kingdom. An Illustrated Guide. Hodder and Stoughton Ltd., The Reader's Digest Association Ltd.
Friday, A. & Ingram, D.S. (1985). The Cambridge Encyclopedia of Life Sciences. Cambridge University Press, Cambridge.
Friedberg, E.C. (1985). DNA Repair. Freeman, New York.
Friedman, D.I. & Gottesman, M. (1983). Lytic mode of lambda development. In: Lambda II (Hendrix, R.W., Roberts, J.W., Stahl, F.W., & Weisberg, R.A., Eds.), pp. 21–51. Cold Spring Harbor Laboratory.
Frisch, K.V. (1938). Noi e la Vita. Biologia Moderna per Tutti. Editore Ulrico Hoepli Milano.
Furth, A.J. (1980). Lipids and Polysaccharides in Biology. The Institute of Biology's Studies in Biology no. 125. Edward Arnold, London.
Galiano, E.F. (1929). Los Fundamentos de la Biología. Editorial Labor, S.A., Barcelona, Buenos Aires.
Garcia-Bellido, A. (1975). Genetic control of wing disk development in *Drosophila*. Ciba Found. Symp. 29, Cell Patterning, 161–182.
Garstang, W. (1922). The theory of recapitulation: a critical re-statement of the biogenetic law. Linn. Soc. J. Zool. XXXV, 81–101.
Garstang, W. (1928a). The morphology of the Tunicata, and its bearings on the phylogeny of the Chordata. Quart. J. Micr. Sci. LXXII, 51–187.
Garstang, W. (1928b). The origin and evolution of larval forms. Presidential Address, Section D., British Ass., Glasgow.
Garstang, W. (1929). The origin and evolution of larval forms. Section D. Zoology. In: Larval Forms and Other Zoological Verses (Garstang, W., Ed.), pp. 77–98. The University of Chicago Press, Chicago and London.
Garstang, W. (1985). Larval Forms and Other Zoological Verses. The University of Chicago Press, Chicago and London.
Gernez, D. (1868). Sur la cristallisation des substances hémiédriques. C. R. Ac. Sc., t. LXVI, p. 353. Paris.
Gernez, D. (1876). Sur les circonstances de la production des deux variétés prismatique et octaédrique du Soufre. C. R. Acad. Sc., t. LXXXVIII, p. 217. Paris.
Gernez, D. (1877). Influence qu'exerce une action mécanique sur la production des divers hydrates dans les solutions sursaturées. C. R. Ac. Sc., t. LXXXIV, p. 1389. Paris.
Ghiselin, M.T. (1988). The origin of molluscs in the light of molecular evidence. In: Oxford Surveys in Evolutionary Biology (Harvey, P.H. & Partridge, L., Eds.), pp. 66–95. Oxford University Press.
Giannoni, G., Padden, F., & Keith, H.D. (1969). Crystallization of DNA from dilute solution. Proc. Natl. Acad. Sci. USA 62, 964–971.
Gierer, A. (1974). Hydra as a model for the development of biological form. Sci. Am. 231(6), 44–54.
Gilbert, W. (1978). Why genes in pieces? Nature 271, 501.
Gilbert, W. (1986). The RNA world—origin of life. Nature 319, 618.

Gilbert, W., Marchionni, M., & McKnight, G. (1986). On the antiquity of introns. Cell 46, 151.

Gitschier, J., Wood, W.I., Goralka, T.M., Wion, K.L., Chen, E.Y., Eaton, D.H., Vehar, G.A., Capon, D.J., & Lawn, R.M. (1984). Characterization of the human factor VIII gene. Nature 312, 326–330.

Glover, D.M. & Hames, B.D. (1989). Genes and Embryos. IRL Press at Oxford University Press, Oxford.

Goebel, K. (1933). Organographie der Pflanzen, Dritter Teil, Samenpflanzen. Gustav Fischer, Jena.

Goethe, J.W. von (1820). Nacharbeiten und Sammlungen. In: I.W. Troll: Goethes Morphologische Schriften, Jena.

Gola, G., Negri, G., & Cappelletti, C. (1943). Tratado de Botánica. Editorial Labor, S.A., Barcelona.

Goldfarb, D.S. (1992). Review: Are the cytosolic components of the nuclear, ER, and mitochondrial import apparatus functionally related? Cell 70, 185–188.

Goodman, M., Czelusniak, J., & Beeber, J.E. (1985). Phylogeny of primates and other eutherian orders: a cladistic analysis using amino acid and nucleotide sequence data. Cladistics 1(2), 171–185.

Gould, J.L., Kirschvink J.L., & Deffeyes, K.S. (1978). Bees have magnetic remanence. Science 201, 1026–1028.

Gould, S.J. (1977). Ontogeny and Phylogeny. The Belknap Press of Harvard University Press, Cambridge, London.

Graham, A., Papalopulu, N., & Krumlauf, R. (1989). The murine and Drosophila homeobox gene complexes have common features of organization and expression. Cell 57, 367–378.

Granick, S. (1965). Evolution of heme and chlorophyll. In: Evolving Genes and Proteins (Bryson, V. & Vogel, H.J., Eds.), pp. 67–88. Academic Press, New York.

Grassé, P.-P. (1948). Traité de Zoologie. Anatomie, Systématique, Biologie. Tome XI. Echinodermes, Stomocordés, Procordés. Masson et Cie, Paris.

Gravis, A. (1920). Eléments de Morphologie végétale. Vaillant-Carmanne, Liège, Vigot Frères, Paris.

Grechushnikov, B.N. (1988). Optical properties of crystals. In: Modern Crystallography IV. Physical Properties of Crystals (Shuvalov, L.A., Ed.), pp. 405–512. Springer-Verlag, Berlin.

Green, M.R. (1986). Pre-mRNA splicing. Ann. Rev. Genet. 20, 671–708.

Greenwood, N.N. & Earnshaw, A. (1989). Chemistry of the Elements. Pergamon Press, Oxford.

Gregory, W.K. (1974). Evolution Emerging. vol. II. Arno Press, New York.

Greulach, V.A. (1973). Plant Function and Structure. Macmillan Publishing Co., Inc., New York.

Grim, R.E. (1968). Clay Mineralogy. 2nd Ed. McGraw-Hill Book Co., New York.

Gustafsson, Å. (1979). Linnaeus' peloria: the history of a monster. Theor. Appl. Genet. 54, 241–248.

Gustafsson, Å. & Mergen, F. (1964). Some principles of tree cytology and genetics. Unasylva 18(2–3), Numbers 73–74, pp. 7–20.

Guyénot, E. (1942). L'Hérédité. G. Doin & Cie, Paris.

Haeckel, E. (1875). Ziele und Wege der heutigen Entwickelungsgeschichte. Dufft, Jena.

Halder, G., Callaerts, P., & Gehring, W.J. (1995). Induction of ectopic eyes by targeted expression of the *eyeless* gene in *Drosophila*. Science 267, 1788–1792.

Halstead, L.B. (1978). The Evolution of the Mammals. Peter Lowe. Eurobook Limited, Oxford.

Hames, B.D. & Glover, D.M. (Eds.) (1990). Gene Rearrangement. IRL Press at Oxford University Press, Oxford.

Hanson, E.D. (1977). The Origin and Early Evolution of Animals. Wesleyan University Press, Middletown, CT.

Hanström, B. & Johnels, A.G. (1962). Benfiskar. In: Djurens Värld, Band 6, Fiskar: 2 (Hanström, B., Ed.), Förlagshuset Norden AB, Malmö.

Hardy, A. (1985). Introduction. In: Larval Forms and Other Zoological Verses (W. Garstang), The University of Chicago Press, Chicago.

Härkönen, T. (1986). Guide to the Otoliths of the Bony Fishes of the Northeast Atlantic. Danbiu ApS. Biological Consultants.

Harvey, R.P. & Melton, D.A. (1988). Microinjection of synthetic Xhox-1A homeobox mRNA disrupts somite formation in developing Xenopus embryos. Cell 53, 687–697.

Haseltine, W. (1983). Ultraviolet light repair and mutagenesis revisited. Cell 33, 13–17.

Hawkins, J.M., Meyer, A., Lewis, T.A., Loren, S., & Hollander, F.J. (1991). Crystal structure of Osmylated C_{60}: Confirmation of the soccer ball framework. Science 252, 312–313.

Heribert-Nilsson, N. (1918). Experimentelle Studien über Variabilität, Spaltung, Artbildung und Evolution in der Gattung *Salix*. Lunds Univ. Arsskr., 14, 1–145.

Hermann, H.R. (1979). Insect sociality—an introduction. In: Social Insects. Volume 1. (Hermann, H.R., Ed.), pp. 1–33. Academic Press, New York.

Herskowitz, I.H. (1977). Principles of Genetics. 2nd Edition. Macmillan Publishing Co., Inc., New York.

Hertwig, O. (1929a). Génesis de los Organismos. Tomo I. Primera Edición. Espasa-Calpe, S.A.

Hertwig, O. (1929b). Génesis de los Organismos. Tomo II. Espasa-Calpe, S.A.

Hertwig, R. (1928). Trattato di Zoologia. Dottor Francesco Vallardi, Milano.

Herzberg, O. & James, M.N.G. (1985). Structure of the calcium regulatory muscle protein troponin-C at 2.8 Å resolution. Nature 313, 653–659.

Heslop-Harrison, J. & Heslop-Harrison, Y. (1958). Studies on flowering-plant growth and organogenesis, III. Leaf shape changes associated with flowering and sex differentiation in *Cannabis sativa*. Proc. Roy. Irish Acad. 59B, 257–283.

Heywood, V.H. (1978). Flowering Plants of the World. Oxford University Press, Oxford.

Hiatt, A., Cafferkey, R., & Bowdish, K. (1989). Production of antibodies in transgenic plants. Nature 342, 76–78.

Hill, A. (1990). Entropy production as the selection rule between different growth morphologies. Nature 348, 426–428.

Hill, J. (1965). Environmental induction of heritable changes in *Nicotiana rustica*. Nature 207, 732–734.

Hill, R.J. & Stollar, B.D. (1983). Dependence of Z DNA antibody binding to polytene chromosomes on acid fixation and DNA torsional strain. Nature 305, 338–340.

Hinchliffe, J.R. & Johnson, D.R. (1980). The Development of the Vertebrate Limb. Clarendon Press, Oxford.

Hinde, R. (1981). Social organization. In: The Oxford Companion to Animal Behaviour (McFarland, D., Ed.), pp. 518–527. Oxford University Press, Oxford.

Hinegardner, R.T. (1975). Morphology and genetics of sea urchin development. Amer. Zool. 15, 679–689.

Hinegardner, R. & Rosen, D.E. (1972). Cellular DNA content and the evolution of teleost fishes. Am. Nat. 106, 621–644.

Holland, P. (1992). Homeobox genes in vertebrate evolution. Bioessays 14, 267–273.

Holmes, S. (1979). Henderson's Dictionary of Biological Terms. Ninth Edition. Longman, London, New York.

Hozumi, N. & Tonegawa, S. (1976). Evidence for somatic rearrangement of immunoglobulin genes coding for variable and constant regions. Proc. Natl. Acad. Sci. USA 73, 3628–3632.

Hutchison, C.A., III, Hardies, S.C., Loeb, D.D., Shehee, W.R., & Edgell, M.H. (1989). LINEs and related retroposons: Long interspersed repeated sequences in the eucaryotic genome. In: Mobile DNA (Berg, D.E. & Howe, M.M., Eds.), pp. 593–617. American Society for Microbiology, Washington, DC.

Hyman, L.H. (1955). The Invertebrates: Echinodermata. The coelomate Bilateria, vol. IV. McGraw-Hill Book Company, Inc., New York.

Iijima, S., Cowley, J.M., & Donnay, G. (1973). High resolution electron microscopy of tourmaline crystals. Tshermaks Mineral. Petrog. Mitt. 20, 216–224.

Imlay, J.A. & Linn, S. (1988). DNA damage and oxygen radical toxicity. Science 240, 1302–1642.

Ingham, P.W. (1988). The molecular genetics of embryonic pattern formation in *Drosophila*. Nature 335, 25–34.

Inoué, S. & Okazaki, K. (1977). Biocrystals. Sci. Am. 236(4), 83–92.

Iyer, S.S. & Xie, Y.-H. (1993). Light emission from silicon. Science 260, 40–46.

Jaffe, H.W. (1988). Introduction to Crystal Chemistry. Student Edition. Cambridge University Press, Cambridge.

Janvier, Ph. (1984). Cladistics: theory, purpose, and evolutionary implications. In: Evolutionary Theory: Paths into the Future (Pollard, J.W., Ed.), pp. 39–75. John Wiley and Sons, New York.

Jeffery, W.R. (1985). The location of maternal mRNA in eggs and embryos. BioEssays 1(5), 196–199.

Jofuku, K.D., den Boer, B.G.W., Van Montagu, M., & Okamuro, J.K. (1994). Control of Arabidopsis flower and seed development by the homeotic gene APETALA2. The Plant Cell 6, 1211–1225.

Jones, W.N. (1934). Plant Chimaeras and Graft Hybrids. Methuen & Co. Ltd., London.

Jussieu, de, M.A. (1873). Cours Elémentaire D'Histoire Naturelle. Botanique. Langlois et Leclercq, Fortin, Masson, Paris.

Katz, B. & Miledi, R. (1966). Input/output relation of a single synapse. Nature 212, 1242–1245.

Kern, R. & Gindt, R. (1958). Contribution à l'étude des accolements réguliers des feldspaths potassiques et des plagioclases. Bull. Soc. franç. Min. Crist. 81, 263–266.

King, J.E. (1964). Seals of the World. The British Museum (Natural History), London.

King, M.C. & Wilson, A.C. (1975). Evolution at two levels in humans and chimpanzees. Science 188, 107–116.

Klein, C. & Hurlbut, C.S., Jr. (1985). Manual of Mineralogy. Twentieth Edition. John Wiley & Sons, New York.

Klug, A. & Rhodes, D. (1987). "Zinc fingers": A novel protein motif for nucleic acid recognition. Trends Biochem. Sci. 12, 464–469.

Knight, C. & Knight, N. (1973). Snow crystals. Sci. Am. 228(1), 100–107.

Kobayashi, I., Stahl, M.M., & Stahl, F.W. (1984). The mechanism of the Chi-*cos* interaction in RecA-RecBC-mediated recombination in phage lambda. Cold Spring Harbor Symposia on Quantitative Biology, vol. XLIX. Recombination at the DNA Level, pp. 497–506. Cold Spring Harbor Laboratory.

Koenig, M., Hoffman, E.P., Bertelson, C.J., Monaco, A.P., Feener, C., & Kunkel, L.M. (1987). Complete cloning of the Duchenne muscular dystrophy (DMD) cDNA and preliminary genomic organization of the DMD gene in normal and affected individuals. Cell 50, 509–517.

Kornberg, A. (1980). DNA Replication. W.H. Freeman, San Francisco.

Kudo, R.R. (1971). Protozoology. Charles C. Thomas, Publisher, Springfield, IL.

Kullinger, B. & Medenbach, O. (1988). Mineralrikets överdåd: Ökenros, agat, svavelkis. Forskning och Framsteg 6: 26–29.

LaBarbera, M. (1985). Foreword. In: Larval Forms and Other Zoological Verses (Garstang, W., Ed.), pp. vii–x. University of Chicago Press, Chicago.

Lagercrantz, H. & Slotkin, A. (1986). The "stress" of being born. Sci. Am. 254(4), 92–102.

Lahue, R.S., Au, K.G., & Modrich, P. (1989). DNA mismatch correction in a defined system. Science 245, 160–164.

Landsmann, J., Dennis, E.S., Higgins, T.J.V., Appleby, C.A., Kortt, A.A., & Peacock, W.J. (1986). Common evolutionary origin of legume and non-legume plant haemoglobins. Nature 324, 166–168.

Langlebert, J. (1901). Histoire Naturelle. 63d Ed. Delalain Frères, Paris.

Langman, J. (1969). Medical Embryology. Human Development—Normal and Abnormal. 2nd ed. The Williams & Wilkins Company, Baltimore.

Lawrence, P.A. (1992). The Making of a Fly. The Genetics of Animal Design. Blackwell Scientific Publications, Oxford.

Lee, T.F. (1991). The Human Genome Project. Cracking the Genetic Code of Life. Plenum Publishing Corporation, New York.

Lehninger, A.L. (1975). Biochemistry. 2nd ed. Worth Publishers Inc., New York.

Leppik, E.E. (1977). The evolution of capitulum types of the Compositae in the light of insect-flower interaction. In: The Biology and Chemistry of the Compositae, vol. I (Heywood, V.H., Harborne, J.B., & Turner, B.L., Eds.), pp. 61–89. Academic Press, London, New York.

Lerner, A.B. (1967). Skin pigmentation. In: The Encyclopedia of Biochemistry (Williams, R.J. & Lansford, E.M., Eds.), pp. 750–752. Reinhold Publishing Corporation, New York.

Levinton, J.S. (1992). The big bang of animal evolution. Sci. Am. 267(5), 52–59.

Levine, M.S. & Harding, K.W. (1989). Drosophila: The zygotic contribution. In: Genes and Embryos (Glover, D.M. & Hames, B.D., Eds.), pp. 39–94. IRL Press at Oxford University Press, Oxford, New York.

Lewin, B. (1983). Genes. John Wiley & Sons, New York.

Lewin, B. (1990). Genes IV. Oxford University Press, Oxford, New York.

Lichtenstein, C. & Draper, J. (1986). Genetic engineering of plants. In: DNA Cloning, Volume II (Glover, D.M., Ed.), pp. 67–119. Oxford IRL Press.

Liebau, F. (1959). Über die Kristallstruktur des Pyroxmangits (Mn, Fe, Ca, Mg) Si O_3. Acta Cryst. 12, 177–181.

Lima-de-Faria, A. (1954). Chromosome gradient and chromosome field in Agapanthus. Chromosoma 6, 330–370.

Lima-de-Faria, A. (1973). Equations defining the position of ribosomal cistrons in the eukaryotic chromosome. Nature New Biology 241, 136–139.

Lima-de-Faria, A. (1980). Classification of genes, rearrangements and chromosomes according to the chromosome field. Hereditas 93, 1–46.

Lima-de-Faria, A. (1983). Molecular Evolution and Organization of the Chromosome. Elsevier, New York.

Lima-de-Faria, A. (1988). Evolution without Selection. Form and Function by Autoevolution. Elsevier, New York.

Lima-de-Faria, A. (1991a). A kit to produce an artificial human chromosome. In: Fundamentals of Medical Cell Biology, vol. 1. Evolutionary Biology (Bittar, E.E., Ed.), pp. 115–161. JAI Press, Greenwich, CT.

Lima-de-Faria, A. (1991b). The autoevolution of the human species. In: Fundamentals of Medical Cell Biology, vol. 1. Evolutionary Biology (Bittar, E.E., Ed.), pp. 273–310. JAI Press, Greenwich, CT.

Lima-de-Faria, A. (1991c). The new ethics in a multibiological society. Polish Bot. Stud. 2, 21–22.

Lima-de-Faria, A., Arnason, U., Widegren, B., Isaksson, M., Essen-Möller, J., & Jaworska, H. (1986). DNA cloning and hybridization in deer species supporting the chromosome field theory. BioSystems 19, 185–212.

Lindahl, T. (1982). DNA repair enzymes. Ann. Rev. Biochem. 51, 61–87.

Line, L. & Reiger, G. (1980). The Audubon Society Book of Marine Wildlife. Harry N. Abrams, Inc., New York.

Lissman, H.W. (1963). Electrical Location by Fishes. Scientific American Inc. New York.

Loomis, W.F. (1988). Four Billion Years: An Essay on the Evolution of Genes and Organisms. Sinauer Associates, Sunderland, MA.

Lowenstam, H.A. (1981). Minerals formed by organisms. Science 211, 1126–1131.

Lund, E.J. (1921). Experimental control of organic polarity by the electrical current. J. Exp. Zool. 54, 471–491.

Lyttle, T.W. (1991). Segregation distorters. Annu. Rev. Genet. 25, 511–557.

Macdonald, D. (Ed.) (1984). The Encyclopaedia of Mammals: Vols. 1 and 2. George Allen & Unwin, London.

Maksymowych, R. (1973). Analysis of Leaf Development. Cambridge University Press, Cambridge.

Mandelbrot, B.B. (1983). The Fractal Geometry of Nature. W.H. Freeman and Company, New York.

Mann, S., Heywood, B.R., Rajam, S., & Walker, J.B.A. (1991). Structural and stereochemical relationships between Langmuir monolayers and calcium carbonate nucleation. J. Phys. D: Appl. Phys. 24, 154–164.

Margoliash, E. & Smith, E.L. (1965). Structural and functional aspects of cytochrome C in relation to evolution. In: Evolving Genes and Proteins (Bryson, V. & Vogel, H.J., Eds.), pp. 221–242. Academic Press, New York.

Margulis, L. & Schwartz, K.V. (1982). Five Kingdoms. An Illustrated Guide to the Phyla of Life on Earth. W.H. Freeman and Company, San Francisco.

Martin, G., Josserand, S.A., Bornman, J.F., & Vogelmann, T.C. (1989). Epidermal focussing and the light microenvironment within leaves of *Medicago sativa*. Physiologia Plantarum 76, 485–492.

Mason, S.F. (1987). Universal dissymmetry and the origin of biomolecular chirality. BioSystems 20, 27–35.

Masterton, W.L. & Hurley, C.N. (1993). Chemistry. Principles and Reactions. 2nd Ed. Saunders College Publishing, Philadelphia.

Matheja, J. & Degens, E.T. (1968). Molekulare Entwicklung mineralisations-fähiger organischer Matrizen. In: Neues Jahrbuch für Geologie und Paläontologie, Monatshefte (Lotze, Fr. & Schindewolf, O.H., Eds.), pp. 215–229. E. Schweizerbart'sche Verlagsbuchhandlung, Stuttgart.

Matthews, L.H. & Carrington, R. (1972). Världens Djurliv. Reader's Digest AB, Stockholm.

Matthews, R.W. & Matthews, J.R. (1978). Insect Behavior. John Wiley & Sons, New York.

Maximov, N.A. (1938). Plant Physiology. (Harvey, R.B. & Murneek, A.E., Eds.), Second English Ed. McGraw-Hill Book Company, Inc., New York.

Mazurs, E.G. (1974). Graphic Representation of the Periodic System during One Hundred Years. University of Alabama Press, Birmingham.

McCarthy, B.J. (1969). The evolution of base sequences in nucleic acids. In: Handbook of Molecular Cytology (Lima-de-Faria, Ed.), pp. 3–20. North Holland Publishing Company, Amsterdam.

McClintock, B. (1950). The origin and behavior of mutable loci in maize. Proceedings of the National Academy of Sciences 36, 347.

McClintock, B. (1978). Mechanisms that rapidly reorganize the genome. In: Stadler Symposium, Vol. 10 (Reder, G.P., Ed.), pp. 25–48. University of Missouri Press, Columbia.

McClintock, B. (1980). Modified gene expressions induced by transposable elements. In: Mobilization and Reassembly of Genetic Information (Scott, W.A. et al., Eds.), pp. 11–19. Academic Press, New York.

McFarland, D. (Ed.) (1981). The Oxford Companion to Animal Behaviour. Oxford University Press, Oxford.

McGinnis, W., & Kuziora, M. (1994). The molecular architects of body design. Sci. Am. 270(2), 36–42.

McMahon, T.A. & Bonner, J.T. (1983). On Size and Life. Scientific American Library, Scientific American Books, Inc., New York.

McNamara, K. (1989). The great evolutionary handicap. New Scientist, 16 September, 47–51.

McPheron, B.A., Smith, D.C., & Berlocher, S.H. (1988). Genetic differences between host races of *Rhagoletis pomonella*. Nature 336, 64–66.

McPherson, A. (1989). Macromolecular crystals. The growth of crystals is now the key to deducing the structure of large molecules. Sci. Am. 260(3), 42–49.

Medenbach, O. & Sussieck-Fornefeld, C. (1983). Minerais (Translation of: Mineralien). Mosaik Verlag, Munich.

Meyerowitz, E.M. (1994). The genetics of flower development. Sci. Am. 271(5), 40–47.

Millar, A.J., Short, S.R., Chua, N.-H., & Kay, S.A. (1992). A novel circadian phenotype based on firefly luciferase expression in transgenic plants. The Plant Cell 4, 1075–1087.

Miller, O.J. & Bakken, A.H. (1972). Gene transcription in reproductive tissues. In: Karolinska Symposia on Research Methods in Reproductive Endocrinology, 5th Symposium, pp. 155–173. Karolinska Institutet, Stockholm.

Miller, O.L. (1981). The nucleolus, chromosomes, and visualization of genetic activity. J. Cell Biol. 91, 15–27.

Mindich, L., Bamford, D., McGraw, T., & Mackenzie, G. (1982). Assembly of bacteriophage PRD1: particle formation with wild-type and mutant viruses. J. Virol. 44(3), 1021–1030.

Mintz, B. (1978). Mutagenized teratocarcinoma cells as probes of mammalian differentiation. Plenary Sessions Symposia, Abstracts, p. 65. XIV International Congress of Genetics, Moscow.

Mintz, B. & Illmensee, K. (1975). Normal genetically mosaic mice produced from malignant teratocarcinoma cells. Proc. Natl. Acad. Sci. USA 72(9), 3585–3589.

Mlodzik, M. & Gehring, W.J. (1987). Expression of the *caudal* gene in the germ line of Drosophila: Formation of an RNA and protein gradient during early embryogenesis. Cell 48, 465–478.

Monod, J., Changeux, J.-P., & Jacob, F. (1963). Allosteric proteins and cellular control systems. J. Mol. Biol. 6, 306–329.

Moret, L. (1940). Manuel de Paléontologie Animale. Masson et Cie, Paris.

Morgan, T.H. (1928). The Theory of the Gene. (2nd ed). Yale University Press, New Haven.

Morton, D.J. (1964). The Human Foot, its Evolution, Physiology and Functional Disorders. Hafner, New York.

Moscona, A.A. (1959). Tissues from dissociated cells. Sci. Am. 200(5), 132–144.

Moscona, A.A. & Hausman, R.E. (1977). Biological and biochemical studies on embryonic cell-cell recognition. In: Cell and Tissue Interactions (Lash, J.W. & Burger, M.M., Eds.), pp. 173–185. Raven Press, New York.

Mossberg, B. & Nilsson, S. (1977). Nordens orkidéer. J.W. Cappelens Förlag A.S., Oslo.

Müller, G.B. (1990). Developmental mechanisms at the origin of morphological novelty: A side-effect hypothesis. In: Evolutionary Innovations (Nitecki, M.H., Ed.), pp. 99–130. The University of Chicago Press, Chicago.

Müller, W.E.G., Müller, I., Kurelec, B., & Zahn, R.K. (1976). Species-specific aggregation factor in sponges. IV. Inactivation of the aggregation factor by mucoid cells from another species. Exp. Cell Res. 98, 31–40.

Mulvey, J. (1979). The new frontier of particle physics. Nature 278, 403–409.

Müntzing, A. (1961). Genetic Research. A Survey of Methods and Main Results. LTs Förlag, Stockholm.

Murray, A.W. & Szostak, J.W. (1987). Artificial chromosomes. Sci. Am. 257(5), 60–70.

Murray, P. & Murray, L. (1963). The Art of the Renaissance. Frederick A. Praeger, New York.

Murzin, A.G. (1993). OB(oligonucleotide/oligosaccharide binding)-fold: Common structural and functional solution for non-homologous sequences. EMBO J. 12(3), 861–867.

Nachtigall, W. (1974). Insects in Flight. A Glimpse Behind the Scenes in Biophysical Research. George Allen & Unwin Ltd., London.

Nagl, W. (1973). The angiosperm suspensor and the mammalian trophoblast: organs with similar cell structure and function? Soc. Bot. Fr. Mém. Coll. Morphol., 1973, 289–302.

Nagl, W. (1976). Zellkern und Zellzyklen. Verlag Eugen Ulmer, Stuttgart.

Nagl, W. (1986). Molecular phylogeny. In: Patterns and Processes in the History of Life (Raup, D.M. & Jablonski, D., Eds.), pp. 223–232. Dahlem Konferenzen 1986. Springer-Verlag, Berlin.

Nakaya, U. (1954). Snow Crystals. Natural and Artificial. Harvard University Press, Cambridge.

Napier, J.R. & Napier, P.H. (1985). The Natural History of the Primates. British Museum (Natural History), London.

Nasmyth, K. & Shore, D. (1987). Transcriptional regulation in the yeast life cycle. Science 237, 1162–1170.

Nassau, K. (1978). The origins of color in minerals. The American Mineralogist, 63, 219–229.

Nassau, K. (1980a). The causes of color. Scientific American 243(4), 106–123.

Nassau, K. (1980b). Gems Made by Man. Chilton Book Co., Radnor, PA.

Needham, A.E. (1965). The Uniqueness of Biological Materials. Pergamon Press, Oxford.

Needham, J. (1950). Biochemistry and Morphogenesis. University Press, Cambridge.

Nelson, R.J., Ziegelhoffer, T., Nicolet, C., Werner-Washburne, M., & Craig, E.A. (1992). The translation machinery and 70 kd heat shock protein cooperate in protein synthesis. Cell 71, 97–105.

Nicolas, A., Treco, D., Schultes, N.P., & Szostak, J.W. (1989). An initiation site for meiotic gene conversion in the yeast *Saccharomyces cerevisiae*. Nature 338, 35–39.

Nielsen, H. (1990). Introns and ribozyme activity in the ciliate Tetrahymena. Licentiatrapport. Biokemisk Institut B, Panum Instituttet, Copenhagen.

Niklitschek, A. (1943). Técnica de la Vida. Iniciación al Estudio de la Biología. Iberia–Joaquin Gil, Barcelona.

Nilsson, D.-E. (1989). Vision optics and evolution. Nature's engineering has produced astonishing diversity in eye design. BioScience 39(5), 298–307.

Nilsson, D.-E. (1990). From cornea to retinal image in invertebrate eyes. TINS 13(2), 55–64.

Nitzelius, T. & Vedel, H. (1966). Skogens Träd och Buskar i Färg. Almqvist och Wiksell, Stockholm.

Noble, E.R. & Noble, G.A. (1976). Parasitology. 4th Ed. Lea & Febiger, Philadelphia.

Nomura, M. (1977). Some remarks on recent studies on the assembly of ribosomes. In: Nucleic Acid-Protein Recognition (Vogel, H.J., Ed.), pp. 443–467. Academic Press, New York.

Nover, L., Hellmund, D., Neumann, D., Scharf, K.-D., & Serfling, E. (1984). The heat shock response of eukaryotic cells. Biol. Zbl. 103, 357–435.

Ny, T., Elgh, F., & Lund, B. (1984). The structure of the human tissue-type plasminogen activator gene: Correlation of intron and exon structures to functional and structural domains. Proc. Natl. Acad. Sci. USA 81, 5355–5359.

O'Brien, E.J. & MacEwan, A.W. (1970). Molecular and crystal structure of polynucleotide complex: Polyinosinic acid plus polydeoxycytidylic acid. J. Mol. Biol. 48, 243–261.

Ohno, S., Muramato, J., Klein, J., & Atkin, N.B. (1969). Diploidtetraploid relationship in Clupeoid and Salmonid fish. Chromosomes Today 2, 139–147.

Ohno, S., Nagai, Y., & Ciccarese, S. (1978). Testicular cells lysostripped of H-Y antigen organize ovarian follicle-like aggregates. Cytogenet. Cell Genet. 20, 351–364.

Okazaki, R., Okazaki, T., Sakabe, K., Sugimoto, K., & Sugino, A. (1968a). Mechanism of DNA chain growth. I. Possible discontinuity and unusual secondary structure of newly synthesized chains. Proc. Natl. Acad. Sci. USA 59, 598–605.

Okazaki, R., Okazaki, T., Sakabe, K., Sugimoto, K., Kainuma, R., Sugino, A., & Iwatsuki, N.T. (1968b). *In vivo* mechanism of DNA chain growth. Cold Spring Harbor Symp. Quant. Biol. 33, 129–143.

Oliver, S.G., van der Aart, Q.J.M., & Agostoni-Carboni, M. et al. (1992). The complete DNA sequence of yeast chromosome III. Nature 357, 38–46.

Olson, W.K. (1977). Spatial configuration of ordered polynucleotide chains: A novel double helix. Proc. Natl. Acad. Sci. USA 74, 1775–1779.

Olsson, V. & Schumann, G. (1945). Lilla Djurboken. AB Lindqvists Förlag, Stockholm.

Orr, R. & Pope, J. (1986). Mammals of Britain & Europe. Peerage Books, London.

Osterud, H.L. (1918). Preliminary observations on the development of *Leptasterias hexactis*. Publ. Puget Sound Biol. Sta. 2, 1–15.

Oudet, P., Gross-Bellard, M., & Chambon, P. (1975). Electron microscopic and biochemical evidence that chromatin structure is a repeating unit. Cell 4, 281–300.

Ow, D.W., Wood, K.V., DeLuca, M., De Wet, J.R., Helinski, D.R., & Howell, S.H. (1986). Transient and stable expression of the firefly luciferase gene in plant cells and transgenic plants. Science 234, 856–859.

Pace, N.R., Olsen, G.J., & Woese, C.R. (1986). Ribosomal RNA phylogeny and the primary lines of evolutionary descent. Minireview. Cell 45, 325–326.

Pagels, H.R. (1982). The Cosmic Code. Quantum Physics as the Language of Nature. Michael Joseph, London.

Palhinha, R.T. & Cunha, A.G. (1939). Curso de Botanica. J. Rodrigues & C., Lisbon.

Palmiter, R.D. & Brinster, R.L. (1986). Germ line transformation of mice. Annu. Rev. Genet. 20, 465–499.

Pasteur, L. (1847). Application de la Polarisation Rotatoire des Liquides à la Solution de Diverses Questions de Chimie.Bachelier, Paris.

Pasteur, L. (1857). Etudes sur les modes d'accroissement des cristaux et sur les causes des variations de leurs formes secondaires. Ann. Ch. et Phys., 3e sér., t. XLIX, p. 5. Paris.

Pasteur, L. (1858). Mémoire sur la fermentation appelé lactique. Ann. Ch. et Phys. 3e sér., t. LII, p. 405. Paris.

Patterson, C. (1987). Molecules and Morphology in Evolution: Conflict or compromise? Cambridge University Press, Cambridge.

Pauling, L. (1949). General Chemistry. W.H. Freeman and Company, San Francisco.

Pauling, L. (1960). The Nature of the Chemical Bond. 3rd ed. Cornell University Press, Ithaca, NY.

Pauling, L. (1987). Determination of ionic radii from cation-anion distances in crystal structures: Discussion. Am. Mineral. 72, 1016.

Pauling, L. & Corey, R.B. (1953). Compound helical configurations of polypeptide chains: Structure of proteins of the alfa-keratin type. Nature 171, 59–61.

Pauling, L. & Hayward, R. (1964). The Architecture of Molecules. W.H. Freeman and Company, San Francisco.

Pays, E., Guyaux, M., Aerts, D., Van Meirvenne, N., & Steinert, M. (1985). Telomeric reciprocal recombination as a possible mechanism for antigenic variation in trypanosomes. Nature (London) 316, 562–564.

Pennycuick, C.J. (1972). Animal Flight. Studies in Biology No. 33, Edward Arnold, London.

Perkin, W.H. & Kipping, F.S. (1932). Organic Chemistry. Lippincott, Philadelphia.

Perrier, R. (1936). Cours Elémentaire de Zoologie. Masson et Cie, Paris.

Perrins, C. (1976). Bird Life. An Introduction to the World of Birds. Elsevier–Phaidon, London.

Petrén, T. (1964). Lärobok i Anatomi. Del I. Rörelseapparaten. Aktiebolaget Nordiska Bokhandelns Förlag, Stockholm.

Pierantoni, U. (1944). Tratado de Zoologia. (Spanish ed.) Editorial Labor, S.A., Barcelona.

Pitt, V.H. (1988). The Penguin Dictionary of Physics. Laurence Urdang Associates Ltd., Penguin Books, New York.

Ponticelli, A.S., Schultz, D.W., Taylor, A.F., & Smith, G.R. (1985). Chi-dependent DNA strand cleavage by RecBC enzyme. Cell 41, 145–151.

Pope, J. (1986). Do Animals Dream? The Natural History Museum London, Michael Joseph, London.

Prenant, M. (1935). Leçons de Zoologie. Annélides. Actualités Scientifiques et Industrielles 196. Hermann et Cie, Paris.

Prenant, M. (1936). Leçons de Zoologie. Prochordés Tuniciers. Actualités Scientifiques et Industrielles 380. Hermann et Cie, Paris.

Press, F. & Siever, R. (1982). Earth. 3rd Ed. W.H. Freeman and Company, New York.

Pringle, J.W.S. (1975). Insect Flight. Oxford Biology Readers: 52 (Head, J.J., Ed.), pp. 1–16. Oxford University Press, London.

Pritchard, D.J. (1986). Foundations of Developmental Genetics. Taylor & Francis, London.

Ptashne, M. (1967a). Isolation of the lambda phage repressor. Proc. Natl. Acad. Sci. USA 57, 306–313.

Ptashne, M. (1967b). Specific binding of the lambda phage repressor to lambda DNA. Nature 214, 232–234.

Ptashne, M. (1978). Lambda repressor function and structure. In: The Operon (Miller, J.H. & Reznikoff, W.S., Eds.), pp. 325–343. Cold Spring Harbor Laboratory.

Ptashne, M. (1986). A Genetic Switch. Cell Press & Blackwell Scientific Publications, San Francisco.

Pugsley, A.P. (1989). Protein Targeting. Academic Press, Inc., New York.

Quiring, R., Walldorf, U., Kloter, U., & Gehring, W.J. (1994). Homology of the *eyeless* gene of *Drosophila* to the *Small eye* gene in mice and *Aniridia* in humans. Science 265, 785–789.

Quirk, S., Bell-Pedersen, D., & Belfort, M. (1989). Intron mobility in the T-even phages: High frequency inheritance of group I introns promoted by intron open reading frames. Cell 56, 455–465.

Rabaud, E. (1932). Zoologie Biologique. Fascicule I Morphologie generale et systeme nerveux. Gauthier-Villars et Cie, Paris.

Rabaud, E. (1934). Zoologie Biologique. Fascicule III: Les Phénomènes de Reproduction. Gauthier-Villars, Paris.

Rabbitts, T.H. (1978). Evidence for splicing of interrupted immunoglobulin variable and constant region sequences in nuclear RNA. Nature 275, 291–296.

Radman, M. & Wagner, R. (1988). The high fidelity of DNA duplication. Sci. Am. 259(2), 24–30.

Rae, P.M.M., Kohorn, B., & Wade, R. (1980). The 10 kb Drosophila virilis 28S rDNA intervening sequence is flanked by a direct repeat of 14 base pairs of coding sequence. Nucl. Acids Res. 8, 3491–3504.

Raff, R.A. & Kaufman, T.C. (1983). Embryos, Genes, and Evolution. The Developmental-Genetic Basis of Evolutionary Change. Macmillan Publishing Co., Inc., New York.

Rich, A., Nordheim, A., & Wang, A.H.-J. (1984). The chemistry and biology of left-handed Z-DNA. Ann. Rev. Biochem. 53, 791–846.

Rimington, C. & Kennedy, G.Y. (1962). Porphyrins: Structure, distribution and metabolism. In: Comparative Biochemistry vol. 4 (Florkin, M. & Mason, H.S., Eds.), pp. 557–607. Academic Press, New York.

Rinne, F. (1922). Das Feinbauliche Wesen der Materie nach dem Vorbilde der Kristalle. 3rd Ed. Verlag von Gebrüder Borntraeger, Berlin.

Roberts, L. (1988a). Academy backs genome project. Science 239, 725–726.

Roberts, L. (1988b). Chromosomes: The ends in view. Science 240, 982–983.

Rockstein, M. (1974). The Physiology of Insecta, vol. III, 2nd Ed. Academic Press, New York.

Röhl, R. & Nierhaus, K.N. (1982). Assembly map of the large subunit (50S) of *Escherichia coli* ribosomes. Proc. Natl. Acad. Sci. USA 79, 729–733.

Röhme, D. (1981). Evidence for a relationship between longevity of mammalian species and lifespans of normal fibroblasts *in vitro* and erythrocytes *in vivo*. Proc. Natl. Acad. Sci. USA 78, 5009–5013.

Romer, A.S. & Parsons, T.S. (1978). The Vertebrate Body. Shorter Version. 5th Ed. W.B. Saunders Company, Philadelphia.

Romoser, W.S. (1973). The Science of Entomology. Macmillan Publishing Co., Inc., New York.

Rose, S.M. (1970a). Regeneration. Appleton-Century-Crofts, New York.

Rose, S.M. (1970b). Differentiation during regeneration caused by repressors in bioelectric fields. Amer. Zool. 10, 91–100.

Ross, H.H. (1962). How to Collect and Preserve Insects. Circular 39, Illinois Natural History Survey, Urbana.

Rugh, R. (1951). The Frog. Its Reproduction and Development. The Blakiston Company, Philadelphia.

Russell-Hunter, W.D. (1979). A Life of Invertebrates. Macmillan Publishing Co., Inc., New York.

Rydén, L.G. & Hunt, L.T. (1993). Evolution of protein complexity: The blue copper-containing oxidases and related proteins. J. Mol. Evol. 36, 41–66.

Saenger, W. (1988). Principles of Nucleic Acid Structure. Springer-Verlag, New York.

Saenger, W., Riecke, J., & Suck, D. (1975). A structural model for the polyadenylic acid single helix. J. Mol. Biol. 93, 529–534.

Sandars, E. (1937). A Beast Book for the Pocket. The Vertebrates of Britain Wild and Domestic other than Birds and Fishes. Oxford University Press, Humphrey Milford, London.

Sanderson, R.T. (1967). Inorganic Chemistry. Reinhold Pub. Co., New York.

Sarich, V.M. (1993). Mammalian systematics: Twenty-five years among their albumins and transferrins. In: Mammal Phylogeny. Placentals (Szalay, F.S., Novacek, M.J., & McKenna, M.C., Eds.), pp. 103–114. Springer-Verlag, New York.

Savage, R.J.G. & Long, M.R. (1986). Mammal Evolution. An Illustrated Guide. British Museum (Natural History), London.

Schenck, H. (1915). Handwörterbuch der Naturwissenschaften. Wasserpflanzen X. Jena.

Scherthan, H., Arnason, U., & Lima-de-Faria, A. (1987). The chromosome field theory tested in muntjac species by DNA cloning and hybridization. Hereditas 107, 175–184.

Scherthan, H., Arnason, U., & Lima-de-Faria, A. (1990). Location of cloned, repetitive DNA sequences in deer species and its implications for maintenance of gene territory. Hereditas 112, 13–20.

Schmalhausen, I.I. (1949). Factors of Evolution. The Theory of Stabilizing Selection. The Blakiston Company, Philadelphia.

Schmidtke, J., Becak, W., & Engel, W. (1976). The reduction of genic activity in the tetraploid amphibian Odontophrynus americanus is not due to loss of ribosomal DNA. Experientia 32, 27–28.

Schopf, J.W. (1978). The evolution of the earliest cells. Sci. Am. 239(3), 85–104.

Schorr, H. (1957). Zur Verhaltensbiologie und Symbiose von *Brachypelta aterrima* Först. (Cydnidae, Heteroptera). Z. Morph. Ökol. Tiere 45, 561–601.

Schumann, W. (1973). Mineral och bergarter. Mineral, ädelstenar, bergarter, malmer. P.A. Norstedt & Söners förlag, Stockholm.

Schwabe, J.W.R. & Travers, A.A. (1993). What is evolution playing at? Curr. Bio. 3(9), 628–630.

Scott, M.P. & Caroll, S.B. (1987). The segmentation and homeotic gene network in early Drosophila development. Cell 51, 689–698.

Segrè, E. (1980). From X-Rays to Quarks. Modern Physicists and their Discoveries. W.H. Freeman and Company, San Francisco.

Shannon, R.D. & Prewitt, C.T. (1969). Effective ionic radii in oxides and fluorides. Acta Crystallogr. Sect. B, 25, 925–945.

Sharp, D.W.A. (1988). The Penguin Dictionary of Chemistry. Penguin Books, London.

Sharp, L.W. (1934). Introduction to Cytology. 3rd Ed. McGraw-Hill Book Company, Inc. New York.

Shi, Y.-B. & Brown, D.D. (1990). Developmental and thyroid hormone-dependent regulation of pancreatic genes in *Xenopus laevis*. Genes and Development 4, 1107–1113.

Shoji, A. & Ozawa, E. (1985). Requirement of Fe ion for activation of RNA polymerase. Proc. Jap. Acad., Ser. B, 61, 494–496.

Shoji, A. & Ozawa, E. (1986). Necessity of transferrin for RNA synthesis in chick myotubes. J. Cell. Physiol. 127, 349–356.

Shore, D. & Nasmyth, K. (1987). Purification and cloning of a DNA binding protein from yeast that binds to both silencer and activator elements. Cell 51, 721.

Shubnikov, A.V. & Koptsik, V.A. (1974). Symmetry in Science and Art (Harker, D., Ed.). Plenum Press, New York.

Shuvalov, L.A., (Ed.) (1988). Modern Crystallography IV. Physical Properties of Crystals. Springer-Verlag, Berlin.

Singer, R.H. (1993). RNA zipcodes for cytoplasmic addresses. Current Biology 3(10), 719–721.

Sinnott, E.W. & Dunn, L.C. (1939). Principles of Genetics. 3rd Ed. McGraw-Hill Book Company, Inc., New York.

Slijper, E.J. (1936). Die Cetaceen. Vergleichend-Anatomisch und Systematisch. Martinus Nijhoff, Haag.

Sluyser, M., Geert, A.B., Brinkmann, A.O., & Blankenstein, R.A. (1993). Preface. In: Zinc-finger Proteins in Oncogenesis. DNA-binding and Gene Regulation. Annals of the New York Academy of Sciences. Vol. 684. The New York Academy of Sciences, New York.

Smalley, R.E. (1991). Great balls of carbon. The story of buckminsterfullerene. The Sciences, March/April 1991, 22–28.

Smith, C.W.J., Patton, J.G., & Nadal-Ginard, B. (1989). Alternative splicing in the control of gene expression. Annu. Rev. Genet. 23, 527–577.

Snodgrass, R.E. (1935). Principles of Insect Morphology. McGraw-Hill, New York.

Sommer, H., Beltrán, J.-P., Huijser, P., Pape, H., Lonnig, W.-E., Saedler, H., & Schwarz-Sommer, Z. (1990). *Deficiens*, a homeotic gene involved in the control of flower morphogenesis in *Antirrhinum majus*: The protein shows homology to transcription factors. EMBO J. 9, 605–613.

Sonneborn, T.M. (1950). *Paramecium* in modern biology. Bios. 21, 31–43.

Spatz, H.C. & Crothers, D.M. (1969). The rate of DNA unwinding. J. Mol. Biol. 42, 191–219.

Spemann, H. (1919). Experimentelle Forschungen zum Determinations- und Individualitäts-Problem. Die Naturwiss. VII, 32.

Spirin, A.S. (1986). Ribosome Structure and Protein Biosynthesis. Benjamin/Cummings Advanced Book Program, Menlo Park, CA.

Springer, M.S. & Kirsch, J.A.W. (1993). A molecular perspective on the phylogeny of placental mammals based on mitochondrial 12S rDNA sequences, with special reference to the problem of the Paenungulata. J. Mam. Evol. 1(2), 149–166.

Stark, G.R., Debatisse, M., Wahl, G.M., & Glover, D.M. (1990). DNA amplification in eukaryotes. In: Gene Rearrangement (Hames, B.D. & Glover, D.M., Eds.), pp. 99–149. IRL Press at Oxford University Press, Oxford.

Stavenga, D.G. (1989). Pigments in compound eyes. In: Facets of Vision (Stavenga, D.G. & Hardie, R.C., Eds.), pp. 152–172. Springer-Verlag, Berlin.

Stavenga, D.G. & Kuiper, J.W. (1977). Insect pupil mechanisms. I. On the pigment migration in the retinula cells of Hymenoptera (suborder Apocrita). J. Comp. Physiol. 113, 55–72.

Stavenga, D.G., Numan, J.A.J., Tinbergen, J., & Kuiper, J.W. (1977). Insect pupil mechanisms. II. Pigment migration in retinula cells of butterflies. J. Comp. Physiol. 113, 73–93.

Stern, H. & Hotta, Y. (1978). Regulatory mechanisms in meiotic crossing-over. Annu. Rev. Plant Physiol. 29, 415–436.

Stix, H., Stix, M., & Abbott, R.T. (1978). The Shell. Abrams, New York.

Strasburger, E. (1943). Tratado de Botánica. Manuel Marín, Barcelona.

Struhl, G. & White, R.A.H. (1985). Regulation of the *Ultrabithorax* gene of *Drosophila* by other bithorax complex genes. Cell 43, 507.

Stryer, L. (1988). Biochemistry. 3rd Ed. W.H. Freeman and Company, New York.

Strynadka, N.C.J. & James, M.N.G. (1989). Crystal structures of the helix-loop-helix calcium-binding proteins. Annu. Rev. Biochem. 58, 951–998.

Stubbe, H. (1966). Genetik und Zytologie von *Antirrhinum* L. sect. *Antirrhinum* G. Fischer, Jena.

Stubbings, H.G. (1975). *Balanus balanoides*. Liverpool University Press, Liverpool.

Sturtevant, A.H. & Beadle, G.W. (1940). An Introduction to Genetics. W.B. Saunders Company, Philadelphia.

Südhof, T.C., Goldstein, J.L., Brown, M.S., & Russel, D.W. (1985a). The LDL receptor gene: A mosaic of exons shared with different proteins. Science 228, 815–822.

Südhof, T.C., Russel, D.W., Goldstein, J.L., & Brown, M.S. (1985b). Cassette of eight exons shared by genes for LDL receptor and EGF precursor. Science 228, 893–895.

Sussman, M. (1964). Growth and Development, 2nd. ed. Prentice-Hall, New York.

Sutherland, G.R. & Richards, R.I. (1994). Dynamic Mutations. Am. Sci. 82(2), 157–163.

Swinburne, T.R. (1986). Stimulation of disease development by siderophores and inhibition by chelated iron. In: Iron, Siderophores, and Plant Diseases (Swinburne, T.R., Ed.), pp. 217–226. Plenum Press, New York.

Tansley, K. (1965). Vision in Vertebrates. Science Paperbacks and Chapman and Hall Ltd., London.

Taylor, G.R. (1983). The Great Evolution Mystery. Secker & Warburg, London.

Terni, T. (1914). Condriosomi, idiozoma e formazioni periidiozomiche nella spermatogenesi degli Anfibii. (Ricerche sul Geotriton fuscus.) Arch. Zellf. 12, 1–96.

Thangue, N.B. La & Rigby, P.W.J. (1988). Trans-acting protein factors and the regulation of eukaryotic transcription. In: Transcription and Splicing (Hames, B.D. & Glover, D.M., Eds.), pp. 1–42. IRL Press, Oxford, WA.

Thompson, D.W. (1952). On Growth and Form, Vol. I. Cambridge University Press, Cambridge.

Thornley, J.H. (1976). Mathematical Models in Plant Physiology.Academic Press, London.

Tieghem, P. van & Costantin, J. (1918a). Eléments de Botanique. Vol. I. Botanique Générale. Masson et Cie, Paris.

Tieghem, P. van & Costantin, J. (1918b). Eléménts de Botanique. Vol. II. Botanique Spéciale. Masson et Cie, Paris.

Töndury, G. (1968). Band I. Bewegungsapparat. Knochen-Gelenke-Muskeln. Rauber/Kopsch. Lehrbuch und Atlas der Anatomie des Menschen. In drei Bänden. Georg Thieme Verlag, Stuttgart.

Tonegawa. S. (1983). Somatic generation of antibody diversity. Nature 302, 575–581.

Turner, C.D. & Bagnara, J.T. (1976). General Endocrinology. 6th Ed. W.B. Saunders Company, Philadelphia.

Tyler, A. (1955). Gametogenesis, fertilization, and parthenogenesis. In: Analysis of Development (Willier, B.H., Weiss, P.A., & Hamburger, V., Eds.), pp. 170–212. W.B. Saunders Company, Philadelphia.

Vainshtein, B.K., Fridkin, V.M., & Indenbom, V.L. (1982). Modern Crystallography II. Structure of Crystals. Springer-Verlag Berlin.

Van der Ploeg, L.H.T. (1990). Antigenic variation in African trypanosomes: Genetic recombination and transcriptional control of VSG genes. In: Gene Rearrangement (Hames, B.D. & Glover, D.M., Eds.), pp. 51–97. Oxford University Press, New York.

Van der Putten, H., Botteri, F.M., Miller, A.D., Rosenfeld, M.G., & Fan, H. (1985). Efficient insertion of genes into the mouse germ line via retroviral vectors. Proc. Natl. Acad. Sci. USA 82, 6148–6152.

Varmus, H. (1988). Retroviruses. Science 240, 1427–1435.

Vavilov, N.I. (1922). The law of homologous series in variation. J. Gen. Vol. XII, 47–89.

Vaynshteyn, B.K. (1988). The theory of symmetry. Symmetrical equality principle in the growth of biological macromolecules and biocrystals. In: Modern Crystallography (Vaynshteyn, B.K. & Chernov, A.A., Eds.), pp. 1–33. Nova Science Publishers, Commack, NY.

Veil, C. (1938). L'influence du milieu sur la couleur des animaux. In: La Biologie (Bonnardel, R., et al., Eds.), pp. 49–56. Palais de la Découverte, Masson et Cie, Paris.

Vogelmann, T.C., Bornman, J.F., & Josserand, S. (1989). Photosynthetic light gradients and spectral régime within leaves of *Medicago sativa*. Phil. Trans. R. Soc. Lond. B 323, 411–421.

Von Baeyer, H.C. (1986). Physics finds a childhood. The Sciences Jul/Aug. 8–10.

Von Baeyer, H.C. (1992). Taming the Atom. The Emergence of the Visible Microworld. Random House, New York.

von Ubisch, L. (1943). Über die Bedeutung der Diminution von Ascaris megalocephala. Acta Biotheoret. 7, 163–182.

Walcott, C., Gould, J.L., & Kirschvink, J.L. (1979). Pigeons have magnets. Science 205, 1027–1028.

Walker, C. (1974). Introduction: The world of birds. In: The World Atlas of Birds, pp. 10–33. Mitchell Beazley Publishers Limited, London.

Wampler, J.E. (1978). Measurements and physical characteristics of luminescence. In: Bioluminescence in Action (Herring, P.J., Ed.), pp. 1–48. Academic Press, New York.

Wang, A.H. -J., Fujii, S., Van Boom, J.H., & Rich, A. (1982). Right-handed and left-handed double-helical DNA: Structural studies. Cold Spring Harbor Symp. Quant. Biol. 47, 33–44.

Wareing, P.F. & Phillips, I.D.J. (1978). The Control of Growth and Differentiation in Plants. 2nd Ed. Pergamon Press, Oxford.

Watson, J.D., Hopkins, N.H., Roberts, J.W., Steitz, J.A., & Weiner, A.M. (1987). Molecular Biology of the Gene, volume I. 4th Ed. The Benjamin/Cummings Publishing Company, Inc., Menlo Park, CA.

Webster, N. (1976). Webster's New Twentieth Century Dictionary of the English Language. Unabridged. 2nd Ed. Collins World.

Wehner, R. & Müller, M. (1993). How do ants acquire their celestial ephemeris function? Naturwissenschaften 80, 331–333.

Weinberg, E.D. (1989). Cellular regulation of iron assimilation. Quart. Rev. Bio. 64(3), 261–290.

Weinberg, S. (1977). The First Three Minutes. A Modern View of the Origin of the Universe. William Collins and Sons, Glasgow.

Weiss, P. (1939). Principles of Development. A Text in Experimental Embryology. Henry Holt and Company, New York.

Wells, M. (1983). Cephalopods do it differently. New Scientist 100, 332–337.

Westheimer, F.H. (1986). Polyribonucleic acids as enzymes. Nature 319, 534–536.

Wettstein, R. (1944). Tratado de Botánica sistemática. Editorial Labor, S.A., Barcelona.

Wharton, R.P., Brown, E.L., & Ptashne, M. (1984). Substituting an alfa-helix switches the sequence-specific DNA interactions of a repressor. Cell 38, 361–369.

Wheeler, H. (Ed.) (1939). The Miracle of Life. Odhams Press Limited, London.

White, B.A. & Nicoll, C.S. (1981). Hormonal control of amphibian metamorphosis. In: Metamorphosis: A Problem in Developmental Biology (Gilbert, L.I. & Frieden, E., Eds.), pp. 363–396. Plenum Press, New York.

Whittaker, E.J.W. & Muntus, R. (1970). Ionic radii for use in geochemistry. Geochim. et Cosmochim. Acta 34, 952–953.

Whitten, D.G.A. & Brooks, J.R.V. (1988). The Penguin Dictionary of Geology. Penguin Books, London.

Wichler, C. (1913). Untersuchungen über den Bastard *Dianthus armeria* x D. *deltoides*, nebst Bemerkungen über einige andere Artkreuzungen der Gattung *Dianthus*. Zeits. ind. Abst. Vererb. 10, 177–232.

Wigglesworth, V.B. (1970). Insect Hormones. Oliver & Boyd, Edinburgh.

Williams, R.J. & Lansford, E.M. (1967). The Encyclopedia of Biochemistry. Reinhold Publishing Corporation, New York.

Willier, B.H. & Rawles, M.E. (1935). Organ-forming areas in the early chick blastoderm. Proc. Soc. Exp. Biol. 32, 1293–1296.

Wilson, A.C. (1976). Gene regulation in evolution. In: Molecular Evolution (Ayala, F.J., Ed.), pp. 225–234. Sinauer Associates, Sunderland, MA.

Wilson, E.B. (1925). The Cell in Development and Heredity. 3rd Ed. Macmillan Publishing Company, New York.

Wineland, D.J., Bergquist, J.C., Itano, W.M., Bollinger, J.J., & Manney, C.H. (1987). Atomic-ion coulomb clusters in an ion trap. Phys. Rev. Lett. 59(26), 2935–2938.

Withers, P.C. (1992). Comparative Animal Physiology. Saunders College Publishing, Philadelphia.

Wolffe, A.P. & Brown, D.D. (1988). Developmental regulation of two 5 S ribosomal RNA genes. Science 241, 1626–1632.

Wondratschek, H. (1987). Determination of ionic radii from cation-anion distances in crystal structures. Am. Mineral. 72, 82.

Woods, H. (1937). Palaeontology. Invertebrate. 7th Ed. Cambridge University Press, Cambridge.

Wright, K. (1989). A breed apart. Finicky flies lend credence to a theory of speciation. Sci. Am. 260(2), 13–14.

Wright, S. (1934a). An analysis of variability in number of digits in an inbred strain of guinea pig. Genetics 19, 506–536.

Wright, S. (1934b). The results of crosses between inbred strains of guinea pigs, differing in numbers of digits. Genetics 19, 537–551.

Yaoita, Y., Shi, Y.-B., & Brown, D.D. (1990). *Xenopus laevis* alpha and beta thyroid hormone receptors. Proc. Natl. Acad. Sci. USA 87, 7090–7094.

Young, J.Z. (1962). The Life of Vertebrates. Clarendon Press, Oxford.

Yung, E. (1920). Traité de Zoologie des Animaux Invertébrés (Achordata). Atar, Genève.

Zambryski, P. (1988). Basic processes underlying *Agrobacterium*-mediated DNA transfer to plant cells. Ann. Rev. Genet. 22, 1–30.

Zambryski, P. (1989). *Agrobacterium*-plant cell DNA transfer. In: Mobile DNA (Berg, D.E. & Howe, M.M., Eds.), pp. 309–334. American Society For Microbiology, Washington, DC.

SOURCES OF ILLUSTRATIONS

All figures, which have no reference to the author at the end of the text, are original graphs and plates.

The sources of the individual pictures used in the plates are given below. The first number of each figure refers to the chapter in which it appears.

Figure 1.1. Sanderson (1967), p. 14.
Figure 1.2. From Greenwood and Earnshaw (1989), p. 28.
Figure 3.1. Period Flight.

Figure 3.2. 1. From Francé (1943), p. 208.
2. After Snodgrass (1935). From Borror, et al. (1976), p. 58.
3. Perrins (1976), p. 15.

Figure 3.3. 1. Brumpt (1936), p. 1356.
2. Boule and Piveteau (1935), p. 493.
3. Beazley (1974), p. 154.
4. Freeman (1972), p. 20.
5. Perrins (1976), p. 28.

Figure 3.4. 1. Pringle (1975), p. 2.
2. Perrins (1976), p. 30.
3. Beazley (1980), p. 166.
4. Pennycuick (1972). From Savage and Long (1986), p. 107.

Figure 3.5. 1. Boulenger (1937), p. 123.
2. Pierantoni (1944), p. 908.

Figure 3.6. 1. After Whinter and Bruun. From Hanström and Johnels (1962), p. 309.
2. After Whinter and Bruun. From Hanström and Johnels (1962), p. 310.
3. From McMahon and Bonner (1983), p. 183.

Figure 3.7. 1. Macdonald (1984), p. 234.
2. Macdonald (1984), p. 183.
3. Macdonald (1984), p. 164.
4. After Goode. From Romer and Parsons (1978), p. 56.
5. Langlebert (1901), p. 267.

Figure 4.1. 1. Niklitschek (1943), p. 320.
2. Rabaud (1932), p. 154.
3. Rabaud (1932), p. 151.
4. Rabaud (1932), p. 158.

Figure 4.2. 1. Vogelmann et al. (1989), p. 419.
2. After Stahl. From Palhinha and Cunha (1939), p. 389.
3. Romoser (1973), p. 137.
4. Stavenga et al. (1977), p. 74.

Figure 4.3. Period Vision.

Figure 4.4. Period Vision.

Figure 4.5. 1. Rabaud (1932), p. 155.
2. Rabaud (1932), p. 156.
3. Rabaud (1932), p. 159.
4. Rabaud (1932), p. 139.

Figure 5.1. 1. Beazley (1980), p. 167.
2. Rabaud (1934), p. 660.
3. Rabaud (1934), p. 659.
4. Frisch (1938), p. 344.

Figure 5.2. Period Placenta.

Figure 5.3. Dorrington (1979), p. 74.

Figure 6.1. 1. Cabrera (1937), p. 145.
2. Cabrera (1937), p. 151.
3. Combes (1938), p. 21.
4. Francé (1943), p. 192.
5. Frisch (1938), p. 223.
6. Bohn (1934b), p. VI-42.

Figure 6.2. Period Luminescence.

Figure 6.3. Withers (1992), p. 70.

Figure 7.1. 1. Frisch (1938), p. 291.
2. Orr and Pope (1986), p. 169.
3. After Noble and Noble (1976). From Barnes (1980), p. 233.
4. After Turner and Bagnara (1976). From Romer and Parsons (1978), p. 300.
5. After Stubbings, H.G. (1975). From Barnes, R.D. (1980).

Figure 7.2. 1. From Pierantoni (1944), p. 530.
2. After Schmarda. From Hertwig (1928), p. 446.

	3. After Belar. From Niklitschek (1943), plate 11.
	4. From Maximov (1938), p. 421.
Figure 7.3.	Period Penis.
Figure 8.1.	1. Rinne (1922), p. 130.
	2. After Schenck (1915). From Schmalhausen (1949), p. 197.
	3. After Young (1962). From Gould (1977), p. 323.
	4. Rinne (1922), p. 130.
	5. After Schenck (1915). From Schmalhausen (1949), p. 197.
	6. After Young (1962). From Gould (1977), p. 323.
Figure 8.2.	1. Pierantoni (1944), p. 694.
	2. Niklitschek (1943), p. 227.
	3. Hertwig (1928), p. 619.
	4. Beazley (1980), p. 163.
	5. Freeman (1972), p. 23.
	6. Beazley (1980), p. 187.
Figure 8.3.	1. Commercial photograph.
	2. Savage and Long (1986), p. 98.
	3. Curry-Lindahl and Tinggaard (1965), p. 57.
	4. Freeman (1972), p. 45.
Figure 8.4.	1. Beazley (1980), p. 118.
	2., 3. Matthews and Carrington (1972), p. 26.
	4. Beazley (1980), p. 187.
	5. Macdonald (1984), p. 139.
	6. Macdonald (1984), p. 139.
	7. Freeman (1972), p. 50.
Figure 8.5.	1. Matthews and Carrington (1972), p. 183.
	2. Freeman (1972), p. 51.
	3. Freeman (1972), p. 51.
Figure 8.6.	1. Freeman (1972), p. 52.
	2. Freeman (1972), p. 52.
	3. Macdonald (1984), p. 510.
Figure 8.7.	Period Return to Aquatic Life.
Figure 9.1.	1. Halstead (1978), p. 79.
	2. Macdonald (1984), p. 824.
	3. Freeman (1972), p. 47.
	4. Macdonald (1984), p. 859.
	5. Sandars (1937), p. 98.
	6. Macdonald (1984), p. 842.
Figure 10.1.	Wells (1983), p. 337.
Figure 10.2.	Pope (1986), p. 53.
Figure 11.1.	1. Vainshtein et al. (1982), p. 6.
	2. After Cruickshank (1961), from Saenger (1988), p. 85.
Figure 11.2.	After Benfy (1964), from Jaffe (1988), p. 11.
Figure 11.3.	1. From Klein and Hurlbut (1985), p. 314.
	2. From Klein and Hurlbut (1985), p. 314.
	3. From Klein and Hurlbut (1985), p. 409.
	4. From Klein and Hurlbut (1985), p. 407. After F. Liebau (1959), *Acta Crystallographica*, vol. 12, p. 180.
Figure 12.1.	1. Pauling (1949), p. 202.
	2. Pauling (1949), p. 202.
	3. Heywood (1978), p. 142.
	4. Heywood (1978), p. 209.
	5. Galiano (1929), p. 214.
	6. Galiano (1929), p. 214.
	7. Curry-Lindahl and Tinggaard (1965), p. 24.
	8. Halstead (1978), p. 98.
Figure 13.1.	From Calvin (1983), p. 8.

Figure 13.2. 1. After Wada and Grenway. From Wilbur, K.M. 1972: in Florkin, M. and Scheer, B.T. (1972), p. 105. Obtained from Barnes, R.D. (1980), p. 393.
2. From Matheja and Degens (1968), p. 225.

Figure 14.1. 1. Mason (1987), p. 31.
2. Furth (1980), p. 15.
3. Shubnikov and Koptsik (1974), Quartz crystals p. 54.
4. Pauling (1949), p. 515.

Figure 14.2. 1. Wang et al. (1982). From Watson et al. (1987), p. 253.
2. After Church. From Denffer et al. (1971), p. 127.
3. From Sturtevant and Beadle (1940), p. 329.
4. From Müntzing (1961), p. 54.

Figure 14.3. 1. From Dyson (1953), p. 20.
2. From Boutaric (1938), p. 864.
3. After Willier and Rawles (1935). From Weiss (1939), p. 396.
4. From Boutaric (1938), p. 867.
5. From Frisch (1938), p. 320.

Figure 14.4. 1. Jaffe (1988), p. 92.
2. Greulach (1973), p. 320.
3. After Hinchliffe and Johnson (1980). From Müller (1990), p. 105.
4. Lima-de-Faria (1954), p. 359.

Figure 14.5. 1. Rinne (1922), p. 128.
2. After Becker (1974). From Berrill and Karp (1976), p. 505.
3. After Waller. From Cluzet and Ponthus (1939), p. 295.
4. After Lissman (1963). From Eckert and Randall (1978), p. 217.

Figure 15.1. 1. After Miller and Bakken (1972). From Bostock and Sumner (1978), p. 160.
2. Herskowitz (1977), p. 88.
3. After Goebel (1933). From Hertwig (1929a), p. 363.
4. Burnie (1988), p. 10.
5. Bentley and Humphreys (1962), p. 214.
6. Russell-Hunter (1979), p. 353.

Figure 15.2. 1. After Iijima et al. (1973). From Klein and Hurlbut (1985), p. 9.
2., 3. After Auber (1969). From Alberts et al. (1983), p. 551.
4. Klein and Hurlbut (1985), p. 427.
5. After Craig, R. From Alberts et al. (1983), p. 551.

Figure 15.3. 1. From Rinne (1922), p. 32.
2. After Lutman. From Palhinha and Cunha (1939), p. 481.
3. After Flemming (1881). From Wilson (1925), p. 125.
4. From Dana (1955).
5. From Palhinha and Cunha (1939), p. 396.
6. After Terni (1914). From Sharp (1934), p. 87.

Figure 15.4. 1. Klein and Hurlbut (1985), p. 34.
2. After Schorr (1957). From Eisenbeis and Wichard (1987), p. 276.
3. After Stratz. From Petrén (1964), p. 17.

Figure 15.5. 1. Rinne (1922), p. 105.
2. After Maksymowych (1973). From Dale (1982), p. 23.
3. McMahon and Bonner (1983), p. 52.

Figure 15.6. 1. Bentley and Humphreys (1962), pp. 184, 209.
2. After Heslop-Harrison and Heslop-Harrison (1958). From Dale (1982), p. 46.
3. After Glücksohn. From Weiss (1939), p. 65.

Figure 15.7. 1. Cabrera (1937), p. 122.
2. Medenbach and Sussieck-Fornefeld (1983), p. 132.
3. Raff and Kaufman (1983), p. 225.
4. After Spemann (1919). From Wilson (1925), p. 1054.
5. Frisch (1938), p. 321.
6. After Bateson (1894). From Darlington (1953), p. 259.

Figure 15.8. 1. After Pasteur (1857). From Blaringhem (1923), p. 53.
2. After Child. From Buchsbaum (1951), p. 146.

	3. After Richters. From Aron and Grassé (1939), p. 233.
	4. Weiss (1939), p. 472.
Figure 15.9.	1., 2. After Kern and Gindt (1958). From Klein and Hurlbut (1985), p. 98.
	3. Jones (1934), p. 21.
	4. Dale (1982), p. 30.
	5. From Mintz and Illmensee (1975), p. 3586.
Figure 15.10.	1.–4. Klein and Hurlbut (1985), p. 376.
	5.–7. Sinnott and Dunn (1939), p. 88.
	8.–10. After Morgan (1928). From Sinnott and Dunn (1939), p. 48.
Figure 15.11.	1.–3. Klein and Hurlbut (1985), p. 330, p. 85.
	4. Medenbach and Sussieck-Fornefeld (1983), pp. 128, 130.
	5.–7. After Heribert-Nilsson (1918).
	8.–10. After Wichler (1913). From Guyénot (1942), p. 518.
Figure 18.1.	1. After Escherich. From Bohn (1935), pp. IV-81.
	2. After Regan. From Bohn (1934b), pp. VI-44.
	3. After Baltzer. From Prenant (1935), p. 50.
Figure 20.1.	1. After Hemplelmann. From Pierantoni (1944), p. 513.
	2. After Fraipont. From Barnes (1980), p. 494.
	3. After Baltzer. From Dawydoff (1928), p. 173.
	4. After Greef. From Yung (1920), p. 262.
	5. After Gerould. From Dawydoff (1928), p. 317.
	6. After Keferstein. From Hertwig (1928), p. 311.
	7. After Wilson. From Dawydoff (1928), p. 696.
	8. Pierantoni (1944), p. 695.
Figure 20.2.	1. After Berlese. From Aron and Grassé (1939), p. 817.
	2. Perrier (1936), p. 487.
	3. Perrier (1936), p. 487.
	4. Romoser (1973), p. 349.
	5. Romoser (1973), p. 349.
	6. Romoser (1973), p. 354.
	7. Romoser (1973), p. 354.
Figure 20.3.	1. Perrier (1936), p. 522.
	2. Aron and Grassé (1939), p. 818.
	3. Romoser (1973), p. 361.
	4. Aron and Grassé (1939), p. 820.
	5. Romoser (1973), p. 371.
	6. Romoser (1973), p. 363.
	7. Romoser (1973), p. 363.
Figure 20.4.	1. After Dawydoff (1928). From Aron and Grassé (1939), p. 710.
	2. After Dawydoff (1928). From Aron and Grassé (1939), p. 710.
	3. Yung (1920), p. 146.
	4., 6. After Dawydoff (1928). From Aron and Grassé (1939), p. 710.
	5. Yung (1920), p. 140.
	7. Yung (1920), p. 139.
Figure 21.1.	1. Perrier (1936), p. 498.
	2. Perrier (1936), p. 498.
	3. Perrier (1936), p. 490.
	4. Perrier (1936), p. 513.
	5. Romoser (1973), p. 351.
	6. Pierantoni (1944), p. 643.
Figure 22.1.	1. After Hartmeyer. From Aron and Grassé (1939), p. 852.
	2. Perrier (1936), p. 611.
	3. After Delsman. From Prenant (1936), p. 41.
	4. After Lohmann. From Pierantoni (1944), p. 766.
	5. Rugh (1951), p. 158.
	6. Olsson and Schumann (1945), p. 42.
Figure 22.2.	From Curry-Lindahl and Tinggaard (1965), pp. 178, 179.

Figure 24.1. 1. Rinne (1922), p. 62.
2. After Nathansohn. From Maximov (1938), p. 128.
3. Cuénot (1932), p. 207.
4. Sandars (1937), p. 82.
5. After Schmalhausen (1949). From Pritchard (1986), p. 63.

Figure 24.2. 1. Klein and Hurlbut (1985), p. 267.
2. Wilson (1925), p. 1060.
3. After Bonnier. From Palhinha and Cunha (1939), p. 362.
4. After Brauer. From Cuénot (1932), p. 593.

Figure 24.3. 1. After Schnorr. From Rinne (1922), p. 117.
2. After Abonyi. From Bohn (1935), pp. IV-49.
3. After Stockard. From Buchsbaum (1951), p. 148.
4. After Keller. From Maximov (1938), p. 381.

Figure 24.4. 1. After Nassau (1980a and b). From Klein and Hurlbut (1985), p. 212.
2., 3. From Atkins (1987). Figure cis-retinal on page 147 and Figure of trans-retinal on page 148.
4., 5. Wigglesworth (1970), p. 97.
6. After Mast. From Veil (1938), p. 51.

Figure 25.1. Alberts et al. (1983), p. 474.

Figure 26.1. 1. Bentley and Humphreys (1962), p. 36.
2. Bentley and Humphreys (1962), p. 37.
3. Bentley and Humphreys (1962), p. 37.

Figure 26.2. 1. From Smalley (1991), p. 22.
2. From Hawkins et al. (1991), p. 313.
3. From Hawkins et al. (1991), p. 312.

Figure 26.3. Greenwood and Earnshaw (1989), p. 403.

Figure 26.4. 1. From Saenger (1988), p. 286.
2. After Arnott et al. (1974). From Saenger (1988), p. 278.
3. After Saenger et al. (1975). From Saenger (1988), p. 303.
4. After Arnott et al. (1981). From Saenger (1988), p. 262.
5. After Arnott and Hukins (1972). From Saenger (1988), p. 257.
6. After O'Brien and MacEwan (1970). From Saenger (1988), p. 281.
7. After Olson (1977). From Saenger (1988), p. 329.

Figure 28.1. 1. Cabrera (1937), p. 118.
2. Gola et al. (1943), p. 924.
3. Cabrera (1937), p. 118.
4. Heywood (1978), p. 270.
5. Cabrera (1937), p. 198.
6. Heywood (1978), p. 182.

Figure 28.2. 1. Dana (1955), p. 191.
2. Wettstein (1944), p. 867.
3. Dana (1955), p. 191.
4. Jussieu (1873), p. 577.
5. Cabrera (1937), p. 45.
6. Heywood (1978), p. 196.

Figure 28.3. 1. After Mann et al. (1991), p. 158.
2. From Bowman et al. (1992).
3. From Desautels (1968), p. 163.
4. From Heywood (1978), p. 99.
5. From Kullinger and Medenbach (1988), p. 26.
6. From Heywood (1978), p. 103.

Figure 28.4. 1. Bentley and Humphreys (1962), p. 206.
2. Eames and MacDaniels (1925), p. 234.
3. Bentley and Humphreys (1962), p. 202.
4. Eames and MacDaniels (1925), p. 234.
5. Rinne (1922), p. 90.
6. Tieghem and Costantin (1918a), p. 105.

Figure 28.5. 1. Wettstein (1944), p. 467.
2. Strasburger (1943), p. 647.
3. Palhinha and Cunha (1939), p. 561.
4. Heywood (1978), p. 47.
5. Jussieu (1873), p. 115.
6. Tieghem and Costantin (1918a), p. 359.
7. Heywood (1978), p. 212.
8. Tieghem and Costantin (1918a), p. 361.

Figure 28.6. 1. Jussieu (1873), p. 399.
2. Wettstein (1944), p. 691.
3. Wettstein (1944), p. 788.
4. Wettstein (1944), p. 739.
5. Heywood (1978), p. 270.
6. Heywood (1978), p. 72.

Figure 28.7. 1. Wettstein (1944), p. 642.
2. Heywood (1978), p. 88.
3. Strasburger (1943), p. 604.
4. Wettstein (1944), p. 632.
5. Wettstein (1944), p. 694.
6. Wettstein (1944), p. 830.

Figure 28.8. 1. Dana (1955), p. 180.
2. Dana (1955), p. 185.
3. Leppik (1977), p. 69.
4. Leppik (1977), p. 69.
5. Leppik (1977), p. 69.
6. Heywood (1978), p. 266.
7. Tieghem and Costantin (1918b), p. 590.
8. Elvers (1965), p. 289.

Figure 29.1. 1. Perrier (1936), p. 88.
2. After Beaver (1967). From Raff and Kaufman (1983), p. 169.
3. After Beaver (1967). From Raff and Kaufman (1983), p. 169.
4. After Beaver (1967). From Raff and Kaufman (1983), p. 169.
5. Perrier (1936), p. 310.
6. After Osterud (1918). From Raff and Kaufman (1983), p. 168.

Figure 29.2. 1. Pierantoni (1944), p. 734.
2. Perrier (1936), p. 311.
3. After Gregory (1900). From Hyman (1955), p. 246.
4. From Hyman (1955), p. 250.
5. From Hyman (1955), p. 252.
6. After Perrier. From Grassé (1948), p. 241.

Figure 30.1. 1. Slijper (1936), p. 93.
2. After Hamann. From Grassé (1948), p. 133.
3. After Blanchard. From Galiano (1929), p. 270.
4. Woods (1937), p. 144.
5. Buchsbaum (1951), p. 93.
6. Hanson (1977), p. 295.

Figure 30.2. 1. After Stempell. From Galiano (1929), p. 201.
2. Buschbaum (1951), p. 122.
3. Rabaud (1934), p. 573.
4. Grassé (1948), p. 664.
5. Rabaud (1932), p. 41.
6. Carroll (1987), p. 76.

Figure 30.3. 1. Carroll (1987), p. 252.
2. Colbert (1980), p. 117.
3. Carroll (1987), p. 257.
4. Carroll (1987), p. 297.

Figure 30.4. 1. Dana (1955), p. 185.

	2. Böhmer (1974), p. 29.
	3. Goebel (1933), p. 1657.
Figure 30.5.	1., 2. Klein and Hurlbut (1985), p. 264.
	3., 4. After Von Goethe (1820). From Gustafsson (1979), p. 243.
	5. After Müller. From Dawydoff (1928), p. 723.
	6. After Wilson, D.P. From Barnes (1980), p. 929.
	7. After Kyle. From Cuenot (1932), p. 488.
	8. After Mast. From Cuenot (1932), p. 506.
Figure 31.1.	A.P. Pugsley (1989), p. 2.
Figure 31.2.	B. Lewin (1983), pp. 321, 322.
Figure 31.3.	Van der Ploeg (1990), pp. 75–77.
Figure 31.4.	Blake et al. (1987), p. 928.
Figure 32.1.	de Duve, C. (1984b), p. 232.
Figure 32.2.	Blake et al. (1987), p. 926.
Figure 32.3.	C. de Duve (1984b), p. 275.

INDEX